RENEWELS 691-4574

DATE DUE

OCT 2 2			
APR 1 2			
WITHDRAWN UTSA LIBRARIES			

Demco, Inc. 38-293

Kurt Faber

Biotransformations in Organic Chemistry

With 31 Figures, 238 Schemes, and 18 Tables

Springer-Verlag
Berlin Heidelberg New York
London Paris Tokyo
HongKong Barcelona Budapest

Univ.-Doz. Dr. Kurt Faber
Institut für Organische Chemie
Technische Univesität Graz
Stremayrgasse 16
A - 8010 Graz

Library
University of Texas
at San Antonio

ISBN 3-540-55762-8 Springer-Verlag Berlin Heidelberg New York
ISBN 0-387-55762-8 Springer-Verlag New York Berlin Heidelberg

This work is subject to copyright. All rights are reserved, whether the whole or part of the material is concerned, specifically the rights of translation, reprinting, reuse of illustrations, recitation, broadcasting, reproduction on microfilm or in other ways, and storage in data banks. Duplication of this publication or parts thereof is permitted only under the provisions of the German Copyright Law of September 9, 1965, in its current version, and permission for use must always be obtained from Springer-Verlag. Violations are liable for prosecution act under German Copyright Law.

© Springer-Verlag Berlin Heidelberg 1992
Printed in Germany

The use of general descriptive names, registered names, trademarks, etc. in this publication does not imply, even in the absence of a specific statement, that such names are exempt from the relevant protective laws and regulations and therefore free for general use.

Typesetting: Camera ready by author
Printing: Color-Druck Dorfi GmbH, Berlin; Binding: Lüderitz & Bauer, Berlin
02/3020-5 4 3 2 1 0 - Printed on acid -free paper

Preface

The use of natural catalysts - enzymes - for the transformation of non-natural man-made organic compounds is not at all new: they have been used for more than one hundred years, employed either as whole cells, cell organelles or isolated enzymes [1]. Certainly, the object of most of the early research was totally different from that of the present day. Thus the elucidation of biochemical pathways and enzyme mechanisms was in the foreground of the reasearch some decades ago. It was mainly during the 1980s that the enormous potential of applying natural catalysts to transform non-natural organic compounds was recognized. What started as a trend in the late 1970s could almost be called a fashion in synthetic organic chemistry in the 1990s. Although the early euphoria during the 'gold rush' in this field seems to have eased somewhat, there is still no limit to be seen for the future development of such methods. As a result of this extensive, recent research, there have been an estimated 5000 papers published on the subject [2]. To collate these data as a kind of 'super-review' would clearly be an impossible task and, furthermore, such a hypothetical book would be unpalatable for the non-expert.

The point of this volume is to provide a condensed introduction to this field: The text is written from an organic chemist's viewpoint in order to encourage more 'pure' organic chemists of any level, to take a deep breath and leap over the gap between the 'biochemical' sciences and 'classic organic chemistry' and to make them consider biochemical methods as an additional tool when they are planning the synthesis of an important target molecule. This book may serve as a guide for updating a dusty organic-chemistry curriculum into which biochemical methods should be incorporated. The wide arsenal of classic synthetic methods has not changed radically but it has been significantly widened and enriched due to the appearance of biochemical methods. Certainly, biochemical methods are not superior in a general sense - they are no panacea -, but they definitely represent a powerful synthetic tool to complement other methodology in modern synthetic organic chemistry.

In this book, the main stream of novel developments in biotransformations, which in many cases have already had some impact in organic chemistry, are put to the fore. Other cases, possessing great potential but still having to show their reliability, are mentioned more briefly. The literature covered by this volume extends to the end of 1991. Special credit, however, is given to some very old' papers as well to acknowledge the appearance of novel concepts. References are selected according to the philosophy that 'more is not always better'. Generally, I have attempted to sort out the most useful references from the pack, in order to avoid writing a book with the charm of a telephone directory! Thus special emphasis is placed on reviews and books, which are often mentioned during the early paragraphs of each chapter to facilitate rapid access to a specific field if desired.

I wish to express my deep gratitude to Stanley M. Roberts for undergoing the laborious task of correcting the manuscript of this book and for raising numerous helpful questions and hints. Special thanks also go to Kurt Königsberger for his untiring support in setting up and maintaining a biotransformation database, which served as an indispensable tool for keeping abreast with the literature.

I shall certainly be pleased to receive comments, suggestions and criticism from readers for incorporation in future editions.

Graz, 1992　　　　　　　　　　　　　　　　　　　　　　　　　　　　K. Faber

References

1. Neidleman SG (1990) The Archeology of Enzymology. In: Abramowicz D (ed) Biocatalysis, Van Nostrand Reinhold, New York, pp 1-24
2. For books and conference proceedings see: Jones JB, Sih CJ, Perlman (eds) (1976) Applications of Biochemical Systems in Organic Chemistry, part I and II, Wiley, New York; Porter R, Clark S (eds) (1984) Enzymes in Organic Synthesis, Ciba Foundation Symposium 111, Pitman, London; Kieslich K (1984) Biotransformations. In: Rehm HJ, Reed G (eds) Biotechnology, volume 6a, Verlag Chemie, Weinheim; Tramper J, van der Plas HC, Linko P (ed) (1985) Biocatalysis in Organic Synthesis, Elsevier, Amsterdam; Schneider MP (ed) (1986) Enzymes as Catalysts in Organic Synthesis, NATO ASI Series C, volume 178, Reidel Publ. Co., Dordrecht; Laane C, Tramper J, Lilly MD (eds) (1987) Biocatalysis in Organic Media, Elsevier, Amsterdam; Davies HG, Green RH, Kelly DR, Roberts SM (1989) Biotransformations in Preparative Organic Chemistry, Academic Press, London; Whitaker JR, Sonnet PE (eds) (1989) Biocatalysis in Agricultural Biotechnology, ACS Symposium Series, volume 389, Washington; Copping LG, Martin R, Pickett JA, Bucke C, Bunch AW (eds) (1990) Opportunities in Biotransformations, Elsevier, London; Abramowicz D (ed) (1990) Biocatalysis, Van Nostrand Reinhold, New York.

Contents

Preface

1 Introduction and Background Information
1.1 Introduction 1
1.2 Common Prejudices Against Enzymes 1
1.3 Advantages and Disadvantages of Biocatalysts 2
    1.3.1 Advantages of Biocatalysts 2
    1.3.2 Disadvantages of Biocatalysts 6
    1.3.3 Isolated Enzymes versus Whole Cells 7
1.4 Enzyme Properties and Nomenclature 8
    1.4.1 Mechanistic Aspects 10
    1.4.2 Classification and Nomenclature 17
    1.4.3 Coenzymes 19
    1.4.4 Enzyme Sources 20
References 21

2 Biocatalytic Applications
2.1 Hydrolytic Reactions 23
    2.1.1 Mechanistic and Kinetic Aspects 23
    2.1.2 Hydrolysis of the Amide Bond 40
    2.1.3 Ester Hydrolysis 48
        2.1.3.1 Esterases and Proteases 48
        2.1.3.2 Lipases 72
    2.1.4 Hydrolysis and Formation of Phosphate Esters ... 97
    2.1.5 Hydrolysis of Epoxides 107
    2.1.6 Hydrolysis of Nitriles 112
References 125
2.2 Reduction Reactions 135
    2.2.1 Recycling of Cofactors 135

- 2.2.2 Reduction of Aldehydes and Ketones Using Isolated Enzymes 141
- 2.2.3 Reduction of Aldehydes and Ketones Using Whole Cells 149
- 2.2.4 Reduction of C=C Bonds Using Whole Cells 157
- References 163
- 2.3 Oxidation Reactions 169
  - 2.3.1 Oxidation of Alcohols and Aldehydes 169
  - 2.3.2 Oxygenation Reactions 174
    - 2.3.2.1 Hydroxylation of Alkanes 178
    - 2.3.2.2 Hydroxylation of Aromatic Compounds 181
    - 2.3.2.3 Epoxidation of Alkenes 183
    - 2.3.2.4 Sulphoxidation Reactions 186
    - 2.3.2.5 Baeyer-Villiger Reactions 189
    - 2.3.2.6 Formation of Peroxides 194
    - 2.3.2.7 Dihydroxylation of Aromatic Compounds 196
  - References 199
- 2.4 Formation of Carbon-Carbon Bonds 204
  - 2.4.1 Aldol Reactions 204
  - 2.4.2 Acyloin Reactions 214
  - 2.4.3 Michael-Type Additions 216
  - References 218
- 2.5 Addition and Elimination Reactions 221
  - 2.5.1 Cyanohydrin Formation 221
  - 2.5.2 Addition of Water and Ammonia 224
  - References 226
- 2.6 Glycosyl-Transfer Reactions 228
  - 2.6.1 Glycosyl Transferases 228
  - 2.6.2 Glycosidases 232
  - References 236
- 2.7 Halogenation and Dehalogenation Reactions 238
  - 2.7.1 Halogenation 238
  - 2.7.2 Dehalogenation 243
  - References 245

3 Special Techniques

- 3.1 Enzymes in Organic Solvents 248
  - 3.1.1 Ester Synthesis 255
  - 3.1.2 Lactone Synthesis 272
  - 3.1.3 Amide Synthesis 273

  3.1.4 Peptide Synthesis.. 274
  3.1.5 Peracid Synthesis.. 279
  3.1.6 Redox Reactions.. 280
 3.2 Immobilization... 283
 3.3 Modified and Artificial Enzymes................................... 293
  3.3.1 Polyethylene Glycol Modified Enzymes.................... 293
  3.3.2 Semisynthetic Enzymes... 295
  3.3.3 Catalytic Antibodies... 297
 References... 301

4 State of the Art and Outlook.. 308

5 Appendix
 5.1 Abbreviations.. 311
 5.2 Suppliers of Enzymes.. 312

Subject Index... 313

1 Introduction and Background Information

1.1 Introduction

Any exponents of classical organic chemistry will probably hesitate to consider a biochemical solution for one of their synthetic problems. This would be due, very often, to the fact, that biological systems would have to be handled. When growth and maintainance of whole microorganisms is concerned, such hesitation is probably justified: In order to save endless frustrations a close collaboration with a biochemist is highly recommended to set up and use fermentation systems [1]. On the other hand isolated enzymes (which may be obtained increasingly easily from commercial sources either in a crude or partially purified form) can be handled like any other chemical catalyst [2]. Due to the enormous complexity of biochemical reactions compared to the repertoire of classical organic reactions, it follows that most of the methods described will have a strong empirical aspect. This 'black box' approach may not entirely satisfy the scientific purists, but as organic chemists are prone to be rather pragmatists, they may accept that the understanding of a biochemical reaction mechanism is not a *conditio sine qua non* for the success of a biotransformation. In other words, a lack of understanding of biochemical reactions should never deter us from using them, if their usefulness has been established. Notwithstanding, it is undoubtedly an advantage to have an acquaintance with basic biochemistry, and with enzymology, in particular.

1.2 Common Prejudices Against Enzymes

If one uses enzymes for the transformation of non-natural organic compounds, the following prejudices are frequently encountered:
- *'Enzymes are sensitive'.*

This is certainly true for most enzymes if one thinks of applying boiling water, but that also holds for most organic reagents, e.g. butyl lithium. If certain precautions are met, enzymes can be remarkably stable. Some candidates even can tolerate hostile environments such as temperatures greater than 100 °C and pressures beyond 200 bar [3].

- *'Enzymes are expensive'*.

Some are, but others can be very cheap if they are produced on a reasonable scale. Considering the higher catalytic power of enzymes compared to chemical catalysts, the overall efficiency of a process may be better even if a rather expensive enzyme is required. Moreover, enzymes can be reused if they are immobilized. It should be emphasized that for most chemical reactions relatively crude and thus reasonably priced enzyme preparations are adequate.

- *'Enzymes are only active on their natural substrates'*.

This statement is certainly true for some enzymes, but it is definetely false for the majority of them. The fact, that nature has now developed its own peculiar catalysts over 3×10^9 years does not necessarily imply that they are designed to work only on their natural target molecules. As a matter of fact, many enzymes are capable of accepting non-natural substrates of an unrelated structural type and can convert them often exhibiting the same high specificities as for the natural counterparts. It seems to be a general trend, that the more complex the enzyme's mechanism, the narrower the limit for the acceptability of 'foreign' substrates. It is a remarkable paradox that many enzymes display high specificities for a specific type of reaction while accepting a wide variety of substrate structures.

- *'Enzymes work only in their natural environment'*.

It is generally true that an enzyme displays its highest catalytic power in water, which in turn represents something of a nightmare for the organic chemist if it is there the solvent of choice. Quite recently, some noteworthy rules for conducting biotransformations in organic media have been delineated. Although the activity is usually lower in such an environment, many other advantages can be accrued thus making many processes more effective by using biocatalysts in organic solvents [4].

1.3 Advantages and Disadvantages of Biocatalysts

1.3.1 Advantages of Biocatalysts

- *Enzymes are very efficient catalysts.*

Typically the rates of enzyme-mediated processes are accelerated, compared to those of the corresponding nonenzymatic reactions, by a factor of 10^8. The acceleration may even exceed a value of 10^{12}, which is far above what chemical catalysts are capable of achieving. Generally chemical catalysts are employed in

1.3 Advantages and Disadvantages of Biocatalysts

concentrations of a mole fraction of 0.1 - 1% whereas most enzymatic reactions can be performed at sufficient rates with a mole fraction of $10^{-3} - 10^{-4}\%$ of catalyst, which clearly makes them more effective by some orders of magnitude.

- *Enzymes are environmentally acceptable.*

Unlike heavy metals, for instance, biocatalysts are completely degraded in the environment.

- *Enzymes act under mild conditions.*

Enzymes act in a pH range of about 5-8, typically around 7, and in a temperature range of 20 - 40 °C, preferably at around 30 °C. This minimizes problems of undesired side-reactions such as decomposition, isomerisation, racemisation and rearrangement, which often plague traditional methodology.

- *Enzymes are not bound to their natural rôle.*

They exhibit a high substrate tolerance by accepting a large variety of man-made unnatural substances and they are not specifically required to work in water. If advantageous for a process, the aqueous medium can often be replaced by an organic solvent.

- *Enzymes can catalyse a broad spectrum of reactions.*

Like all catalysts, enzymes only accelerate a reaction, but they have no impact on the position of the thermodynamic equilibrium of the reaction. Thus, in principle, some enzyme-catalysed reactions can be run in both directions.

There is an enzyme-catalysed process equivalent to almost every type of organic reaction [5]: for example

Hydrolysis-synthesis of esters [6], amides [7], lactones [8], acid anhydrides [9], epoxides [10] and nitriles [11].

Oxidation-reduction of alkanes [12], alkenes [13], aromatics [14], alcohols [15], aldehydes and ketones [16, 17], sulphides and sulphoxides [18].

Addition-elimination of water [19], ammonia [20], hydrogen cyanide [21].

Halogenation and dehalogenation [22], alkylation and dealkylation [23], isomerisation [24], acyloin- [25] and aldol reactions [26]. Even Michael-additions have been reported [27].

Some major exceptions, where equivalent reaction types cannot be found in nature are the Diels-Alder reaction and the Cope-rearrangement, although [3,3]-sigmatropic rearrangements such as the Claisen-rearrangement are known. On the other hand, some biocatalysts can accomplish reactions impossible to emulate in organic chemistry e.g. the selective functionalization of ostensibly non-activated positions in organic molecules.

Enzymes display three major types of selectivities:

- *Chemoselectivity*

Since the purpose of an enzyme is to act on a single type of functional group, other sensitive functionalities, which would normally react to a certain extent under chemical catalysis, survive. For instance, enzymatic ester hydrolysis does not show any propensity for acetal-cleavage.

- *Regioselectivity and Diastereoselectivity*

Due to their complex three-dimensional structure, enzymes may distinguish between functional groups, which are chemically situated in different regions of the same substrate molecule [28].

- *Enantioselectivity*

Last but not least, all enzymes are made from L-amino acids and thus are chiral catalysts. As a consequence, any type of chirality present in the substrate molecule is 'recognized' upon the formation of the enzyme-substrate complex. Thus a prochiral substrate may be transformed into an optically active product and both enantiomers of a racemic substrate may react at different rates, affording a kinetic resolution. These latter properties collectively constitute the 'specificity' of an enzyme and represent its most important feature for selective and asymmetric exploitation [29].

All the major biochemical events taking place in an organism are governed by enzymes. Since the majority of them are highly selective with respect to the chirality of a substrate, it is obvious that the enantiomers of a given bioactive compound such as a pharmaceutical or an agrochemical cause different biological effects. Consequently, they must be regarded as two distinct species. The isomer with the highest activity is denoted as the 'eutomer', whereas its enantiomeric counterpart possessing less or even undesired activities is termed as the 'distomer'. The range of effects derived from the distomer can extend from lower (although positive) activity, none or toxic events. The ratio of the activities of both enantiomers is defined as the 'eudismic ratio'. Some representative examples of different biological effects are given in Scheme 1.1.

As a consequence, racemates of pharmaceuticals and agrochemicals should be regarded with suspicion. Quite astonishingly, 88% of the 480 chiral synthetic drugs on the market were sold in racemic form in 1982 while the respective situation in the field of pesticides was even worse (92% racemic out of 480 chiral agents) [30]. Although at present most bioactive agents are still applied as racemates mainly due to economic reasons, this situation will definitely change due to the legislation pressure. This leads to an increased need for enantiomerically pure compounds [31].

Scheme 1.1. Biological effects of enantiomers.

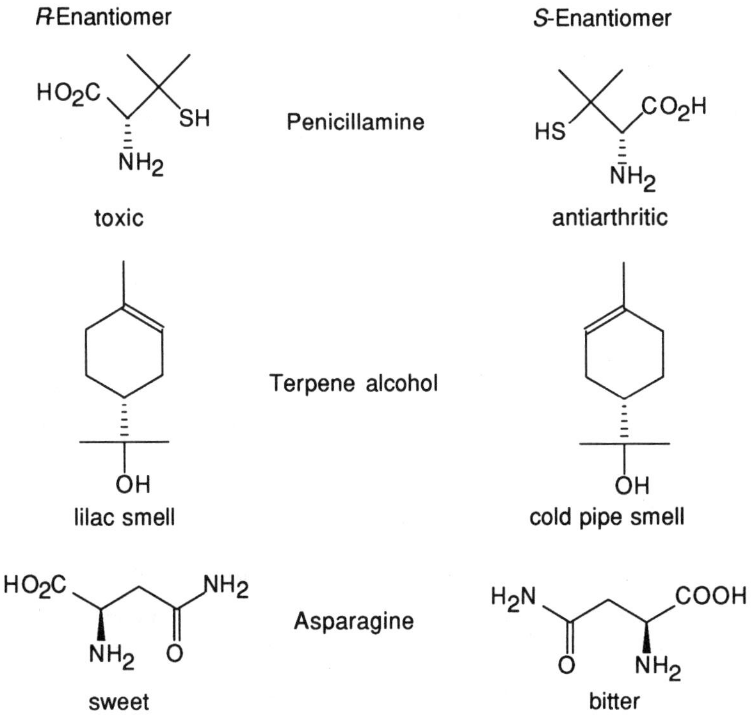

R-Enantiomer		S-Enantiomer
Penicillamine (HO₂C, SH, NH₂)	Penicillamine	Penicillamine (HS, CO₂H, NH₂)
toxic		antiarthritic
Terpene alcohol (OH)	Terpene alcohol	Terpene alcohol (OH)
lilac smell		cold pipe smell
Asparagine (HO₂C, NH₂, NH₂, O)	Asparagine	Asparagine (H₂N, COOH, NH₂, O)
sweet		bitter

Unfortunately, less than 10% of organic compounds crystallize as a conglomerate (the remainder form racemic crystals) offering the possibility of separating their enantiomers by simple crystallisation techniques - such as by seeding a supersaturated solution of the racemate with crystals of one pure enantiomer.

The principle of asymmetric synthesis [32] makes use of enantiomerically pure auxiliary reagents which are used in catalytic or sometimes even stoicheometric amounts. They are often expensive and cannot be recovered in many cases.

Likewise, starting a synthesis with an enantiomerically pure compound which has been selected from the large stock of enantiomerically pure natural compounds [33] such as carbohydrates, amino acids, terpenes or steroids - the so-called 'chiral pool' -, has its limitations. According to a survey from 1984 [34] only about 10 - 20% of compounds are available from the chiral pool at an affordable price (in the range of 100 - 250 US$ per kg). Considering the above mentioned problems with the alternative ways of obtaining enantiomerically pure compounds, it is obvious that enzymatic methods represent a valuable kit for the already existing toolbox available for the asymmetric synthesis of fine chemicals.

1.3.2 Disadvantages of Biocatalysts

There are certainly some drawbacks worthy of mention for a chemist using biocatalysts:
- *Enzymes are provided by Nature in only one enantiomeric form.*

Since there is no way of creating mirror-imaged enzymes from D-amino acids, it is impossible to invert an undesired direction of chirality of a given enzymatic reaction by choosing the 'other enantiomer' of the biocatalyst, a strategy which is possible if chiral chemical catalysts are involved. To gain access to the other enantiomeric product, one has to walk a long path in search of an enzyme with exactly the opposite stereochemical selectivities. However, this is sometimes possible.
- *Enzymes require narrow operation parameters.*

The obvious advantage of working under mild reaction conditions can sometimes turn into a drawback: If a reaction proceeds only slowly under given parameters of temperature or pH, there is only a narrow scope for alteration. Elevated temperatures as well as extreme pH lead to deactivation of the protein as do high salt concentrations. The usual technique of lowering the reaction temperature in order to gain an increase in selectivity is of limited use with enzymatic transformations. The narrow temperature range for the operation of enzymes prevents radical changes, although positive effects from small changes have been reported [35]. Quite astonishingly, some of them remain catalytically active even in ice [36].
- *Enzymes display their highest catalytic activity in water.*

Due to its high boiling point and high heat of vaporization, water is usually the least desired solvent of choice for most organic reactions. Furthermore, the majority of organic compounds are only poorly soluble in aqueous media. Thus shifting enzymatic reactions from an aqueous to an organic medium would be highly desired, but the unavoidable price one has to pay is usually some loss of activity [37].
- *Enzymes are prone to inhibition phenomena.*

Many enzymatic reactions are prone to substrate or product inhibition, which causes the enzyme to cease to work at higher substrate and/or product

concentrations, a factor which limits the efficiency of the process. Whereas substrate inhibition can be circumvented comparatively easily by keeping the substrate concentration at a low level through continuous addition, product inhibition is a more complicated problem. The gradual removal of product by physical means is usually difficult as is the engagement of another step to the reaction sequence in order to effect a chemical removal of the product.
- *Enzymes may cause allergies.*

Enzymes may cause allergic reactions. However, this may be minimized if enzymes are regarded as chemicals and handled with the same care.

1.3.3 Isolated Enzymes versus Whole Cell Systems

The physical state of biocatalysts which are used for biotransformations can be very different. The final decision, as to whether one should use isolated, more or less purified enzymes or whole microorganisms - either in a free or immobilized form - depends on many factors, such as (i) the type of reaction, (ii) if there are cofactors to be recycled and (iii) the scale in which the biotransformation has to be performed. The general pros and cons of the situation are outlined in Table 1.1.

A whole section of biochemistry, microbiology and biochemical engineering - biotechnology- has led to the development of routes to a lot of speciality chemicals (ranging from amino acids to penicillins) starting from cheap carbon sources (such as carbohydrates), cocktail of salts and using viable whole cells. Such syntheses requiring a multitude of biochemical steps are usually referred to as 'fermentation' processes since they constitute de novo syntheses in a biological sense. In contrast, the majority of microbially mediated biotransformations, often starting from relatively complex organic molecules, makes use of only a single (or a few) biochemical synthetic step(s) by using (or rather 'abusing'!) the microbe´s enzymatic potential to convert a non-natural organic compound into a desired product. To distinguish these latter processes from typical fermentations where a multitude of enzymes are involved, the term 'enzymation' is often used.

Table 1.1. Pros and cons of using isolated enzymes and whole cell systems.

System	Form	Pros	Cons
isolated enzymes	any	simple apparatus, simple work-up, better productivity due to higher concentration tolerance	cofactor recycling necessary
	dissolved in water	high enzyme activities	side reactions possible, lipophilic substrates insoluble, workup requires extraction
	suspended in organic solvents	easy to perform, easy work-up, lipophilic substrates soluble, enzyme recovery easy	low activities
	immobilized	enzyme recovery easy	loss of activity during immobilization
whole cells	any	expensive equipment, tedious work-up due to large volumes, low productivity due to lower concentration tolerance, low tolerance of organic solvents, side reactions likely due to metabolism	no cofactor recycling necessary
	growing culture	higher activities	large biomass, more by-products
	resting cells	workup easier, fewer by-products	lower activities
	immobilized cells	cell reuse possible	lower activities

1.4 Enzyme Properties and Nomenclature

The polyamide chain of an enzyme is kept in a three-dimensional structure - the one with the lowest ΔG [38] - believed to be determined by its primary sequence and called the 'primary structure'. For an organic chemist it may roughly be compared with a ball of yarn. Due to the natural aqueous environment, the hydrophilic polar groups such as -COOH, -OH, -NH$_2$, -SH,

1.4 Enzyme Properties and Nomenclature

-CONH$_2$ - are mainly located on the outer surface of the enzyme in order to become hydrated, with the lipophilic substituents - the aryl and alkyl chains - buried inside. The surface of an enzyme is covered by a layer of water, which is tightly bound and which generally cannot be removed by lyophilization. Thus, this residual water, accounting for about 5 - 10% of the total dry weight of a freeze-dried enzyme, is typically called the 'structural water' [39]. It is a distinctive part of the enzyme, necessary to retain the enzyme's three-dimensional structure (and thus its activity) and it differs significantly in its physical state from the 'bulk water' of the surrounding solution. There are very restricted rotation movements in the 'bound water' and it cannot freely reorientate upon freezing. Exhaustive drying of an enzyme (e.g. by chemical means) would force the molecule to change its conformation resulting in a loss of activity. The whole structure is stabilized by a large number of relatively weak binding forces such as Van-der-Waals interaction of aliphatic chains, π-π stacking of aromatics, or salt bridges between charged parts of the molecule. The only covalent bonds, besides the main polyamide backbone, are -S-S- disulfide bridges. Consequently, enzymes are intrinsically unstable in solution and can be deactivated by denaturation, caused by increased temperature, an extreme pH or an unfavourable dielectric environment such as a high salt concentration. The types of reaction leading to an enzyme's deactivation are listed below [40]:

- Rearrangement of peptide chains due to partial unfolding starts at around 40 - 50 °C. Most of these rearrangements are reversible and therefore relatively harmless.
- Hydrolysis of peptide bonds in the backbone, in particular at asparagine units, occurs at more elevated temperatures. Functional groups of amino acids can be hydrolytically cleaved, again especially at asparagine and glutamine to give aspartic and glutamic acid, respectively. Both reaction mechanisms are favoured by the presence of neighbouring groups such as glycine which facilitate the formation of a cyclic intermediate. Thus, a negative charge (-COO$^-$) is created from a neutral part (-CONH$_2$) of the enzyme molecule; this moiety drives a rearrangement of the enzyme's structure in order to become hydrated. These reactions are associated with an irreversible deactivation.
- Thiol groups may interchange the -S-S- disulfide bridges, leading to a modification of covalent bonds within the enzyme.
- Elimination and oxidation reactions (often involving cystein residues) cause the final destruction of the protein.

Thermostable enzymes from thermophilic microorganisms show an astonishing upper operation limit of 60 - 70 °C and differ from their mesophilic counterparts by only small changes in primary structure. The three-dimensional structure of such enzymes is often the same as those derived from mesophilics [41] but generally they possess less asparagine residues and more salt- or disulfide bridges.

1.4.1 Mechanistic Aspects

Among the numerous theories and rationales which have been developed in order to understand enzyme catalysis, the most illustrative models for the organic chemist are briefly discussed here [42-44].

'Lock and Key' Mechanism
The first proposal for a general mechanism of enzymatic action was developed by E. Fischer in 1894 [45]. It assumes that an enzyme and its substrate mechanistically interact like a lock and a key, respectively. Although this assumption was quite sophisticated at that time, it assumes a completely rigid enzyme structure.

Figure 1.1. Schematic representation of the Lock and Key mechanism.

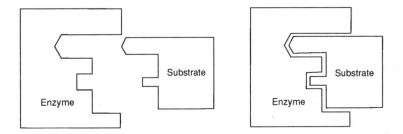

Thus, it cannot explain why many enzymes do act on large substrates, while they are inactive on smaller similar counterparts. Given Fischer's rationale, small substrates should be transformed at higher rates than larger substrates since the access to the active site would be easier. Furthermore, the hypothesis cannot explain, why many enzymes can convert not only their natural substrates but also numerous non-natural compounds possessing different structural features. Thus, a more sophisticated model had to be developed.

Induced-Fit Mechanism
This rationale, which takes into account that enzymes are not entirely rigid, was developed by Koshland jr. in the late 1960s [46]. It assumes, that upon

approach of a substrate during the formation of the enzyme-substrate complex the enzyme can change its conformation under the influence of the substrate structure so as to wrap around its guest. A similar picture is given by the interaction of a hand (the substrate) and a glove (the enzyme). This model can indeed explain why in many cases several structural features on a substrate are required. These structural features may be located quite a distance from the actual site of the reaction. The most typical 'induced-fit' enzymes are the lipases. They can convert an amazing large variety of artificial substrates which possess structures which do not have much in common with the natural substrates, triglycerides.

Figure 1.2. Schematic representation of the Induced-Fit mechanism.

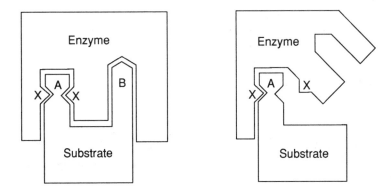

A represents the reactive group of the substrate, X is the respective reactive group(s) of the enzyme, the 'chemical operator'. Substrate part B forces the enzyme to adapt a different (active) conformation. Both of the 'active' groups X of the enzyme - the 'chemical operators' - are then positioned in the right way to effect catalysis. If part B is missing, no conformational change (the 'induced fit') takes place and thus the chemical operators stay in their inactive state.

Desolvation and Solvation-Substitution Theory

Quite recently, M.J.S. Dewar developed a different rationale [47] in attempting to explain the high conversion rates of enzymatic reactions, which are often substantially faster than the chemically-catalysed equivalent processes [48]. The theory (called the 'desolvation-theory') assumes, that the kinetics of enzyme reactions have much in common with those of gas-phase reactions. If a substrate enters the active site of the enzyme, it replaces all of the water molecules from the active site of the enzyme. Then, a formal gas phase reaction can take place which mimics two reaction partners interacting without disturbing solvent. In solution, the water molecules impede the approach of the partners, hence the reaction rate is reduced. This theory would inter alia explain why small substrate molecules are often more slowly converted than larger analogues, since the former would not be able to replace all the water molecules from the active site. However, there is still much debate about this theory.

This 'desolvation'-theory has recently been substituted by a 'solvation-substitution' theory [49]. It is based on the assumption that the enzyme would not be able to strip off the water which is surrounding the substrate to effect a 'desolvation', which would be energetically unfavoured. Instead, the solvent is displaced by another polar environment by a so-called 'solvation substitution'. Thus the (often) hydrophobic substrates are replacing the water in the (often) hydrophobic site of the enzyme.

In any case it is clear that a maximum change in entropy is only obtained upon a 'tight and close fit' of a substrate into the pocket of an active site of an enzyme [50].

Three-Point Attachment Rule

This widely used rationale to explain the enantioselectivity of enzymes was suggested by A. G. Ogston [51]. To get a high degree of enantioselection, a substrate must be held firmly in three dimensional space. Therefore, there must be at least three different points of attachment of the substrate onto the active site [52].

Scheme 1.2. Examples for central, axial and planar chirality.

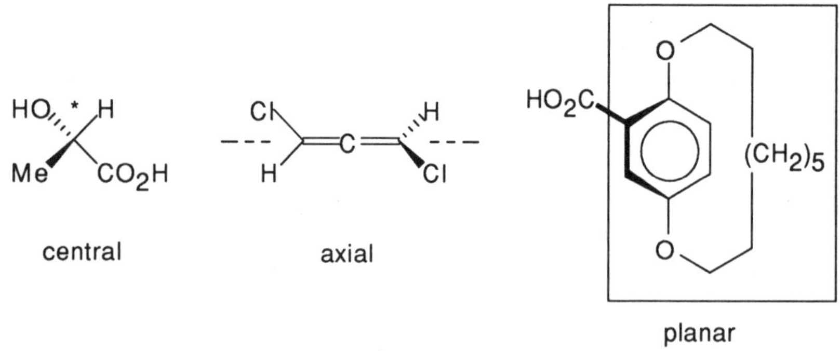

This is exemplified for the discrimination of the enantiomers of a racemic substrate (A and B, see below) with its chirality located on a sp^3-carbon atom.

1.4 Enzyme Properties and Nomenclature

For compounds possessing an axial or planar chirality involving sp^2- or sp-carbon atoms, respectively, analogous pictures can be created.

Figure 1.3. Schematic representation of enzymatic enantiomer discrimination.

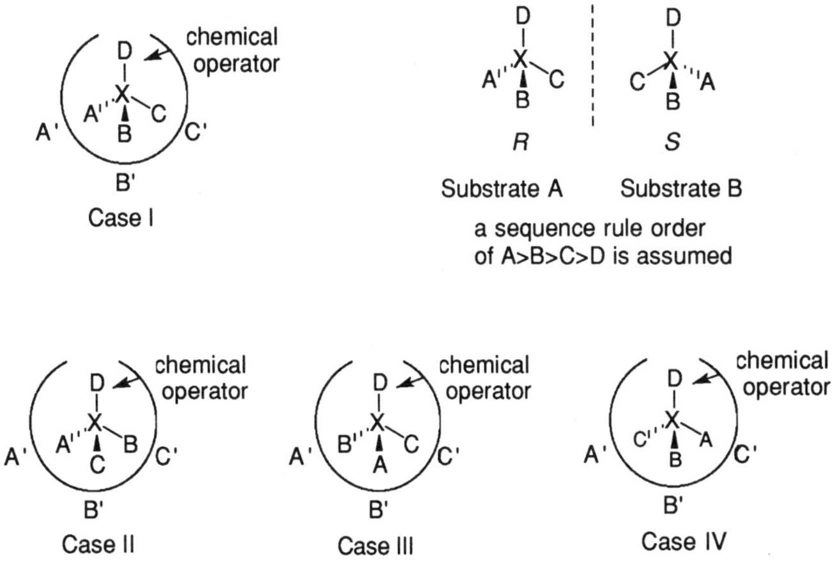

Case I

Substrate A (R)   Substrate B (S)
a sequence rule order of A>B>C>D is assumed

Case II   Case III   Case IV

Case I: Enantiomer A is a good substrate by allowing an optimal interaction of its groups (A,B,C) with their complementary binding site areas of the enzyme (A', B', C'). It ensures an optimal orientation of the reactive group (D) towards the chemical operator which is required for a successful transformation.

Cases II through IV: Enantiomer B is a poor substrate because optimal binding and orientation of the reactive group D is not possible. Thus poor catalysis will be observed.

If a prochiral substrate (C), bearing two chemically identical but stereochemically different enantiotopic groups (A), is involved the same model can be applied to rationalize the favoured transformation of one of the two leading to an 'enantiotopos differentiation'.

Scheme 1.3. Enantiotopos and -face nomenclature.

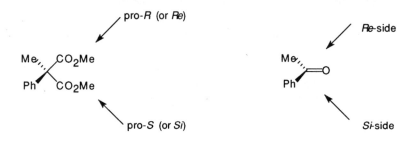

Figure 1.4. Schematic representation of enzymatic enantiotopos discrimination.

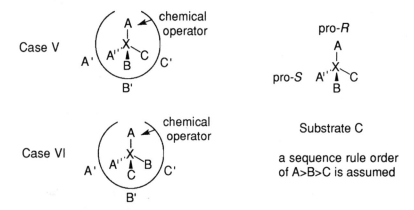

Case V shows good binding of a prochiral substrate (C) to the complementary enzyme's binding sites with the pro-(R) group out of the two reactive groups being positioned to the chemical operator.

Case VI: Positioning of the pro-(S) reactive group towards the chemical operator results in poor orientation of the other functions to their complementary sites, resulting in poor catalysis. As a consequence, the pro-(R) group is cleaved preferentially to its pro-(S) counterpart.

The ability of enzymes, to distinguish between two enantiomeric faces of a prochiral substrate (D) - an 'enantioface differentiation' - is illustrated in Figure 1.5.

1.4 Enzyme Properties and Nomenclature

Figure 1.5. Schematic representation of enzymatic enantioface discrimination.

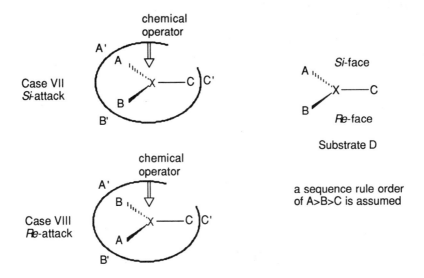

Case VII: An optimal match between the functional groups of substrate D leads to an attack of the chemical operator to the central atom X from the *Si*-side.

Case VIII: The mirror image orientation (also called 'alternative fit') of substrate D in the active site of the enzyme leads to a mismatch in the binding of the functional groups, thus an attack of the chemical operator, which would come from the *Re*-side in this case, is unfavoured.

Generally, many functional groups - and sometimes also coordinated metal ions - have to work together in the active site of the enzyme to effect catalysis. Individual enzyme mechanisms have been elucidated in certain cases where the exact three-dimensional structure is known. For most of the other enzymes used for the biotransformation of non-natural organic compounds, assumptions are made about their molecular action.

Kinetic Reasons for Selectivity

As in every other catalytic reaction, an enzyme (E) accelerates the reaction by lowering the energy barrier between substrate (S) and product (P) - the activation energy (E_a). The origin of this catalytic power - the rate acceleration - has generally been attributed to transition-state stabilisation of the reaction by the enzyme [53], assuming that the catalyst binds more strongly to the transition state than to the ground state of the substrate, by a factor approximately equal to the rate acceleration [54].

Figure 1.6. Energy diagram of catalysed vs. uncatalysed reaction.

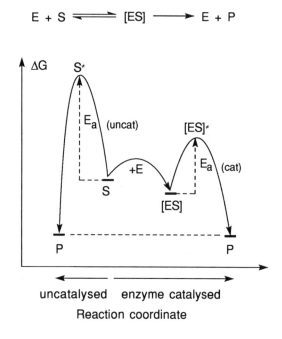

S = substrate, P = product, E = enzyme, [ES] = enzyme-substrate complex, ≠ denotes a transition state, E_a = activation energy.

Virtually all stereoselectivities of enzymes originate from the energy difference in the enzyme-transition-state complex [ES]≠. For instance, in an enantioselective reaction both of the enantiomeric substrates (A and B, see Figure 1.3) or the two forms of mirror-image orientation of a prochiral substrate involving its enantiotopic groups or -faces (see Figures 1.4 and 1.5) compete for the active site of the enzyme. Due to the chiral environment of the active site of the enzyme, *diastereomeric* enzyme-substrate complexes [EA] and [EB] are formed, which possess different values of free energy (ΔG) for their respective transition states [EA]≠ and [EB]≠. The result is a difference in activation energy for both of the enantiomeric substrates or the 'enantiomeric orientations', respectively, and as a consequence, one enantiomer (or orientation) will be transformed faster than the other.

1.4 Enzyme Properties and Nomenclature

Figure 1.7. Energy diagram for an enzyme-catalysed enantioselective reaction.

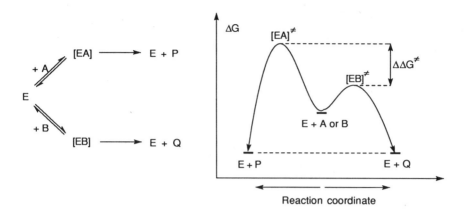

E = enzyme, A and B = enantiomeric substrates, P and Q = enantiomeric products, [EA] and [EB] = enzyme-substrate complexes, $\neq$ denotes a transition state, $\Delta\Delta G^{\neq}$ = difference in free energy.

The value of this difference in free energy, expressed as $\Delta\Delta G^{\neq}$, is a direct measure for the selectivity of the reaction which in turn governs the ultimate ratio of the enantiomers P and Q, i.e. the optical purity of the product. Some representative values of enantiomeric excess of product (e.e.) corresponding to a given $\Delta\Delta G^{\neq}$ of the reaction are presented in Table 1.2.

Table 1.2. Energy values $\Delta\Delta G^{\neq}$ for representative optical purities of product.

$\Delta\Delta G^{\neq}$ (kcal/mol)	e.e. (%)
0.118	10
0.651	50
1.74	90
2.17	95
3.14	99
4.50	99.9

$$\text{e.e. (\%)} = \frac{P - Q}{P + Q} \times 100$$

1.4.2 Classification and Nomenclature

At present more than 2100 enzymes have been recognized by the International Union of Biochemistry [55] and if the speculation that there are about 25 000 enzymes existing in nature is true [56], about 90% of this vast reservoir of biocatalysts remains still to be discovered and used. However, only a minor fraction of the enzymes already investigated (roughly 300, ~15%) is commercially available.

For its identification every enzyme has got a 4-digit number EC A.B.C.D where EC stands for 'Enzyme Commission' with the following properties encoded:

A   denotes the main type of reaction (see Table 1.3).
B   stands for the subtype, indicating the substrate type or the type of transferred molecule.
C   indicates mainly the co-substrate allocation.
D   is the individual enzyme number.

As depicted in Table 1.3, enzymes have been classified into six categories according to the type of reaction they can catalyse. On a first glimpse it would be advantageous, to keep this classification throughout this book, since organic chemists are used to think in reaction principles. Unfortunately, this does not work in practice for the following reasons:

The importance of practical applications for organic synthesis is not at all evenly distributed amongst the different enzyme classes, as may be seen from the 'utility' column shown below. Furthermore, due to the widespread use of crude enzyme preparations (consisting of more than one active biocatalyst), often one does not know which enzyme is actually responsible for the biotransformation. Last but not least, there are many useful reactions which are performed with whole microbial cells, where it only can be speculated as to which of the numerous enzymes in the cell is actually involved in the transformation. An alternative classification of enzymes according the the ease they can be used with in biotransformations has been proposed recently [57].

Another warning should be given concerning catalytic activities which are measured in several different systems. The standard unit system is the 'International Unit' (1 I.U. = 1 µmol of substrate transformed per min) but other units such as nmol/min or nmol/hour are also common. Another system of units is based on the katal (1 kat = 1 mol s^{-1} of substrate transformed) but it has not yet been widely accepted. The activities which are observed when non-natural organic compounds are used as substrates are often significantly below the values which could be expected with natural substrates.

1.4 Enzyme Properties and Nomenclature

Table 1.3. Classification of enzymes

Enzyme class	Number classified	available	Reaction type	Utility[§]
1. Oxidoreductases	650	90	Oxidation-reduction: oxygenation of C-H, C-C, C=C, C=O bonds, or overall removal or addition of hydrogen atom equivalents.	++ 25%
2. Transferases	720	90	Transfer of groups: aldehydic, ketonic, acyl, sugar, phosphoryl or methyl.	+ <5%
3. Hydrolases	636	125	Hydrolysis-formation of esters, amides, epoxides, nitriles, anhydrides.	+++ 65%
4. Lyases	255	35	Addition-elimination of small molecules on C=C, C=N, C=O bonds.	+ <5%
5. Isomerases	120	6	Isomerisations such as racemisation, epimerisation.	± <5%
6. Ligases	80	5	Formation-cleavage of C-O, C-S, C-N, C-C bonds with concomitant triphosphate cleavage	± <5%

[§] The estimated 'utility' of an enzyme class for the transformation of non-natural substrates ranges from +++ (very useful) to ± (little use). The values (%) indicate the percentage of research performed with enzymes from a given class for the 1987-88 period [58].

1.4.3 Coenzymes

A remarkable fraction of interesting enzyme-catalysed reactions require cofactors (coenzymes) [59]. These are compounds of relatively low molecular weight compared to the enzyme (few hundred Da, in contrast to the general range of 15 000 to 1 000 000 for enzymes) which provide either 'chemical reagents' such as redox- (hydrogen, oxygen, electrons) and carbon units, or 'chemical energy' stored in energy-rich functional groups, such as acid anhydrides, etc. Some of the cofactors are gradually destroyed due to undesired side-reactions occurring in the medium. These cofactors are too expensive to be used in the stoichiometric amounts formally required. Accordingly, when coenzyme-dependent enzymes are employed, the corresponding coenzymes are used in catalytic amounts in conjunction with an efficient and inexpensive system for their regeneration in situ. Some methods for cofactor-recycling are already well developed, but others are still problematic, as depicted in Table 1.4.

Some coenzymes are tightly bound to their respective enzymes such that external recycling is not required. Many enzymes require coordinated metals such as Fe, Ni, Cu, Zn, Mg or Mn. In nearly all cases, chemists do not have to worry about these metals since they are tightly bound to the enzyme.

Table 1.4. Common coenzymes required for biotransformations.

Coenzyme	Reaction type	Recycling[a]
NAD$^+$/NADH	removal or addition of	+ [++]
NADP$^+$/NADPH	hydrogen	+ [+]
ATP[b]	phosphorylation	+ [++]
SAM	C_1-alkylation	+ [±]
Acetyl-CoA	C_2-alkylation	+ [±]
Flavines[c]	oxygenation	-
Pyridoxal-phosphate	transamination	-
Biotin	carboxylation	-
Metal-porphyrin complexes	peroxidation	-

[a] Recycling of a cofactor is necessary (+) or not required (-), the feasibility of which is indicated in square brackets as 'feasible' [++] to 'complicated' [±].
[b] For other triphosphates such as GTP, CTP and UTP the situation is similar.
[c] Many flavine-dependent mono- or dioxygenases require NADPH as an indirect reducing agent.

1.4.4 Enzyme Sources

The large majority of enzymes used for biotransformations in organic chemistry are employed in a crude form and are thus affordable. The preparations typically contain only about 1 - 30% of actual enzyme, the remainder being inactive proteins, buffer salts or carbohydrates from the fermentation broth from which they have been isolated. Interestingly, crude preparations are often more stable than purified enzymes.

The main sources for enzymes are as follows:

- The detergent industry produces many proteases in huge amounts. These are largely used as additives for detergents to effect hydrolysis of protein.
- The food industry uses proteases for meat and cheese processing and numerous lipases for the amelioration of fats and oils [60].

- Numerous enzymes can be isolated from slaughter waste or cheap mammalian organs such as kidney or liver. A minor fraction of enzymes is obtained from plant sources.
- Pure enzymes are usually very expensive and thus are mostly sold by the unit, while crude preparations are often shipped by the kg. Since the techniques for protein purification are becoming easier, thus making their isolation more economically feasible, the use of pure enzymes in biotransformations is steadily increasing.

References

1. Goodhue CT (1982) Microb. Transform. Bioact. Compd. 1: 9
2. For a list of enzyme suppliers see the appendix
3. Baross JA, Deming JW (1983) Nature 303: 423
4. Laane C, Boeren S, Vos K, Veeger C (1987) Biotechnol. Bioeng. 30: 81
5. Sih CJ, Abushanab E, Jones JB (1977) Ann. Rep. Med. Chem. 12: 298
6. Ohno M, Otsuka M (1989) Org. Reactions 37: 1
7. Schmidt-Kastner G, Egerer P (1984) Amino Acids and Peptides. In: Kieslich K (ed) Biotechnology, Verlag Chemie, Weinheim, volume 6a, pp 387-419
8. Gutman AL, Zuobi K, Guibe-Jampel E (1990) Tetrahedron Lett. 2037
9. Yamamoto Y, Yamamoto K, Nishioka T, Oda J (1988) Agric. Biol. Chem. 52: 3087
10. Weijers CAGM, de Haan A, de Bont JAM (1988) Microbiol. Sci. 5: 156
11. Nagasawa T, Yamada H (1989) Trends Biotechnol. 7: 153
12. Mansuy D, Battoni P (1989) Alkane Functionalization by Cytochromes P-450 and by Model Systems Using O_2 or H_2O_2. In: Hill CL (ed) Activation and Functionalization of Alkanes, Wiley, New York
13. May SW (1979) Enzyme Microb. Technol. 1: 15
14. Boyd DR, Dorrity MRJ, Hand MV, Malone JF, Sharma ND, Dalton H, Gray DJ, Sheldrake GN (1991) J. Am. Chem. Soc. 113: 667
15. Lemiere GL, Lepoivre JA, Alderweireldt FC (1985) Tetrahedron Lett. 4527
16. Walsh CT, Chen Y-C J (1988) Angew. Chem., Int. Ed. Engl. 27: 333
17. Servi S (1990) Synthesis 1
18. Phillips RS, May SW (1981) Enzyme Microb. Technol. 3: 9
19. Findeis MH, Whitesides GM (1987) J. Org. Chem. 52: 2838
20. Akhtar M, Botting NB, Cohen MA, Gani D (1987) Tetrahedron 43: 5899
21. Effenberger F, Ziegler Th (1987) Angew. Chem., Int. Ed. Engl. 26: 458
22. Morrison M, Schonbaum GR (1976) Ann. Rev. Biochem. 45: 861
23. Buist PH, Dimnik GP (1986) Tetrahedron Lett. 1457
24. Schwab JM, Henderson BS (1990) Chem. Rev. 90: 1203
25. Fuganti C, Grasselli P (1988) Baker's Yeast Mediated Synthesis of Natural Products. In: Whitaker JR, Sonnet PE (eds) Biocatalysis in Agricultural Biotechnology, ACS Symposium Series, volume 389, pp 359-370
26. Toone EJ, Simon ES, Bednarski MD, Whitesides GM (1989) Tetrahedron 45: 5365
27. Kitazume T, Ikeya T, Murata K (1986) J. Chem. Soc., Chem. Commun. 1331
28. Sweers HM, Wong C-H (1986) J. Am. Chem. Soc. 108: 6421
29. Chen C-S, Sih CJ (1989) Angew. Chem., Int. Ed. Engl. 28: 695
30. Ariens EJ (1988) Stereospecificity of Bioactive Agents. In: Ariens EJ, van Rensen JJS, Welling W (eds) Stereoselectivity of Pesticides, Elsevier, Amsterdam, pp 39-108
31. Morrison JD (1983) A summary of ways to obtain optically active compounds, Morrison JD (ed) in Asymmetric Synthesis, vol 1, p 1, Academic Press, Orlando
32. Morrison JD (ed) (1985) Chiral Catalysis. In: Asymmetric Synthesis, volume 5, Academic Press, London

33. Hanessian S (1983) Total Synthesis of Natural Products: the 'Chiron' Approach, Pergamon Press, Oxford
34. Scott JW (1984) Readily Available Chiral Carbon Fragments and their Use in Synthesis. In: Morrison JD, Scott JW (eds) Asymmetric Synthesis, Academic Press, New York, volume 4, pp 1-226
35. Pham VT, Phillips RS, Ljungdahl LG (1989) J. Am. Chem. Soc. 111: 1935
36. Schuster M, Aaviksaar A, Jakubke H-D (1990) Tetrahedron 46: 8093
37. Klibanov AM (1990) Acc. Chem. Res. 23: 114
38. Anfinsen CB (1973) Science 181: 223
39. Cooke R, Kuntz ID (1974) Ann. Rev. Biophys. Bioeng. 3: 95
40. Ahern TJ, Klibanov AM (1985) Science 228: 1280
41. Mozhaev VV, Martinek K (1984) Enzyme Microb. Technol. 6: 50
42. Jencks WP (1969) Catalysis in Chemistry and Enzymology, McGraw-Hill, New York
43. Fersht A (1985) Enzyme Structure and Mechanism, 2nd edition, Freeman, New York
44. Walsh C (ed) (1979) Enzymatic Reaction Mechanism, Freeman, San Francisco
45. Fischer E (1894) Ber. dtsch. chem. Ges. 24: 2683
46. Koshland DE, Neet KE (1968) Ann. Rev. Biochem. 37: 359
47. Dewar MJS (1986) Enzyme 36: 8
48. A 'record' of rate acceleration factor of 10^{14} has been reported. See: Lipscomb WN (1982) Acc. Chem. Res. 15: 232
49. Warshel A, Aqvist J, Creighton S (1989) Proc. Natl. Acad. Sci. 86: 5820
50. Johnson LN (1984) Inclusion Compds. 3: 509
51. Ogston AG (1948) Nature 162: 963
52. The following rationale was adapted from: Jones JB (1976) Biochemical Systems in Organic Chemistry: Concepts, Principles and Opportunities. In: Jones JB, Sih CJ, Perlman D (eds) Applications of Biochemical Systems in Organic Chemistry, part I, Wiley, New York, pp 1-46
53. Kraut J (1988) Science 242: 533
54. Wong C-H (1989) Science 244: 1145
55. International Union of Biochemistry (1979) Enzyme Nomenclature, Academic Press, New York
56. Kindel S (1981) Technology 1: 62
57. Whitesides GM, Wong C-H (1983) Aldrichimica Acta 16: 27
58. Crout DHG, Christen M (1989) Biotransformations in Organic Synthesis. In: Scheffold R (ed) Modern Synthetic Methods, volume 5, pp 1-114
59. A 'cofactor' is tightly bound to an enzyme (*e.g.* FAD), whereas a 'coenzyme' can dissociate into the medium (*e.g.* NADH). In practice, however, this distinction is not always made in a consequent manner
60. Spradlin JE (1989) Tailoring enzymes for food processing, Whitaker JR, Sonnet PE (eds) ACS Symposium Series, vol 389, p 24, J. Am. Chem. Soc., Washington

2 Biocatalytic Applications

2.1 Hydrolytic Reactions

Among all the types of enzyme-catalysed reactions, hydrolytic transformations involving amide- and ester-bonds are most easy to perform by using proteases, esterases or lipases. A lack of sensitive cofactors which would have to be recycled, and a large number of readily available enzymes possessing relaxed substrate specificities to choose from are the main features which made hydrolases the favourite class of enzyme for organic chemists during the past decade. About $2/3$ of the total research in the field of biotransformations has been performed using hydrolytic enzymes of this type. The reversal of the reaction, giving rise to ester- or amide-*synthesis* has been particularly well investigated using enzymes in solvent systems of low water activity. The special methodologies involved in this latter type of reaction are described in Section 3.1.

Other types of application of hydrolase enzymes involving the formation and/or cleavage of phosphate esters, nitriles and epoxides are generally more complicated to perform and are therefore described in a separate section. In contrast to the group of proteases, esterases and lipases, they have caused less impact in organic chemistry, although their synthetic potential should not be underestimated.

2.1.1 Mechanistic and Kinetic Aspects

The mechanism of amide- and ester-hydrolysing enzymes is very similar to that describing chemical hydrolysis by base. A nucleophilic group from the active site of the enzyme attacks the carbonyl group of the substrate ester or amide. This nucleophilic 'chemical operator' can be either an hydroxy group of a serine (e.g. pig liver esterase, subtilisin and lipases from porcine pancreas or *Mucor* sp.), a carboxy group of an aspartic acid (e.g. pepsin) or a thiol-functionality of a cysteine (e.g. papain) [1].

The mechanism which has been elucidated in detail is that of the serine-hydrolases [2] (see Scheme 2.1). Two additional groups (Asp and His) located together with the serine residue (which is the actual reacting chemical operator in the active site) form the so called 'catalytic triade' [3]. The special arrangement of these three groups effects a decrease of the pK-value of the serine hydroxy group thus enabling it to perform a nucleophilic attack on the carbonyl group of the substrate R^1-CO-OR^2 (step 1). Thus the acyl moiety of the substrate is covalently linked onto the enzyme, forming the 'acyl-enzyme intermediate' by liberating the leaving group (R^2-OH). Then, a nucleophile (Nu) - usually water - can in turn attack the acyl-enzyme intermediate, regenerating the enzyme and releasing a carboxylic acid R^1-CO-OH (step 2).

Scheme 2.1. The serine hydrolase mechanism.

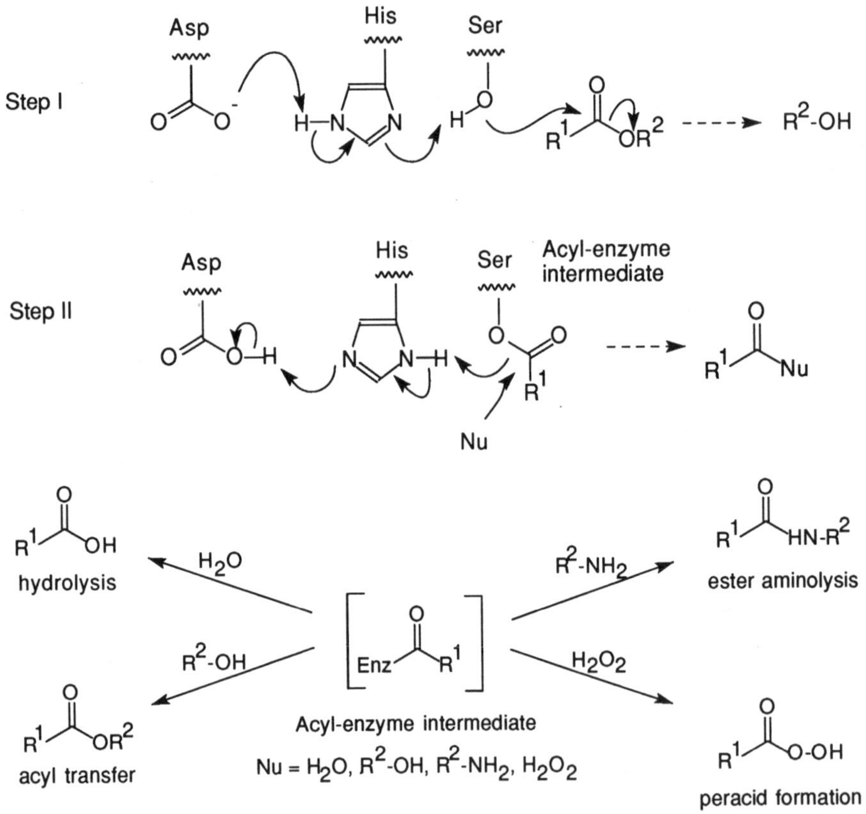

When the enzyme is operating in an environment of low water activity - in other words at low water concentrations - any other nucleophile can compete

with the water for the acyl-enzyme intermediate thus leading to a number of synthetically useful transformations:
- attack of another alcohol R^2-OH leads to another ester R^1-CO-OR^2, an interesterification reaction, called enzymatic 'acyl transfer' [4],
- an incoming amine R^2-NH_2 results in the formation of an amide R^1-CO-NH-R^2, yielding an enzymatic aminolysis of esters [5,6], and
- peracids of type R^1-CO-OOH are formed when hydrogen peroxide is acting as the nucleophile [7].
- Hydrazinolysis provides access to hydrazides [8, 9], and the action of hydroxylamine results in the formation of hydroxamic acid derivatives [10]. Both of the latter transformations, however, have not created a significant impact.

During the course of all of these reactions, any type of chirality in the substrate is 'recognized' by the enzyme and this causes a preference for one of the two possible stereochemical directions in a reaction. The value of this discrimination is a crucial parameter since it stands for the 'selectivity' of the reaction. In turn this is governed by the reaction kinetics. It should be noted, that the following section is not an elaborate discussion on enzyme kinetics but a short summary of the most important conclusions which are the necessary considerations for an optimisation of stereoselective enzymatic transformations.

Since representative examples for the different types of chiral recognition can be found within the class of enzymes called hydrolases, the underlying principles of this recognition process and the corresponding kinetic implications are discussed here [11]. However, some of these types of transformations can be found within other groups of enzymes as well, and the corresponding rules can be applied.

Enantioface Differentiation

Hydrolases can distinguish between the two enantiomeric faces of achiral substrates such as enol esters possessing a plane of symmetry within the molecule. The attack of the enzyme´s nucleophilic chemical operator preferably occurs from one side, leading to a side-directed enolisation of the unstable free enol towards one preferred side within the chiral environment of the enzyme´s active site [12, 13]. During the course of the reaction a new center of chirality is created in the product.

Scheme 2.2. Enantioface differentiation.

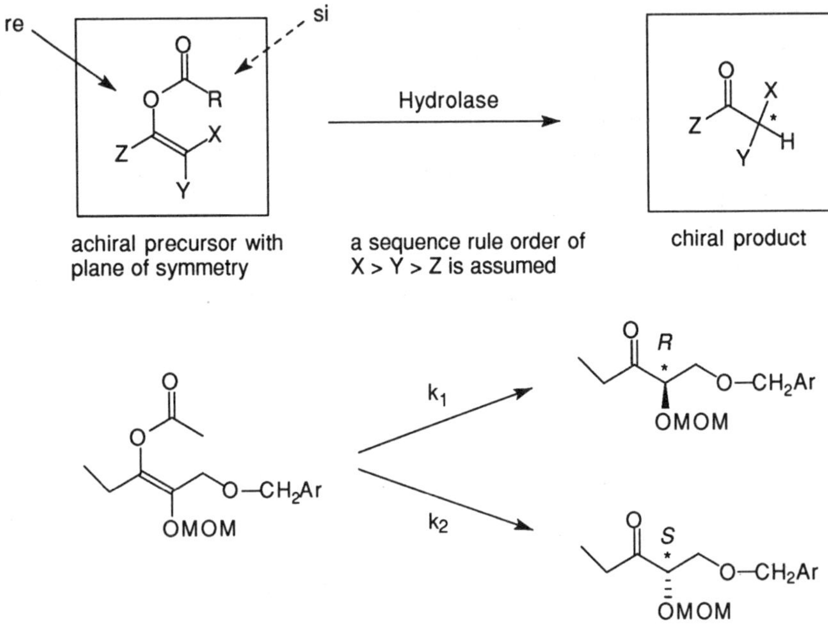

achiral precursor with plane of symmetry → a sequence rule order of X > Y > Z is assumed → chiral product

Enantiotopos Differentiation

If prochiral substrates possessing two chemically identical but enantiotopic reactive groups X designated as pro-R and pro-S are subjected to an enzymatic transformation such as hydrolysis, a chiral discrimination between them occurs during the transformation of group X into Y, thus leading to a chiral product (see Scheme 2.3). During the course of the reaction the plane of symmetry within the substrate is broken. The single-step asymmetric hydrolysis of a prochiral α,α-disubstituted malonic diester by pig liver esterase or α-chymotrypsin is a representative example [14]. Here, the reaction terminates at the monoester stage since highly polar compounds of such type are generally not accepted by hydrolases [15].

On the other hand, when the substrate is a diacetate, the resulting monoesters usually undergo further cleavage in a second step to yield an achiral diol [16].

2.1 Hydrolytic Reactions

Scheme 2.3. Enantiotopos differentiation (prochiral substrates).

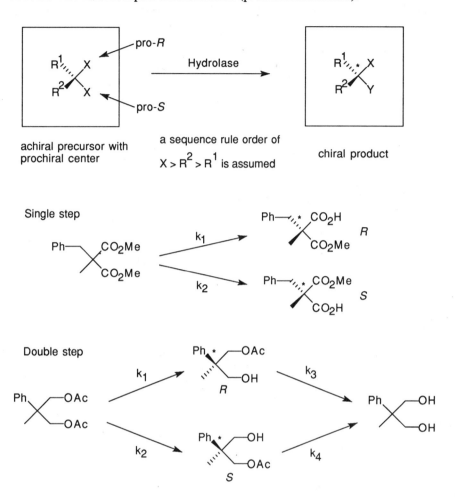

Similarly, the two chemically identical groups X, positioned on carbon atoms of opposite configuration of a *meso*-substrate (R, S), can react at different rates in a hydrolase-catalysed reaction. As a result, the optically inactive substrate is transformed into an optically active product due to the transformation of one of the reactive groups from X into Y going in hand with the destruction of the plane of symmetry within the substrate. Numerous open-chain or cyclic *cis-meso*-diesters have been transformed into chiral monoesters by this technique [17]. Again, with dicarboxylates the reaction usually stops after the first step, whereas two hydrolytic steps are observed with diacetoxy esters [18]. In contrast to the kinetic resolution of a racemate, the theoretical yield of these conversions is always 100%.

Scheme 2.4. Asymmetrisation of *meso*-substrates.

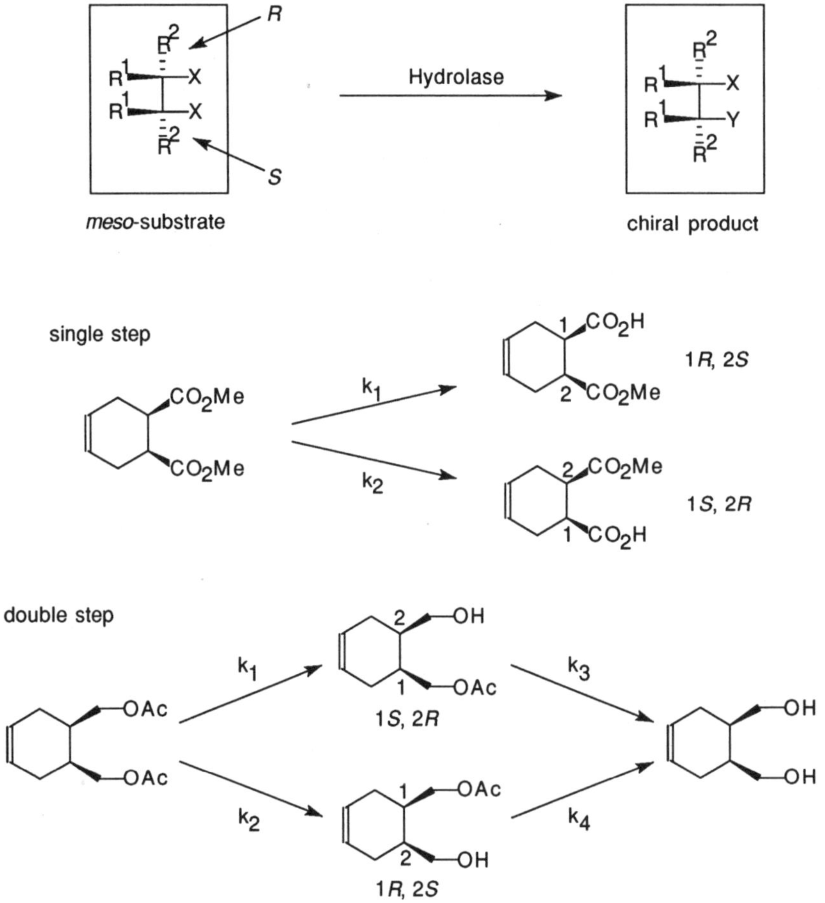

If required, the interconversion of a given product into its enantiomer can be achieved by a simple two-step protection-deprotection sequence. Thus, regardless of the enantiomeric preference of the enzyme which is used to perform the asymmetrisation of the substrate, both enantiomers of the product are available and no 'unwanted' enantiomer is produced. This technique has no generally applicable counterpart in conventional organic chemistry and is often referred to as the 'meso-trick'.

Since hydrolytic reactions are performed in an aqueous environment, they are completely irreversible. The kinetics of all of the single-step reactions shown above is very simple [19]: a prochiral or a *meso*-substrate S is transformed into two enantiomeric products P and Q at different rates

2.1 Hydrolytic Reactions

determined by the apparent first-order rate constants k_1 and k_2, respectively. The selectivity of the reaction α is only governed by the ratio of k_1/k_2, which remains constant throughout the reaction. As a consequence, the optical purity of the product (e.e.) is not dependent on the extent of the conversion. The selectivity observed in such a reaction can only be improved therefore by changing the 'environment' of the system (e.g. substrate modification, choice of another enzyme, the addition of organic cosolvents, and variations in temperature or pH) but not by stopping the reaction at various rates of conversion. For a more detailed presentation of different techniques to improve the selectivity of enzymatic reactions by variations in the 'environment' see below.

Figure 2.1. Single-step kinetics.

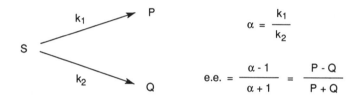

On the other hand, in some cases a second successive reaction step cannot be avoided with bifunctional prochiral or *meso*-diesters as shown above (see Schemes 2.3 and 2.4). For such types of substrates the reaction does not terminate at the chiral monoester stage to give products P and Q (step 1), but rather proceeds via a second step at usually slower reaction rates to yield achiral products (R) again. Here, the reaction kinetics become more complicated.

As depicted in Figure 2.2, the ratio of P and Q - in other words the optical purity of the product (e.e.) - depends now on all four of the rate constants k_1, k_2, k_3 and k_4, since the second hydrolytic step cannot be neglected as in the former case. From the fact that enzymes usually show a continuous preference for reactive groups with the same chirality, one may conclude that if S is transformed more quickly into P, Q will be hydrolysed faster (into diol R) than P. Thus, the rate constants governing the selectivity of the reaction are in an order of $k_1 > k_2$ and $k_4 > k_3$. As an important consequence, the optical purity of the product hemiester (e.e.) becomes a function of the conversion of the reaction, and generally follows a bell-shaped curve. In the early stages of the reaction the optical purity of the product is mainly determined by the selectivity of the first reaction step, which constitutes an enantiotopos or -face differentiation depending on the type of substrate. As the reaction proceeds, the second hydrolytic step, which actually constitutes per se a kinetic resolution, starts to take place to a more significant extent, and its 'opposite' selectivity compared to that of the first step (remember that $k_1 > k_2$, $k_4 > k_3$) leads to an enhancement of optical purity. After having reached a maximum at a certain conversion the product concentration (P+Q) finally drops off again, when most of the substrate S is consumed and the second hydrolytic step to form R constitutes the main reaction. The same analogous considerations are pertinent for a reversed situation - an esterification reaction.

Figure 2.2. Double-step kinetics.

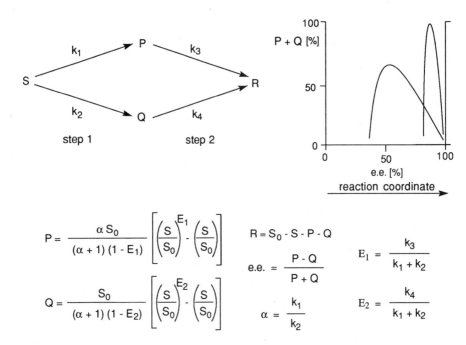

$$P = \frac{\alpha S_0}{(\alpha + 1)(1 - E_1)} \left[\left(\frac{S}{S_0}\right)^{E_1} - \left(\frac{S}{S_0}\right) \right]$$

$$Q = \frac{S_0}{(\alpha + 1)(1 - E_2)} \left[\left(\frac{S}{S_0}\right)^{E_2} - \left(\frac{S}{S_0}\right) \right]$$

$$R = S_0 - S - P - Q$$

$$e.e. = \frac{P - Q}{P + Q}$$

$$\alpha = \frac{k_1}{k_2}$$

$$E_1 = \frac{k_3}{k_1 + k_2}$$

$$E_2 = \frac{k_4}{k_1 + k_2}$$

Besides trial-and-error experiments, i.e. by stopping such a reaction at varous intervals and checking the optical purity of the product, the e.e.-conversion dependence may also be calculated. Determination of the amounts of substrate S and hemiester P and Q and the optical purities at various intervals can be used to determine the kinetic constants α, E_1 and E_2 for a given reaction by using the equations listed above. Thus, the enantiomeric excess of the monoester may be predicted as a function of its percentage present in the reaction mixture. The validity of this method has been verified by the asymmetrisation of a prochiral *meso*-diacetate using pig liver esterase (PLE) and porcine pancreatic lipase (PPL) as shown below [20].

2.1 Hydrolytic Reactions

Scheme 2.5. Asymmetrisation of a *meso*-diacetate.

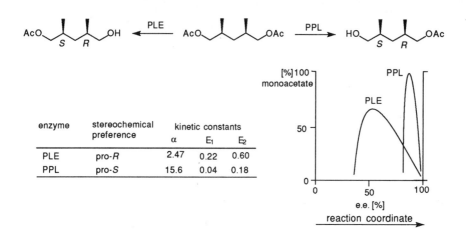

enzyme	stereochemical preference	kinetic constants		
		α	E_1	E_2
PLE	pro-R	2.47	0.22	0.60
PPL	pro-S	15.6	0.04	0.18

Enantiomer Differentiation

When a racemic substrate is subjected to enzymatic hydrolysis, chiral discrimination of the enantiomers occurs [21]. It should be noted that the chirality does not necessarily have to be of a central type, but also can be axial or planar to be 'recognized' by enzymes. Due to the chirality of the active site of the enzyme, one enantiomer fits better into the active site than its counterpart and is therefore converted at a higher rate, resulting in a kinetic resolution of the racemate. The vast majority of enzymatic transformations occur in a stereoselective manner and interestingly, this particular synthetic potential of hydrolytic enzymes was realized as early as 1903 [22]!

Of course, in contrast to the above-mentioned types of stereoselective transformations showing a theoretical yield of 100%, each of the enantiomers from a kinetic resolution of a racemate can be obtained in only 50% yield. To overcome the occurrence of the undesired 'wrong' enantiomer, two strategies are possible:

- The unwanted enantiomer may be racemised after its recovery from the resolution procedure thus making recycling a possibility or
- the resolution is carried out under conditions, under which the enantiomers of the substrate are in a rapid equilibrium but those of the product are not. Thus, as the 'well-fitting' substrate-enantiomer is depleted by the enzyme, the equilibrium provides more of it by racemising the 'bad-fitting' counterpart. It is understandable that the above mentioned criteria are very difficult to be met experimentally and hence examples of this type of biotransformation are rare [23, 24].

Scheme 2.6. Enantiomer differentiation.

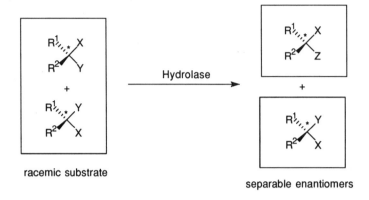

racemic substrate

separable enantiomers

In some ideal cases the difference in the reaction rates of both enantiomers is so extreme, that the 'good' enantiomer is transformed quickly and the other is not converted at all. Then, the enzymatic reaction will cease automatically at 50% conversion, when there is nothing left of the more reactive enantiomer [25].

In practice, however, most cases of enzymatic resolution of a racemic substrate do not show this ideal situation, i.e. in which one enantiomer is rapidly converted and the other not at all. The difference in - or better: the ratio of - the rates of conversion of the enantiomers is not infinite, but measurable. The thermodynamic reasons for this have been discussed in Chapter 1. What one observes in these cases, is not a complete standstill of the reaction at 50% conversion but a marked decrease in reaction rate at around this point. In these numerous cases one encounters some crucial dependencies:
- The velocity of the transformation of each enantiomer varies with the degree of conversion, since the ratio of the two enantiomers does not remain constant during the reaction.
- Therefore, the optical purity of both substrate (e.e.$_S$) and product (e.e.$_P$) is a function of the extent of conversion.

An very useful treatment of the kinetics of enzymatic resolution, describing the dependency of the conversion (c) and the enantiomeric excess of substrate and product (e.e.$_S$, e.e.$_P$), was developed by C. J. Sih in 1982 [26] on a theoretical basis laid by K. B. Sharpless [27] and K. Fajans [28]. The parameter describing the selectivity of a resolution was introduced as the 'Enantiomeric Ratio' (E), which remains constant throughout the reaction and is only determined by the 'environment' of the system. A related alternative method has been proposed recently [29, 30].

2.1 Hydrolytic Reactions

Irreversible reaction. Hydrolytic reactions are considered as completely irreversible due to the high 'concentration' of water in the aqueous environment (55.5 mol/l). Assuming the absence of enzyme inhibition so that both enantiomers of the substrate compete freely for the active site of the enzyme, Michaelis-Menten kinetics effectively describe the reaction in which two enantiomeric substrates (A and B) are transformed by an enzyme (Enz) into the corresponding enantiomeric products (P and Q).

Figure 2.3. Enzymatic kinetic resolution (irreversible reaction).

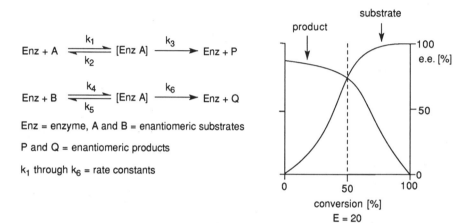

Instead of determining all individual rate constants for each of the enantiomers (a wearysome task for synthetic organic chemists), to get access to the corresponding relative rates which govern the selectivity of the reaction - expressed as the Enantiomeric ratio (E) - , the ratio of the initial reaction rates can be mathematically linked to the conversion (c) of the reaction, and the optical purities of substrate (e.e.$_S$) and product (e.e.$_P$). These parameters are usually much easier to determine.

The dependence of the selectivity and the conversion of the reaction is

for the product: $\qquad$ for the substrate:

$$E = \frac{\ln[1-c(1+e.e._P)]}{\ln[1-c(1-e.e._P)]} \qquad E = \frac{\ln[(1-c)(1-e.e._S)]}{\ln[(1-c)(1+e.e._S)]}$$

c = conversion, e.e. = enantiomeric excess of substrate (S) or product (P); E = Enantiomeric Ratio.

An example of an enzymatic resolution showing a selectivity of E = 20 is depicted in Figure 2.3. As may be seen from the 'product'-curve, the product can be obtained in its highest optical purities before 50% conversion, where the enzyme can freely chose the 'good-fitting' enantiomer from the racemic mixture. As a consequence, the 'good-fitting' enantiomer is gradually depleted from the reaction mixture during the course of the reaction, leaving behind the 'bad-fitting' counterpart. Beyond 50% of conversion, the high relative concentration of the 'bad-fitting' counterpart leads to its increased transformation by the enzyme. Thus after 50% conversion the e.e.p rapidly decreases.

Analogous considerations are true for the optical purity of the residual enantiomer of the substrate (e.e.$_S$, as shown in the 'substrate'-curve). Its optical purity remains low before 40%, then climbs significantly at around 50% and reaches its maximum beyond 60% conversion.

Using these equations the expected optical purity of substrate and product can be calculated at given conversions and the enantiomeric ratio (E) can be determined as a convenient constant value for the 'selectivity' of an enzymatic resolution. As a rule of thumb, enantiomeric ratios below 15 are insufficient for practical purposes. They can be regarded as moderate to good from 15-30, and above this value they are excellent. It must be emphasized that values of E >200 cannot be accurately determined due to the inaccuracies emerging from the determination of the enantiomeric excess (e.g. by NMR or GLC) due to the fact that in this range even a very small variation of e.e.$_S$ or e.e.p cause a significant change in the numerical value of E.

In order to accomplish resolutions of the numerous racemic substrates which exhibit moderate selectivities (E values ca. 20), one can proceed as follows (see Figure 2.4): The reaction is terminated at a conversion of 40%, where the 'product'-curve reaches its optimum in chemical and optical yield being closest to the point X. The product is isolated and the remaining substrate - showing a low optical purity at this stage of conversion - is subjected to a second hydrolytic step, until an overall conversion of about 60% is reached, where the 'substrate'-curve is closest to X. Now, the substrate is harvested with an optimal chemical and optical yield and the 20% of product from the second step is sacrificed or recycled. This two-step process [31] can be used to make use of numerous enzyme-catalysed hydrolysis reactions which show moderate selectivities. On the other hand, it is possible to repeat a resolution by starting the second step with a substrate which is already moderately optically enriched.

2.1 Hydrolytic Reactions

Figure 2.4. Two-step enzymatic resolution.

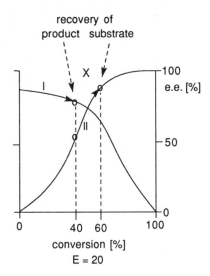

Reversible reaction. The situation becomes more complicated when the reaction is reversible [32, 33]. This is the case if the concentration of the nucleophile which attacks the acyl-enzyme intermediate is limited and is not in excess (cf. water in a hydrolytic reaction). In this situation, the equilibrium constant of the reaction, neglected in the irreversible type of reaction, plays an important rôle and therefore has to be determined.

The equations linking the selectivity of the reaction [the Enantiomeric ratio (E)], the conversion (c), the optical purities of substrate (e.e.$_S$) and product (e.e.$_P$) and also the equilibrium constant K are as follows:

for the product:

$$E = \frac{\ln[1-(1+K)c(1+e.e._P)]}{\ln[1-(1+K)c(1-e.e._P)]}$$

for the substrate:

$$E = \frac{\ln[1-(1+K)(c+e.e._S\{1-c\})]}{\ln[1-(1+K)(c-e.e._S\{1-c\})]}$$

c = conversion, e.e. = enantiomeric excess of substrate (S) or product (P), E = Enantiomeric Ratio, K = equilibrium constant of the reaction.

As shown in Figure 2.5, the 'product'-curve of an enzymatic resolution following a reversible reaction type stays quite the same. However, a significant difference as compared to the irreversible case is found in the 'substrate'-curve: particularly at higher levels of conversion (beyond 70 %) the reverse reaction (i.e. esterification instead of a hydrolysis) starts to take place to a significant extent. Since the main steric requirements and hence the preferred chirality of the substrate stays the same, it is clear that the same enantiomer from the substrate and the product react preferentially in both the forward and the reverse reaction, respectively. Assuming that A is the better substrate than B, accumulation of product P and unreacted B will occur as well as minor amounts of A and Q. For the reverse reaction, however, P is a better substrate than Q, because it is of the same chirality as A and it will therefore progressively be transformed back into P at a faster rate than that of the corresponding reaction of Q, as the extent of the conversion increases. As a consequence, the optical purity of the remaining substrate enantiomers A and B is depleted. In other words, the reverse reaction, taking place at higher rates of conversion, constitutes a second - and in this case an undesired - selection of chirality which causes a depletion of the e.e.s of the remaining substrate.

All the tricks to improve the optical purity of substrate and product of reversible enzymatic resolutions consist in shifting the reaction out of the equilibrium to obtain an irreversible type. The easiest way to achieve this is the use of an excess of cosubstrate: to obtain an equilibration constant of K >10, about 20 molar equivalents of nucleophile vs. substrate are considered as being sufficient for most cases to obtain an irreversible type of reaction. Other techniques, using special cosubstrates which cause an irreversible type of reaction, are discussed in Section 3.1.1.

Figure 2.5. Enzymatic kinetic resolution (reversible reaction).

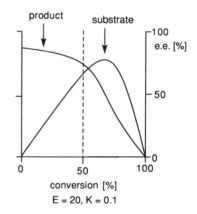

Enz + A $\rightleftarrows^{k_1}_{k_2}$ [Enz A] $\rightleftarrows^{k_3}_{k_4}$ Enz + P

Enz + B $\rightleftarrows^{k_5}_{k_6}$ [Enz A] $\rightleftarrows^{k_7}_{k_8}$ Enz + Q

Enz = enzyme, A and B = enantiomeric substrates

P and Q = enantiomeric products

k_1 through k_8 = rate constants

K = equilibrium constant

Sequential biocatalytic resolutions. In the case wherein a racemic substrate has *two* chemically identical reactive groups of opposite stereochemical configuration, an enzymatic resolution proceeds through two consecutive steps via an intermediate monoester stage. During the course of such a reaction the substrate is forced to enter the active site of the enzyme twice, it is therefore 'double-selected'. Since each of the selectivities of both of the sequential steps determine the final optical purity of the product, exceptionally high selectivities can be achieved by using such a 'double-sieving' procedure.

Figure 2.6. Sequential enzymatic kinetic resolution.

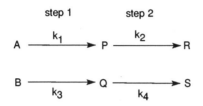

A, B = enantiomeric starting material (diesters)
P, Q = enantiomeric intermediate products (monoesters)
R, S = enantiomeric final products (diols)
k_1 through k_4 = relative rate constants

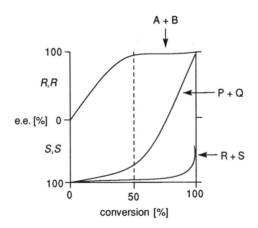

As depicted in Figure 2.6 a racemic substrate consisting of its enantiomers A and B is enzymatically resolved via a first step to give the intermediate enantiomeric products P and Q. The selectivity of this step is governed by the constants k_1 and k_3. Then, both of the intermediate monoester products (P, Q) undergo a second reaction step, the selectivity of which is determined by k_2 and k_4, to form the enantiomeric products R and S. As a consequence, the optical purity of all of the substrates (R, S), the intermediate monoesters (P, Q) and the final products (R, S) are a function of the conversion of the reaction, as exemplified in the graph of Figure 2.6. The selectivities of each of the steps can be determined experimentally and the optical purities of the three species e.e. $_{A/B}$, e.e. $_{P/Q}$ and e.e. $_{R/S}$ can be calcultated [34].

This technique has been proven to be highly flexible. It was shown to work successfully not only in a hydrolytic reaction using cholesterol esterase [35], microbial cells [36] but also in the reverse esterification direction in an organic solvent catalysed by a *Pseudomonas* sp. lipase (see Scheme 2.7).

A special type of sequential enzymatic resolution involving a hydrolysis-esterification [37] or an alcoholysis-esterification sequence [38] has been reported recently (see Figure 2.7). In view of the mechanistic symmetry of enzymatic acyl-transfer reactions, the resolution of a racemic alcohol can be effected by enantioselective hydrolysis of the corresponding ester or by its esterification counterpart. As the biocatalyst displays the same stereochemical preference in both reactions, the desired product can be obtained with higher optical yields, if the two steps are coupled in a sequence. The basis of this approach parallels that of product-recycling in hydrolytic reactions. However, tedious chromatographic separation of the intermediates and accompagnied re-esterification is omitted. Thus, in an organic medium containing a minimum amount of water, the racemic starting acetate A and B is hydrolysed to give alcohols P and Q, which in turn, by the action of the lipase, are re-esterified with cyclohexanoic acid present in the medium.

As a consequence, the alcohol moiety of the substrate has to enter the active site of the lipase twice during the course of its transformation into the final product cyclohexanoyl ester R and S (see Scheme 2.8). An apparent selectivity of $E = 400$ was achieved by this way, whereas the coresponding isolated single-step resolutions of this process, which were carried out independently for reason of comparison, were $E = 8$ for the hydrolysis of the acetate A/B, and $E = 97$ for the independent esterification of alcohol P/Q with cyclohexanoic acid.

2.1 Hydrolytic Reactions

Scheme 2.7. Sequential enzymatic resolution by hydrolysis and esterification.

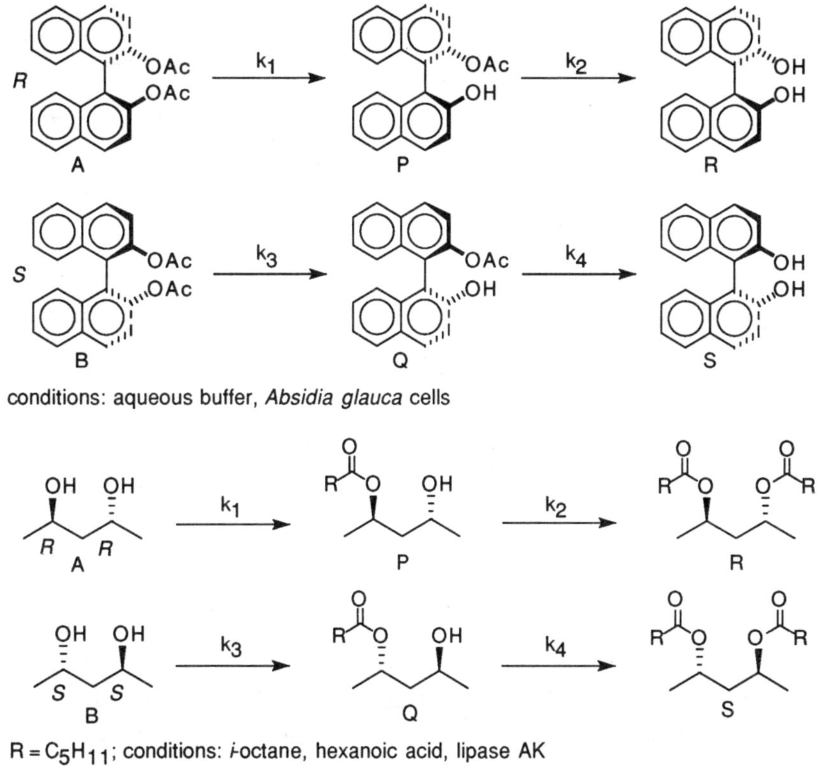

conditions: aqueous buffer, *Absidia glauca* cells

R = C_5H_{11}; conditions: *i*-octane, hexanoic acid, lipase AK

Figure 2.7. Mechanism of sequential enzymatic kinetic resolution via hydrolysis-esterification.

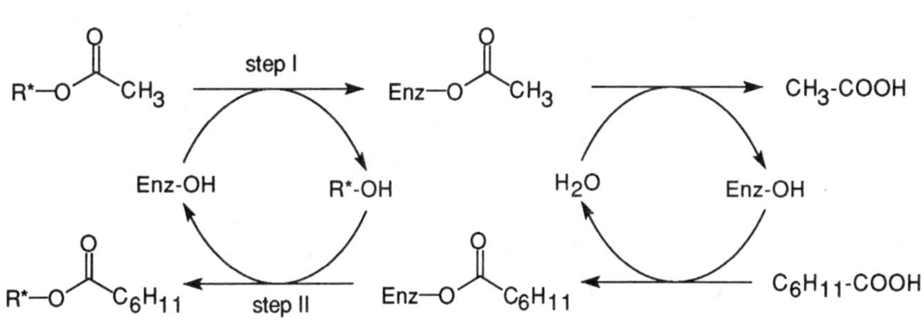

Scheme 2.8. Sequential enzymatic kinetic resolution via hydrolysis-esterification.

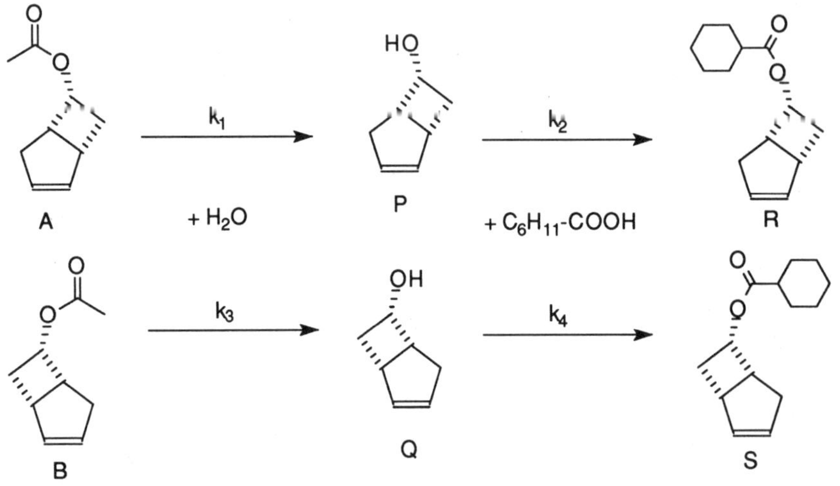

2.1.2 Hydrolysis of the Amide Bond

The enzymatic hydrolysis of the carboxamide bond is naturally linked to the chemistry of amino acids and peptides [39]. The world production of enantiomerically pure amino acids accounted for more than 0.5 million tons of material and a market of ca. 2 billion US$ [40] per annum in 1980. The three amino acids dominating this area with respect to output and value, L-glutamic acid, L-lysine and D,L-methionine, are produced by fermentation and synthetic methods. However, a considerable number of optically pure L-amino acids are prepared by using one of the available enzymatic methods. The L-amino acids are increasingly used as additives for infusion solutions and as optically pure starting materials for the synthesis of pharma- and agrochemicals or artificial sweeteners. Quite recently, some selected amino acids possessing the unnatural D-configuration have gained an increasing importance as bioactive compounds or components of such agents. For instance, D-phenylglycine and its p-hydroxy derivative are used for the synthesis of antibiotics such as ampicillin and amoxicillin, respectively, and D-valine is an essential component of the synthetic pyrethroid fluvalinate.

2.1 Hydrolytic Reactions

Table 2.1. World production of amino acids using enzymatic processes (1980).

Amino acid	Amount [t/a]
L-alanine	130
L-aspartic acid	450
L-2,4-dihydroxyphenylalanine	200
L-methionine	240*
L-phenylalanine	240*
L-tryptophane	200
L-valine	150

* 1990 value.

Among the principal methods for the enzymatic synthesis of enantiomerically pure amino acids depicted in Scheme 2.9, the resolution of racemic starting material (synthetically prepared from inexpensive bulk chemicals) making use of easy-to-use hydrolytic enzymes such as proteases and esterases represents the most widely applied strategy. In contrast, transformations by means of lyases and dehydrogenases are generally more complex. All of the methods described below which have been selected from the numerous strategies of enzymatic amino acid synthesis [41-47] are of a more or less general applicability. They also represent useful tools for the preparation of enantiomerically pure non-natural amino acids [48], not only for industrial needs but also for reasearch on a laboratory scale.

The majority of the reactions shows a common pattern: the substrate enantiomer possessing the 'natural' L-configuration is accepted by the enzyme, the 'unnatural' D-counterpart is not and thus can be recovered from the reaction medium. For some selected processes such as the hydantoinase-method (see below) different enzymes possessing the opposite enantiospecificity are available for the transformation of the desired enantiomer. Using these other enzyme systems, additional synthetic protection and/or deprotection steps are required in cases, where the unnatural D-amino acid constitutes the desired product. The work-up procedure is usually easy in such processes. Thus the difference in solubility of the product and the remaining substrate at different pH-values in the aqueous medium facilitates their separation by conventional ion-exchange or extraction methods. Alternatively, highly insoluble Schiff-bases can be prepared from aldehydes by condensation with *N*-unprotected amino acid derivatives in order to facilitate their isolation. The free amino acids are readily obtained from these derivatives by mild acid hydrolysis without loss of optical purity.

However, there is a common limitation to all of these methods. The α-carbon atom bearing the amino group must not be fully substituted since such bulky substrates are generally not accepted by hydrolases. Thus, optically pure α-methyl or α-ethyl amino acids are generally not accessible by these methods, although some rare exceptions are known [49].

The recycling of the undesired enantiomer from the enzymatic resolution is of crucial importance particularly on industrial scale applications [50]. The classic chemical method consists of the thermal racemisation of an amino acid ester at about 150-170 °C. Milder conditions can be applied for the racemisation of the corresponding amides via intermediate formation of Schiff-bases with benzaldehyde. Unfortunately, the use of isomerase enzymes (such as amino acid racemases [51]) for this purpose has not yet become widespread.

Scheme 2.9. Important enzymatic routes to enantiomerically pure α-amino acids.

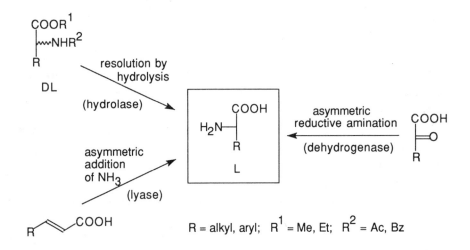

Esterase-Method

A racemic amino acid ester can be enzymatically resolved by the action of a protease or (in selected cases) an esterase. The first resolution of this type using a crude porcine pancreatic extract was reported in 1906 [52]. The enzymatic action of a protease when it is cleaving a carboxylic ester bond has frequently been denoted as 'esterase-activity', although the mechanism of action involved does not differ in principle from that of an amide-hydrolysis. Bearing in mind

2.1 Hydrolytic Reactions

the greater stability of an amide bond as compared to that of an ester, it is reasonable that a protease which is able to cleave a much stronger amide bond by a nucleophilic attack, is capable of performing a similar action on a carboxylic ester. Esterases, on the other hand, are generally unable to cleave amide bonds, although they can catalyse their formation (see Scheme 2.1). The report of the hydrolysis of a highly strained β-lactam derivative by pig liver esterase is an exception [53].

Scheme 2.10. Enzymatic resolution of amino acid esters via the esterase-method.

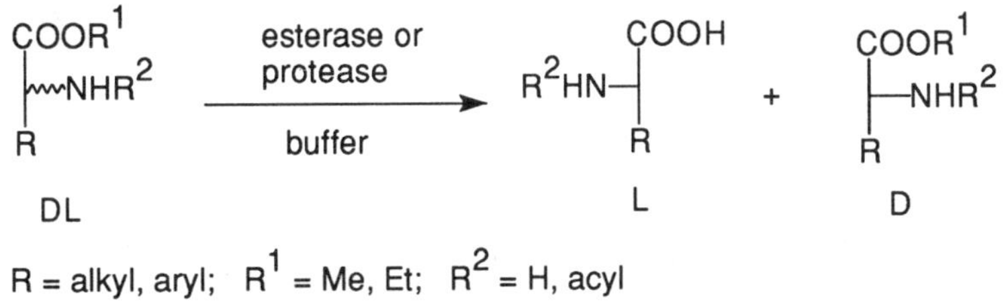

$$\underset{DL}{\overset{COOR^1}{\underset{R}{\overset{|}{\text{---}}}\text{NHR}^2}} \quad \xrightarrow[\text{buffer}]{\text{esterase or protease}} \quad \underset{L}{\overset{COOH}{\underset{R}{R^2HN\text{---}|}}} \quad + \quad \underset{D}{\overset{COOR^1}{\underset{R}{|\text{---NHR}^2}}}$$

R = alkyl, aryl; R^1 = Me, Et; R^2 = H, acyl

The *N*-amino group of the substrate may be either free or (better) protected by an acyl functionality, preferably an acetyl-, benzoyl-, or the *tert*-butyloxy-carbonyl-(Z)-group in order to avoid possible side-reactions such as ring-closure with the formation of diketopiperazines. The ester moiety should constitute of a short-chain aliphatic alcohol such as methyl or ethyl to ensure a reasonable reaction rate. When carboxyl ester hydrolases such as lipases are used in this process, more lipophilic alcoholic groups bearing electron withdrawing substituents such as chloroethyl- [54] or trifluoroethyl alcohols [55] are recommended.

Numerous enzymes have been used to hydrolyse *N*-acyl amino acid esters, the most versatile and thus very popular is α-chymotrypsin (Scheme 2.11) [56-58] isolated from bovine pancreas. It is one of the very few examples of a pure enzyme which can be used readily for biotransformations and furthermore it is one of the few biocatalysts whose mode of action is well understood. A useful and quite reliable model of its active site has been proposed in order to rationalize the stereochemical outcome of resolutions performed with α-chymotrypsin [59, 60].

Other proteases, *e.g.* subtilisin [61, 62], thermolysin [63] and alkaline protease [64] have been reported for the resolution of aminoacid esters, but these have been used less often. Even whole microorganisms such as baker's yeast, possessing unspecific proteases, can be used as biocatalysts for this type of transformation [65].

Scheme 2.11. Resolution of *N*-acetyl aminoacid esters by α-chymotrypsin [66, 67].

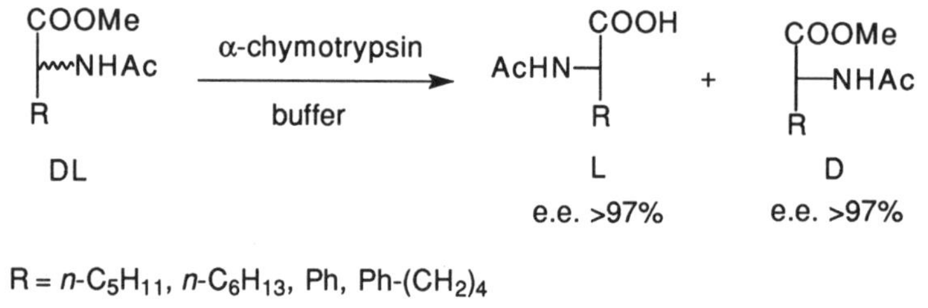

$$\underset{DL}{\underset{R}{\overset{COOMe}{\wedge\wedge\wedge NHAc}}} \xrightarrow[\text{buffer}]{\alpha\text{-chymotrypsin}} \underset{\substack{L \\ e.e. >97\%}}{\underset{R}{\overset{COOH}{AcHN-}}} + \underset{\substack{D \\ e.e. >97\%}}{\underset{R}{\overset{COOMe}{-NHAc}}}$$

R = *n*-C$_5$H$_{11}$, *n*-C$_6$H$_{13}$, Ph, Ph-(CH$_2$)$_4$

Amidase-Method

L-Aminoacid amides are hydrolysed enantioselectively by amino acid amidases (occasionally termed aminopeptidases) from various sources, such as mammalian kidney and pancreas [68] and from different microorganisms, in particular *Pseudomonas*, *Aspergillus* or *Rhodococcus* sp. [69].

Scheme 2.12. Enzymatic resolution of amino acid amides via the amidase-method.

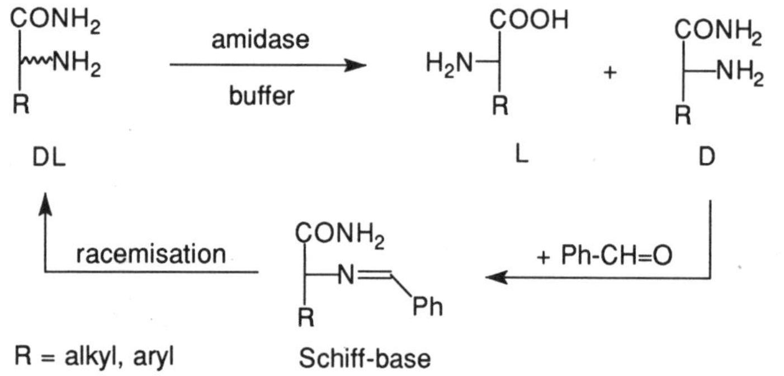

$$\underset{DL}{\underset{R}{\overset{CONH_2}{\wedge\wedge\wedge NH_2}}} \xrightarrow[\text{buffer}]{\text{amidase}} \underset{L}{\underset{R}{\overset{COOH}{H_2N-}}} + \underset{D}{\underset{R}{\overset{CONH_2}{-NH_2}}}$$

racemisation ← Schiff-base + Ph-CH=O

R = alkyl, aryl

Again, the L-amino acids thus formed are separated from the unreacted D-amino acid amide by the difference in solubility in various solvents at various pH. Since amino acid amides are less susceptible to spontaneous chemical hydrolysis in the aqueous environment than the corresponding esters, the products which are obtained by this method are often of higher optical purities.

Acylase-Method

Aminoacylases catalyse the hydrolysis of N-acyl amino acids derivatives, with acyl groups being preferably acetyl, chloroacetyl or propionyl. Most recently, the corresponding N-carbamoyl derivatives have also been used successfully [70]. Enzymes of the amino acylase type have been isolated from hog kidney, and from *Aspergillus* or *Penicillium* sp. [71-73]. The versatility of this type of enzyme has been demonstrated by the resolution of racemic N-acetyl tryptophan and -phenylalanine on an industrial scale using column reactors [74].

Scheme 2.13. Enzymatic resolution of N-acyl amino acids via the acylase-method.

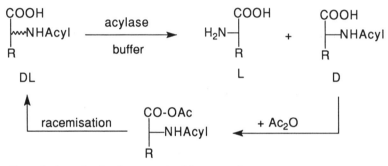

R = alkyl, aryl; Acyl = acetyl, chloroacetyl

On a laboratory scale, the readily available amino acylase from hog kidney is recommended [75]. It seems to be extremely substrate-tolerant, allowing variations of the alkyl- or aryl moiety R while retaining very high enantiospecificities; it has been frequently been used as an aid in natural product synthesis [76, 77]. Unwanted enantiomers of N-acetyl amino acids can be conveniently racemised via the intermediate formation of a mixed anhydride with acetic acid. Interestingly, N-acyl α-aminophosphonic acid derivatives have been resolved using penicillin acylase [78].

Scheme 2.14. Enzymatic resolution of *N*-acyl amino acids for natural product synthesis.

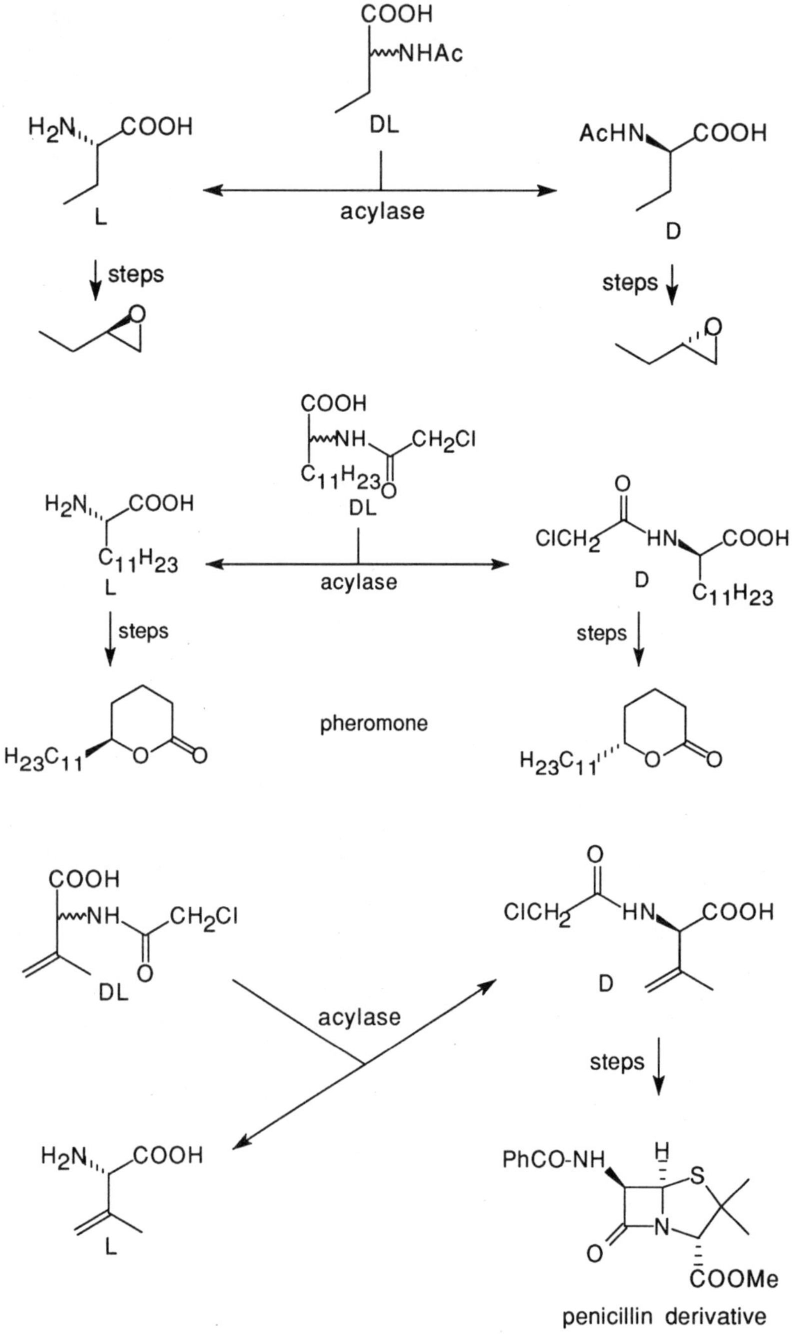

2.1 Hydrolytic Reactions

Hydantoinase-Method

5-Substituted hydantoins may be obtained easily in racemic form from cheap starting materials such as an aldehyde, hydrogen cyanide and ammonium carbonate using the Bücherer-Bergs synthesis [79]. Hydantoinases (occasionally also called 'dihydropyrimidinases') from different microbial origins [80] catalyse the hydrolytic ring-opening to form the corresponding N-carbamoyl-α-amino acids.

Scheme 2.15. Enzymatic resolution of N-acyl amino acids via the hydantoin-method.

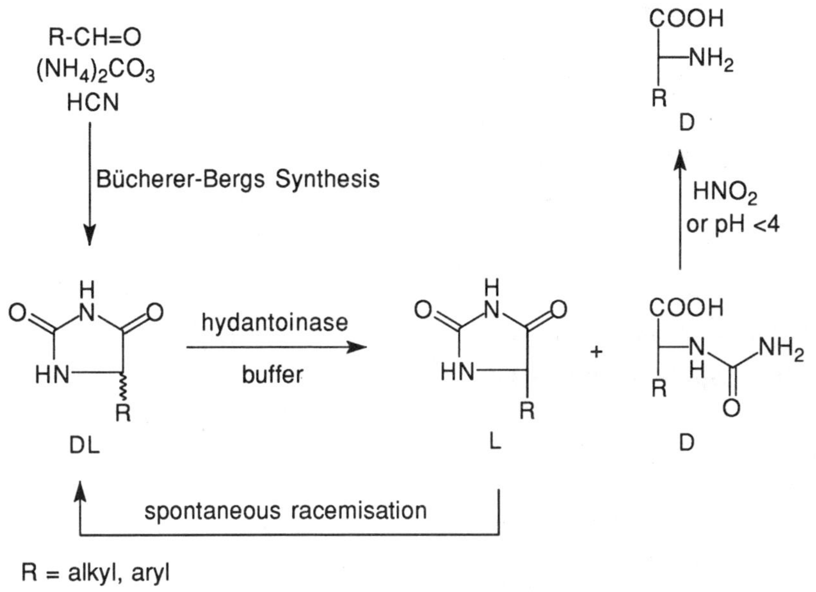

R = alkyl, aryl

Interestingly, and in contrast to the above mentioned amino acid resolution methods involving aminoacid esters, -amides, or N-acyl acids where the natural L-enantiomer is preferably hydrolysed from a racemic mixture, hydantoinases usually convert the opposite enantiomer [81-83], although L-hydantoinases are known [84, 85]. The D-N-carbamoyl derivatives thus derived can be transformed into the corresponding D-amino acids by treatment with nitrous acid or by exposure to an aqueous solution pH < 4 or enzymatically by using an N-carbamoyl D-amino acid amidohydrolase, which is often produced by the same microbial species. One property of 5-substituted hydantoins, which makes them particularly attractive for large-scale resolutions is their ease of racemisation. When R constitutes an aromatic group, the enantiomers of the starting hydantoins are readily equilibrated even at pH 7 - others racemise at somewhat elevated pH. This means a theoretical yield of 100% can be achieved for the resolution process.

2.1.3 Ester Hydrolysis

2.1.3.1 Esterases and Proteases

In contrast to the large number of readily available microbial lipases, less than a handful of true 'esterases' - such as pig and horse liver esterases (PLE [86, 87] and HLE, respectively) - have been used to perform the bulk of a large number of highly selective hydrolytic reactions. As a consequence, the movement to the use of a different esterase is not easy in the case that the hydrolysis of an ester proceeds non-selectively with a standard enzyme such as PLE.

However, another esterase which has been shown to catalyse the hydrolysis of non-natural esters with exceptional high selectivities is acetylcholine esterase (ACE). It would certainly be a valuable enzyme to add to the limited number of available esterases but it has a significant disadvantage since it is isolated from the electric eel. Comparing the natural abundance of this species with the occurrence of horses or pigs, its high price (which is prohibitive for most synthetic applications), is probably justified. Thus the number of ACE applications is limited [88-91]. Additionally, cholesterol esterase is also of limited use, since it seems to work only on relatively bulky substrates which show some structural similarities with steroid esters [92].

To overcome this narrow range of readily available esterases, whole microbial cells are sometimes used to perform the reactions instead of isolated enzyme preparations [93]. Although some highly selective conversions using whole-cell systems have been reported, it is clear that any optimisation by controlling the reaction conditions is very complicated when living organisms are employed. Furthermore, for most cases the nature of the actual active enzyme system remains unknown.

Very recently, novel microbial esterases such as carboxyl-esterase NP have been isolated during the search for biocatalysts with high specificities for certain types of substrates. Although they have been made available in generous amounts by genetic engineering, they still have to enter general use.

Fortunately, as mentioned in the foregoing chapter, a large number of proteases can also selectively hydrolyse carboxylic esters and this effectively compensates for the limited number of esterases [94]. The most frequently used

2.1 Hydrolytic Reactions

members of this group are α-chymotrypsin [95], subtilisin [96] and - to a somewhat lesser extent - trypsin, pepsin [97], papain [98] and penicillin acylase [99, 100]. Since many of the studies on the ester-hydrolysis catalysed by α-chymotrypsin and subtilisin have been performed together with PLE in the same investigation, representative examples are not singled out in a separate chapter but are incorporated into the section dealing with studies on PLE. A recently established member of this group is a protease from *Aspergillus oryzae* which seems to be particularly useful for the selective hydrolysis of bulky esters.

As a rule of thumb, when acting on non-natural carboxylic esters, most proteases seem to retain a preference for the hydrolysis of the enantiomer, which mimics the configuration of an L-amino acid most closely [101].

The structural features of more than 90% of the substrates which have been transformed by esterases and proteases can be reduced to the following general formulae given in Scheme 2.16.

Scheme 2.16. Types of substrates for esterases and proteases.

R^1, R^2 = alkyl, aryl; R^3 = Me, Et; * = centre of (pro)chirality

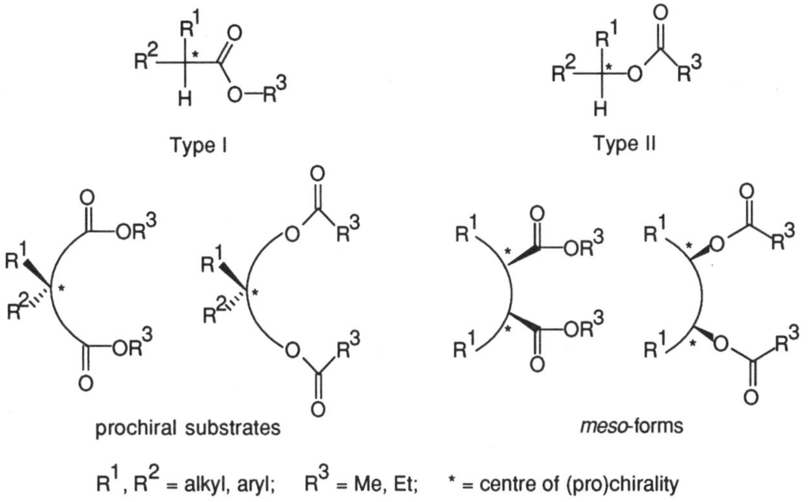

For both esters of the general type I and II, the center of chirality, (as indicated with an asterisk [*]), should be located as close as possible to the site of the reaction, that is the carbonyl group of the ester. Thus, α-substituted carboxylates and esters of secondary alcohols are usually more selectively hydrolysed than their β-substituted counterparts and esters of chiral primary alcohols, respectively. Both substituents R^1 and R^2 can be alkyl- or aryl-groups, but they should differ in size and polarity to aid the chiral recognition process of the enzyme. They may also be joined together to form cyclic structures. Polar groups, such as -COOH, -CONH$_2$ or -NH$_2$, which are heavily hydrated in an aqueous environment should be absent since esterases (and in particular lipases) do not accept highly polar hydrophilic substrates. If such groups are required, they should be protected with a less polar unit. The alcohol moieties (R^3) of type I esters should be as short as possible - preferably methyl, cyanomethyl, ethyl or 2-haloethyl - since carboxylates bearing long-chain alcohols are usually hydrolysed at reduced reaction rates with esterases and proteases. The same is true for acylates of type II, where acetates, haloacetates or propionates are the preferred acyl moieties. Increasing the carbonyl reactivity of the substrate ester by adding electron-withdrawing substituents such as halogen (leading to e.g. haloacetates or 2-haloethyl esters) is a frequently used method to enhance the reaction rate in enzyme-catalysed ester-hydrolysis [102]. One limitation in substrate construction is common for both types: The remaining hydrogen atom at the chiral center must not replaced since α,α,α-trisubstituted carboxylates and esters of tertiary alcohols are usually not accepted by esterases and proteases. This limitation turns them into protective groups for carboxy- and alcoholic functionalities, where an enzymatic hydrolysis is not desired. It is clear that both general types (which themselves would constitute racemic substrates) may be further combined into suitable prochiral or *meso*-substrates.

Pig Liver Esterase and α-Chymotrypsin

Amongst all the esterases, pig liver esterase (PLE) is clearly the champion cosidering its general versatility. This enzyme is constitutionally complex and consists of at least five so-called *iso*-enzymes, which are associated as trimers of three individual proteins [103]. However, from an organic chemist´s viewpoint this crude mixture can be regarded as a single enzyme since all of the isoenzyme subunits usually possess a very similar stereospecificity [104], although some significant differences among them have been reported recently [105]. The biological rôle of PLE is the hydrolysis of various esters occurring in the porcine diet, which would explain its exceptionally wide substrate tolerance. For preparative reactions, it is not absolutely necessary to use the commercially available partially purified enzyme because for many applications a crude acetone powder which can easily be prepared is a cheap and efficient alternative [106].

Whereas an unspecific esterase isolated from horse liver (HLE) has been used to some extent [107], all the respective enzymes from other sources such

2.1 Hydrolytic Reactions

as chicken-, hamster-, guinea pig- or rat-liver were found to be less versatile when compared to PLE.

Mild hydrolysis. Acetates of primary and secondary alcohols such as cyclopropyl acetate [108] and methyl or ethyl carboxylates (such as the labile cyclopentadiene ester [109]) can be selectively hydrolysed under mild conditions using PLE avoiding decomposition reactions which would occur during a chemical hydrolysis under acid- or base-catalysis. For example, this strategy has been used for the final deprotection of the carboxyl moiety of prostaglandin E_1 avoiding the destruction of the delicate molecule [110].

Scheme 2.17. Mild ester hydrolysis by PLE.

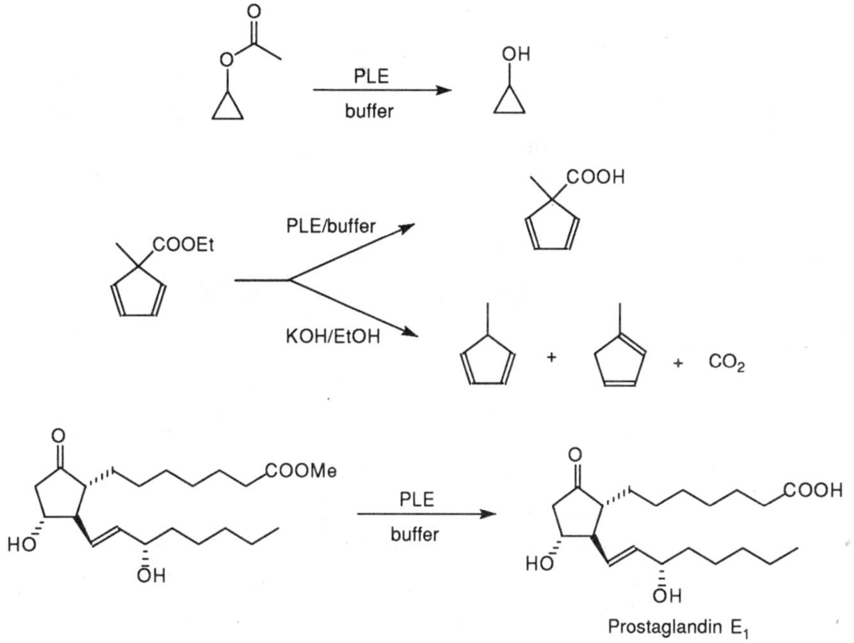

Regioselective hydrolysis. Regiospecific hydrolysis of dimethyl malate at the 1-position can be effected with PLE as the catalyst [111]. Similarly, hydrolysis of an *exo-/endo*-mixture of diethyl dicarboxylates with a bicyclo[2.2.1]heptane framework occurs only on the less hindered *exo*-position [112] by leaving the *endo*-ester untouched, thus allowing a facile separation of the two positional isomers in a diastereomeric mixture.

Scheme 2.18. Regioselective ester hydrolysis by PLE.

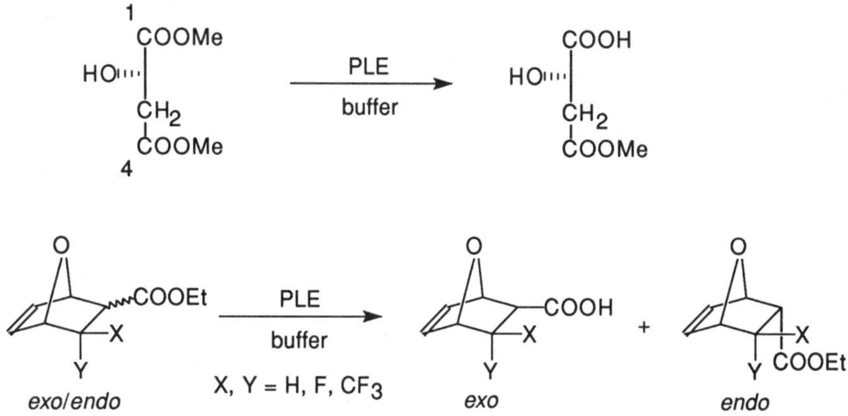

Asymmetrisation of prochiral diesters. It is noteworthy that PLE has been used relatively infrequently for the resolution of racemic esters; α-chymotrypsin had played a more important rôle. Instead, the '*meso*-trick' has been thoroughly used with the former enzyme. Since there is no enzymological reason for this phenomenon, this situation may probably be attributed to the different traditions of the researchers involved.

As depicted in Scheme 2.19, α,α-disubstituted malonic diesters can be selectively transformed by PLE or α-chymotrypsin to give the corresponding chiral monoesters [113-115].

These transformations demonstrate an illustrative example for an 'alternative fit' of substrates with different steric requirements. Whereas PLE performs a selective hydrolysis of the pro-*S* ester group on all substrates possessing α-substituents (R) of a smaller size ranging from ethyl through *n*-butyl to phenyl, an increase of the steric bulkiness of R forces the substrate to enter the enzyme's active site with the opposite stereochemistry. Thus with the more bulky substituents the pro-*R* ester is preferentially cleaved.

The synthetic utility of the chiral hemiesters was demonstrated by the stereoselective degradation of the carboxyl group in the benzyl derivative using a Curtius-rearrangement via an acyl azide intermediate (see Scheme 2.20). Finally, hydrolysis of the remaining ester group led to an optically pure α-methyl amino acid as exemplified for (*S*)-α-methylphenyl alanine. As mentioned in the foregoing chapter, such sterically demanding amino acids normally cannot be obtained by resolution of the appropriate amino acid derivative.

2.1 Hydrolytic Reactions

Scheme 2.19. Asymmetrisation of prochiral malonates by PLE and α-chymotrypsin.

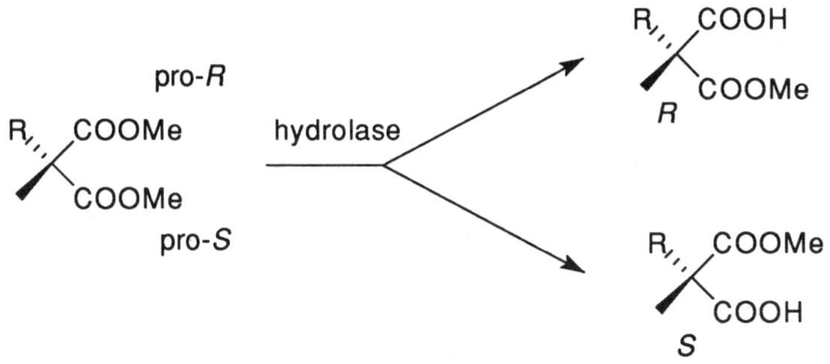

Enzyme	R	Configuration	e.e. [%]
PLE*	Ph-	S	86
PLE	C_2H_5-	S	73
PLE	n-C_3H_7-	S	52
PLE	n-C_4H_9-	S	58
PLE	n-C_5H_{11}-	R	46
PLE	n-C_6H_{13}-	R	87
PLE	n-C_7H_{15}-	R	88
PLE	p-MeO-C_6H_4-CH_2-	R	82
PLE	t-Bu-O-CH_2-	R	96
α-chymotrypsin	Ph-CH_2-	R	~100

* The ethyl ester was used.

Scheme 2.20. Chemo-enzymatic synthesis of α-methyl-L-amino acids.

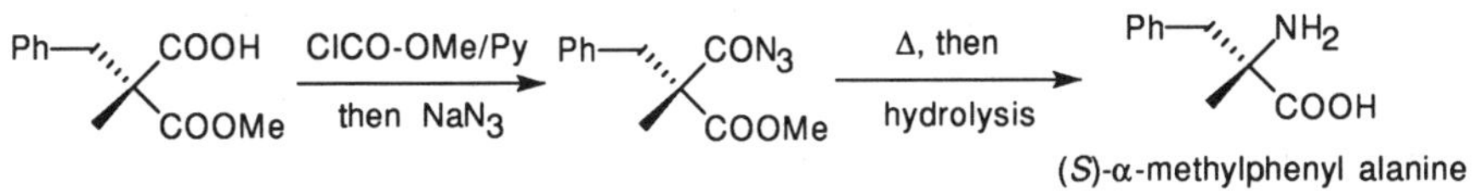

As shown in Scheme 2.21, the prochiral center may be moved from the site of the reaction into the β-position. Thus, a recognition of chirality by PLE [116-120] and α-chymotrypsin [121-124] is retained during the asymmetrisation of prochiral 3-substituted glutaric esters. Whole cells of *Acinetobacter lowfii* and *Arthrobacter* sp. have also been used as a source for esterase activity [125] and, once again, depending on the substitutional pattern on carbon-3, the asymmetrisation can lead to both enantiomeric products.

Scheme 2.21. Asymmetrisation of prochiral glutarates.

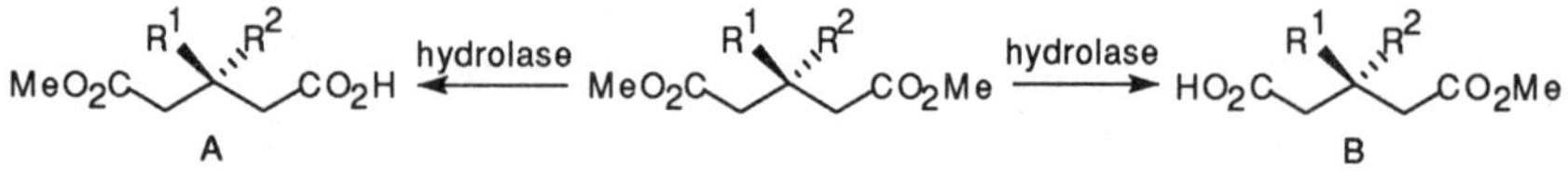

$$MeO_2C-CR^1R^2-CO_2H \xleftarrow{hydrolase} MeO_2C-CR^1R^2-CO_2Me \xrightarrow{hydrolase} HO_2C-CR^1R^2-CO_2Me$$
$$\quad\quad\quad A \quad\quad\quad\quad\quad\quad\quad\quad\quad\quad\quad\quad\quad\quad\quad\quad\quad\quad\quad\quad\quad\quad\quad B$$

Hydrolase	R^1	R^2	Product	e.e. [%]
α-chymotrypsin*	AcNH-	H	B	79
α-chymotrypsin*	OH-	H	B	85
α-chymotrypsin	Ph-CH$_2$-O-	H	B	84
α-chymotrypsin	Ph-CO-O-	H	B	92
α-chymotrypsin	CH$_3$OCH$_2$O-	H	B	93
PLE	CH$_3$-	H	B	90
PLE	AcNH-	H	B	93
PLE	OH-	H	A	12
PLE	t-Bu-CO-NH-	H	A	93
PLE	Ph-CH$_2$-O-CONH-	H	A	93
PLE	Ph-CH$_2$-CH=CH-CH$_2$-	H	A	88
PLE	OH-	CH$_3$	A	99
Acinetobacter sp.*	OH-	H	B	>95
Arthrobacter sp.*	OH-	H	A	>95

* The corresponding ethyl esters were used.

Acyclic *meso*-dicarboxylic esters with a succinic and glutaric acid backbone were also good substrates for PLE [126] and α-chymotrypsin (see Scheme 2.22) [127]. It is interesting to note here that an additional hydroxy group in the substrate led to an enhancement of the chiral recognition process in both cases.

Scheme 2.22. Asymmetrisation of acyclic *meso*-dicarboxylates by α-chymotrypsin and PLE.

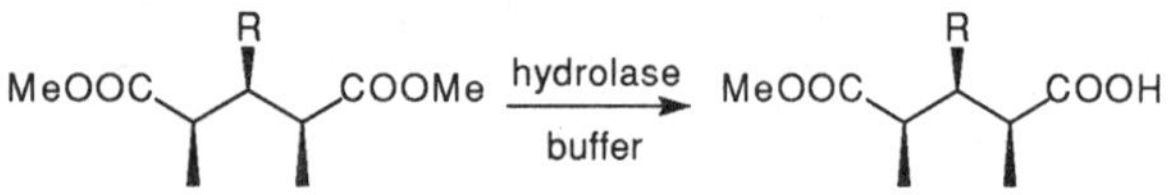

MeOOC-CHR-CHR-COOMe $\xrightarrow{\text{hydrolase, buffer}}$ MeOOC-CHR-CHR-COOH

R	enzyme	e.e. [%]
H	PLE	64
OH	PLE	98
H	α-chymotrypsin	77

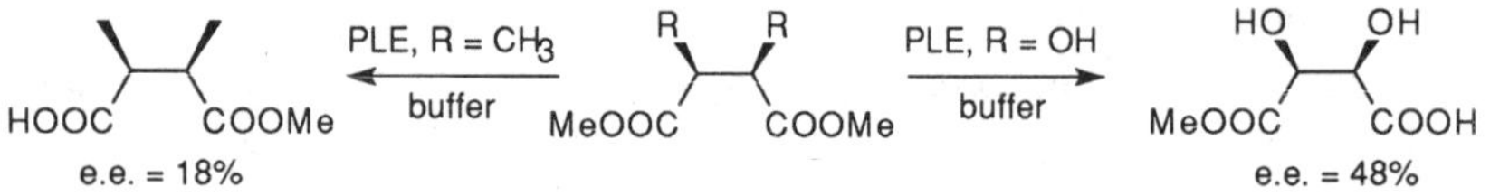

HOOC-C(CH$_3$)-C(CH$_3$)-COOMe (e.e. = 18%) $\xleftarrow{\text{PLE, R = CH}_3\text{, buffer}}$ MeOOC-CHR-CHR-COOMe $\xrightarrow{\text{PLE, R = OH, buffer}}$ MeOOC-C(OH)-C(OH)-COOH (e.e. = 48%)

2.1 Hydrolytic Reactions

The full synthetic potential of the 'meso-trick' has been exemplified by the asymmetrisation of cyclic diesters possessing all kinds of structural patterns. Another striking example of a 'reversal of best fit' for cyclic meso-1,2-dicarboxylates, this time caused by variation of the ring size, is shown in Scheme 2.23 [128]: When the rings are small, the (S)-carboxyl ester is selectively cleaved whereas the (R)-counterpart reacts when the rings are larger. The cyclopentane derivative of moderate ring-size is somewhat in the middle of the range and its chirality is not very well recognized. The fact that the alcohol moiety of such esters can have a significant impact in both the reaction rate and stereochemical outcome of the hydrolysis was shown by the poor chiral recognition of the corresponding diethyl ester of the cyclohexane derivative, which was slowly hydrolysed to give the monoethyl ester of poor optical purity [129].

Scheme 2.23. Asymmetrisation of cyclic meso-1,2-dicarboxylates by PLE.

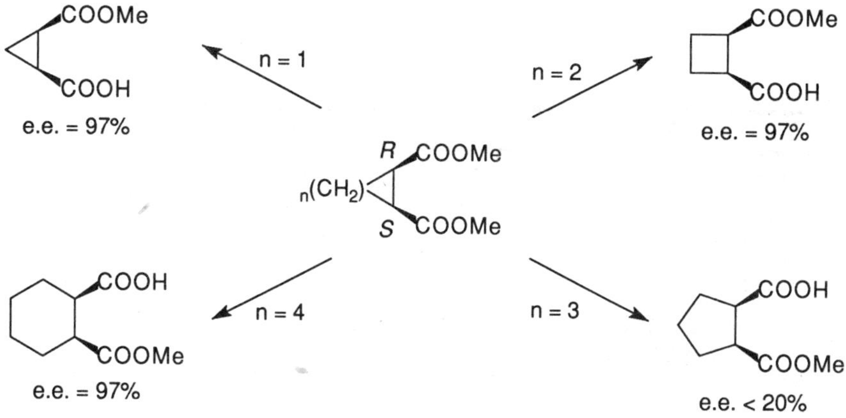

Bulky bicyclic meso-dicarboxylates, which were extensively used as optically pure starting material for the synthesis of bioactive products, can be accepted by PLE [130]. Some selected examples are shown below. During these studies it was shown that the exo-configurated diesters were good substrates, whereas the corresponding more shielded endo-counterparts were less selectively hydrolysed at a significantly reduced reaction rate. The importance of the appropriate choice of the alcohol moiety is also exemplified [131]. Thus whereas the short-chain methyl and ethyl esters were asymmetrically hydrolysed, the propyl ester was not.

Scheme 2.24. Asymmetrisation of polycyclic *meso*-1,2-dicarboxylates by PLE.

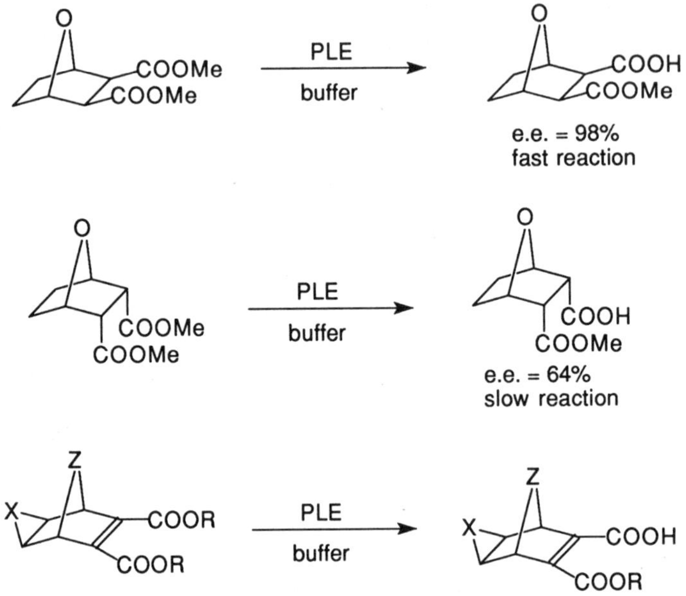

X	Z	R	e.e. [%]
bond	CH_2	Me	41
O	O	Me	77
O-C(CH$_3$)$_2$-O	O	Me	77
O-C(CH$_3$)$_2$-O	CH_2	Me	85
O-C(CH$_3$)$_2$-O	CH_2	Et	100
O-C(CH$_3$)$_2$-O	CH_2	n-Pr	45

Cyclic *meso*-diacetates can be hydrolysed in a similar fashion. As shown in Scheme 2.25, the cyclopentane hemiester [132], which constitutes one of the most important chiral synthon for prostaglandins and derivatives thereof [133], was obtained in high optical purity. In accordance with above mentioned hypotheses for the construction of esterase-substrates, a significant influence of the acyl moiety of the ester was observed: The optical purity of the hemiester gradually declined from 86% to 33% as the acyl chain of the starting substrate ester was changed from acetate to butanoate.

2.1 Hydrolytic Reactions

Scheme 2.25. Asymmetrisation of cyclic *meso*-diacetates by PLE and ACE.

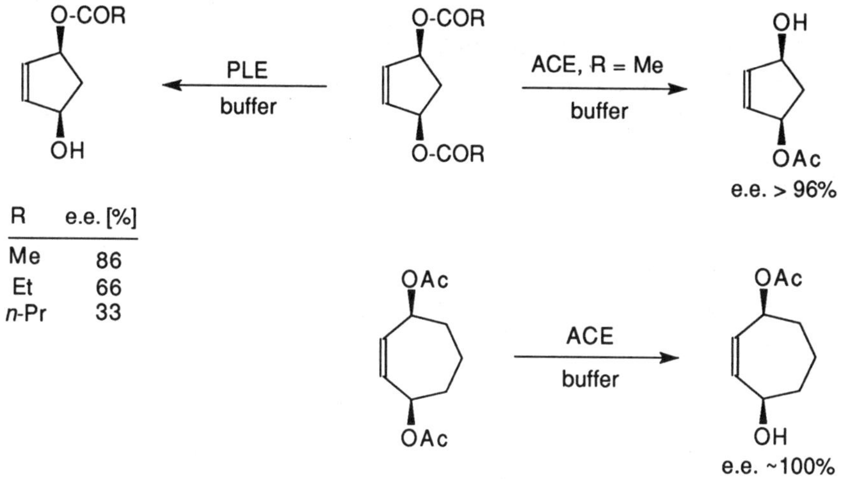

R	e.e. [%]
Me	86
Et	66
n-Pr	33

Scheme 2.26. Asymmetrisation of N-containing cyclic *meso*-diesters by PLE.

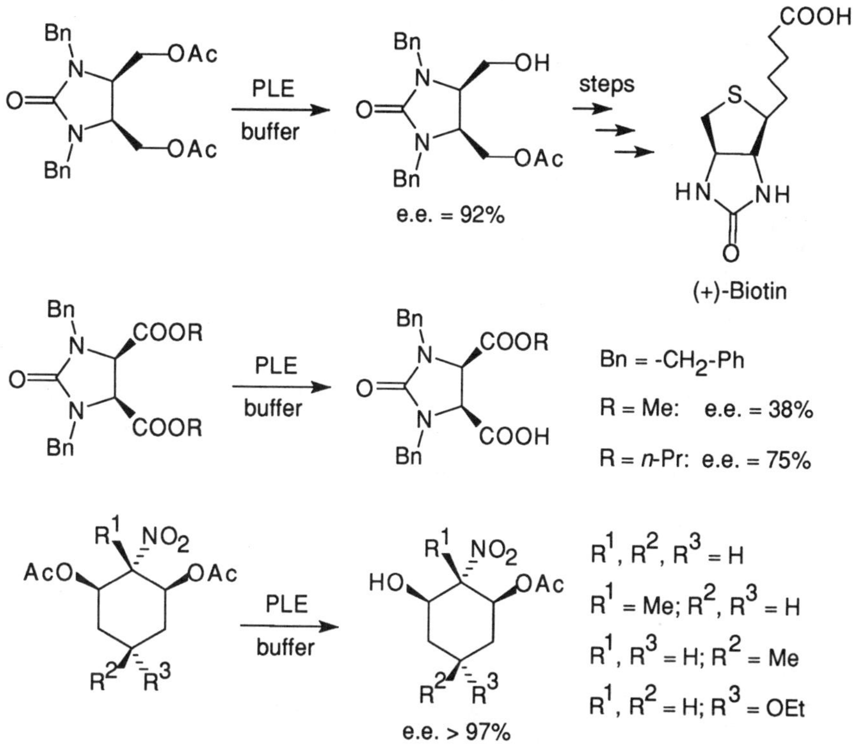

Bn = -CH$_2$-Ph

R = Me: e.e. = 38%
R = n-Pr: e.e. = 75%

$R^1, R^2, R^3 = H$
$R^1 = Me; R^2, R^3 = H$
$R^1, R^3 = H; R^2 = Me$
$R^1, R^2 = H; R^3 = OEt$

In order to avoid recrystallisation of the optically enriched material (86% e.e.), which was obtained from the hydrolysis of *meso*-1,3-diacetoxycyclopentene with PLE, to enantiomeric purity, a search was conducted for another esterase, which would hydrolyse this substrate with an even better stereoselectivity. Acetylcholine esterase (ACE) was shown to be the best choice [134]. It hydrolysed the substrate with complete stereospecificity but with the opposite configuration as compared to PLE.

Other cyclic *meso*-diacetates containing nitrogen functionalities proved to be excellent substrates for PLE (Scheme 2.26). In the benzyl-protected 1,3-imidazolin-2-one system, (which serves as a starting material for the synthesis of the vitamin (+)-biotin), the optical yield of PLE-catalysed hydrolysis of the *cis*-diacetate [135] was much superior to that of the corresponding *cis*-dicarboxylate [136]. Thus, PLE hydrolyses of a dicarboxylate and a diacetate are often complementary to each other in terms of selectivity.

Comparison of an acetone powder prepared from porcine liver and the more expensive partially purified PLE showed comparable selectivity with respect to certain nitro-compounds as shown in Scheme 2.26 [137].

Resolution of racemic esters. As mentioned above, PLE catalysed resolution of racemic esters have been performed less often as compared to the asymmetrisation of prochiral and *meso*-diesters. However, as exemplified below, it can be a valuable technique.

It has been shown that chirality does not neccessarily need to be located on a tetrahedral carbon atom, as in the case of the *trans*-epoxy dicarboxylates (Scheme 2.27) [138]. For example, the helical chirality of the racemic iron-carbonyl complex [139] and of the allenic carboxylic ester shown below [140], was well recognized by PLE.

Resolution of an *N*-acetylaminocyclopentanecarboxylate was used to get access to optically pure starting material for the synthesis of carbocyclic nucleoside analogues with promising antiviral activity [141] (Scheme 2.28). The cyclopentanone dicarboxylate [142] was also required for natural product synthesis, as was the bulky tricyclic monoester shown in Scheme 2.28 [143].

2.1 Hydrolytic Reactions

Scheme 2.27. Resolution of acyclic carboxylic esters by PLE.

[Scheme showing three reactions:

Reaction 1: rac epoxide with two COOMe groups → PLE, buffer → epoxide with COOH and COOMe (e.e. > 95%) + epoxide with COOMe and COOH (e.e. > 95%)

Reaction 2: rac Fe(CO)₃ diene with COOEt → PLE, buffer → Fe(CO)₃ diene with COOH (e.e. = 85%) + Fe(CO)₃ diene with COOEt (e.e. = 85%)

Reaction 3: rac Ph-allene-COOMe → PLE, buffer → Ph-allene-COOH (e.e. = 90%) + Ph-allene-COOMe (e.e. = 61%)]

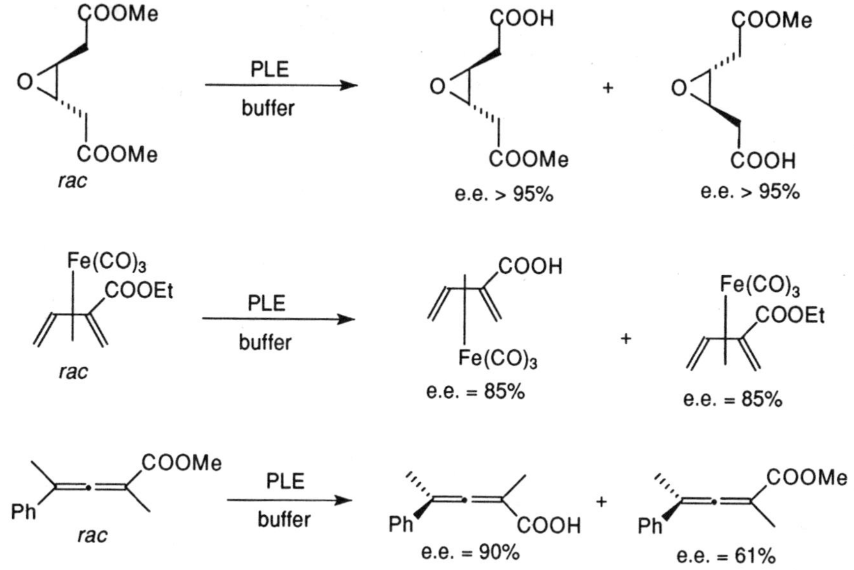

Scheme 2.28. Resolution of cyclic carboxylic esters by PLE.

[Scheme showing three reactions:

Reaction 1: rac cyclopentene with AcHN and CO₂Me → PLE, buffer → cyclopentene with HOOC and NHAc (e.e. = 97%) + cyclopentene with AcHN and CO₂Me (e.e. = 87%)

Reaction 2: rac cyclopentanone with two CO₂Me → PLE, buffer → cyclopentanone with MeO₂C and CO₂H (e.e. = 95%) + cyclopentanone with MeO₂C and CO₂Me (e.e. = 95%)

Reaction 3: rac tricyclic CO₂Et ketone → PLE, buffer, MeCN → tricyclic HO₂C ketone (e.e. = 96%) + tricyclic CO₂Et ketone (e.e. = 83%)]

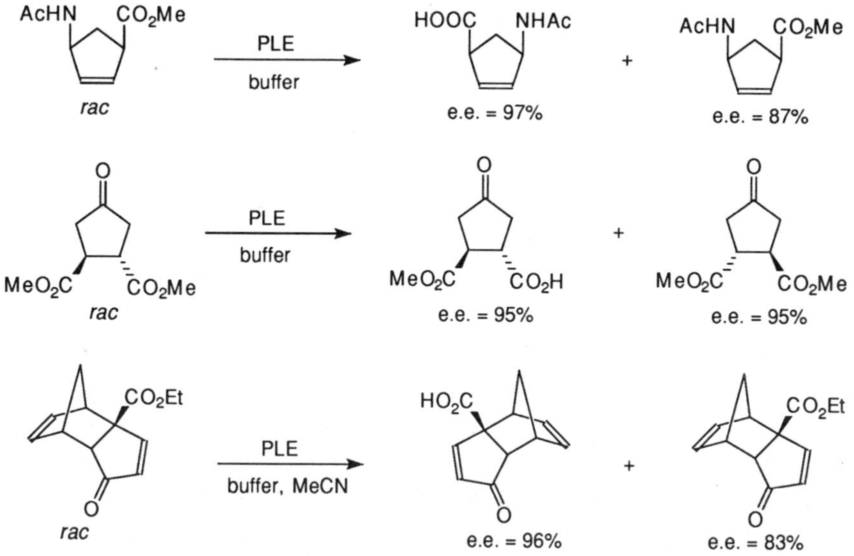

Illustrative examples of the high specificity of hydrolytic enzymes are the kinetic resolution of cyclic *trans*-1,2-diacetates, which occurs with concomitant regioselectivity in case of the cyclobutane derivative [144]. The PLE-catalysed hydrolysis terminates when the (R,R)-diacetate is hydrolysed from the racemic mixture to give the (R,R)-monoacetate leaving the (S,S)-diacetate untouched. Only a minor fraction of (S,S)-diol is formed in the reaction. Again, as observed in the asymmetrisation of cis-meso-1,2-dicarboxylates, the optical purity of the products declines in going from the four- to the five-membered ring system and then increases again on proceeding to the six-membered system with a reversal of stereoselectivity. It should be noted that an asymmetrisation of the correponding cis-meso-1,2-diacetates is restricted a priori by non-enzymic acyl migration which leads to racemisation of any monoester that is formed.

Scheme 2.29. Resolution of cyclic *trans*-1,2-diacetates by PLE.

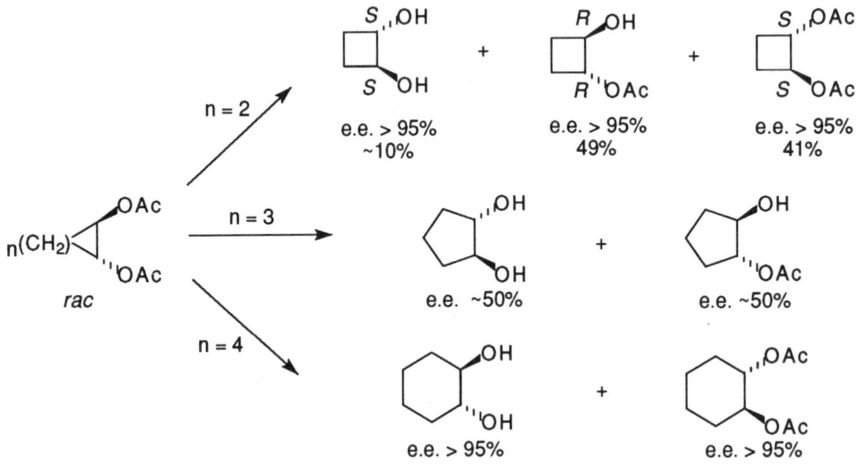

Microbial Esterases

Whole microbial cells have also been used to catalyse some esterolytic reactions. Interesting examples can be found for both bacteria and fungi, for instance *Bacillus subtilis* [145], *Brevibacterium ammoniagenes* [146], *Bacillus coagulans* [147], *Pichia miso* [148] and *Rhizopus nigricans* [149]. Although the reaction control becomes more complex on using fermenting microorganisms, the selectivities achieved are sometimes significantly higher when compared to the use of isolated enzymes. This was shown by the resolution of a secondary alcohol via hydrolysis of its acetate by a *Bacillus* sp. In comparison other biocatalytic methods to obtain the desired masked chiral hydroxyaldehyde failed [150]. Note that the polar terminal carboxyl

2.1 Hydrolytic Reactions

functionality was blocked as its *tert*-butyl ester in order to prevent an undesired ester-hydrolysis in this position.

Scheme 2.30. Microbial resolution of a masked hydroxyaldehyde by *Bacillus* sp.

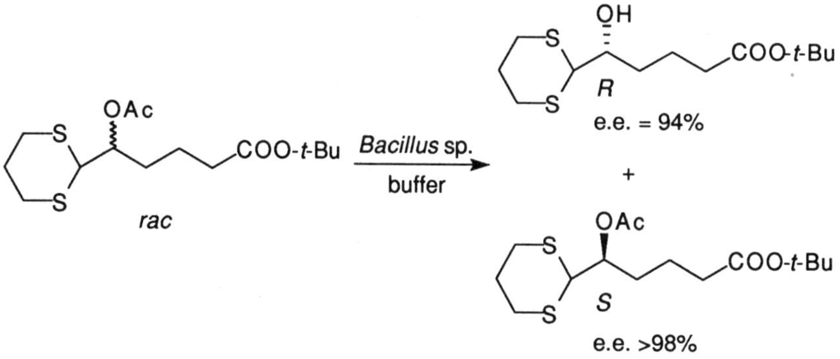

Scheme 2.31. Microbial resolution of acetates by Baker's yeast.

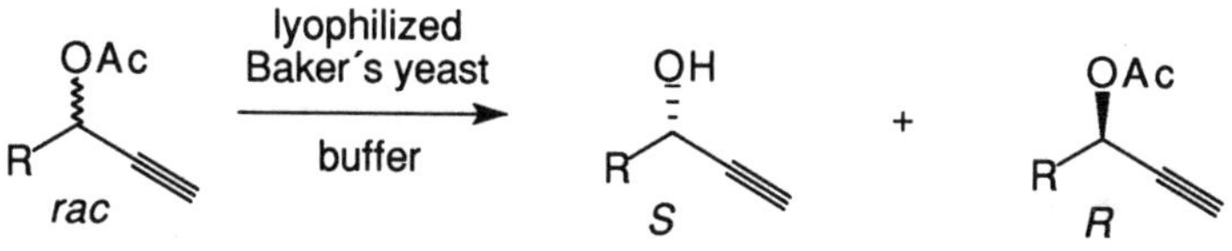

R	e.e. [%]	e.e. [%]
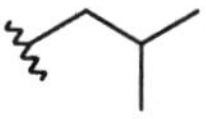 ![isopropyl]	91	72
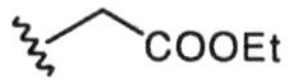 ~~COOEt	91	>97
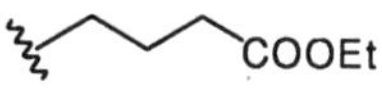 ~~~COOEt	96	59

e.e. = 86% e.e. = 95%

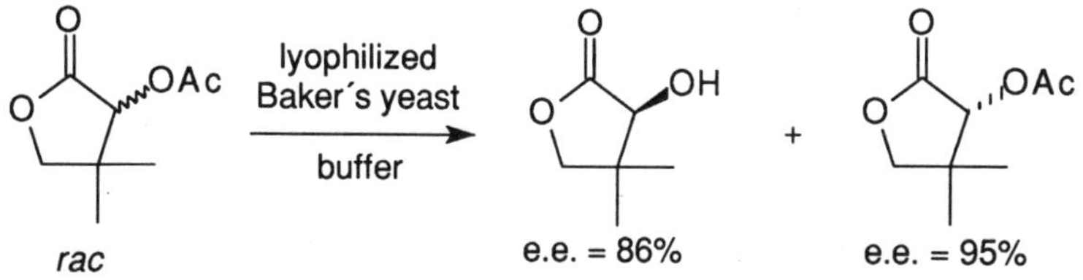

To overcome the problems of reaction control arising from the metabolism of fermenting microorganisms, resting cells of lyophilized Baker's yeast have been proposed as a source of esterase-activity. As shown in Scheme 2.31, 1-alkyn-3-yl acetates [151] and the acetate of pantolactone - a synthetic procursor of pantothenic acid which constitutes a member of the vitamin B group [152] - were well resolved.

Scheme 2.32. Synthetic potential of chiral 1-alkyn-3-ols.

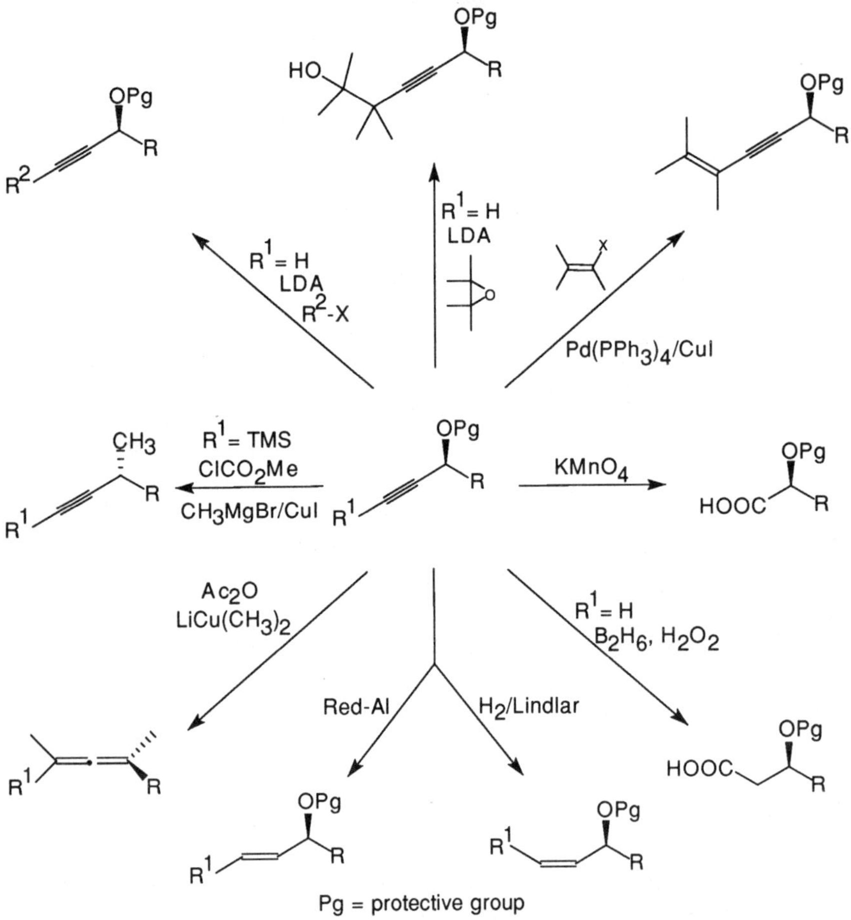

Pg = protective group

The high synthetic value of chiral 1-alkyn-3-ols is summarized in Scheme 2.32. Extension of the carbon chain may be achieved by replacement of the acetylenic hydrogen via alkylation [153] or acylation [154] (R^2 = alkyl, acyl) or via nucleophilic ring-opening of epoxides [155]. Similarly, vinyl halides may

2.1 Hydrolytic Reactions

be coupled using Pd^0- or Cu^I-catalysis [156]. Oxidative cleavage of the triple-bond by permanganate or via hydroboration gives rise to α- or β-hydroxycarboxylic acids, respectively [157]. Stereoselective reduction of the acetylene leads to allylic alcohols with the *E*- or *Z*-configuration [158]. Chiral allenes are obtained by rearrangement reactions [159], and finally, the hydroxyl moiety may be replaced in a stereoselective fashion by alkyl substituents [160].

Due to the importance of α-aryl- and α-aryloxy-substituted propionic acids as anti-inflammatory agents (e.g. Naproxen) and agrochemicals (e.g. Diclofop), respectively, where the majority of the biological activity resides in only one enantiomer (*S* in case of Naproxen), a convenient way for the separation of their enantiomers was sought by biocatalytic methods. As shown in Scheme 2.33, the carboxyl esterase NP, isolated from a *Bacillus* strain was able to resolve the target esters with very high selectivity [161]. It is interesting to note that the stereochemistry of the preferred enantiomer is reversed by going from the α-aryl- to the α-aryloxy derivatives. Very recent studies indicate that this esterase may be useful for the hydrolysis of other non-natural esters [162].

Scheme 2.33. Resolution of α-substituted propionates by carboxylesterase NP.

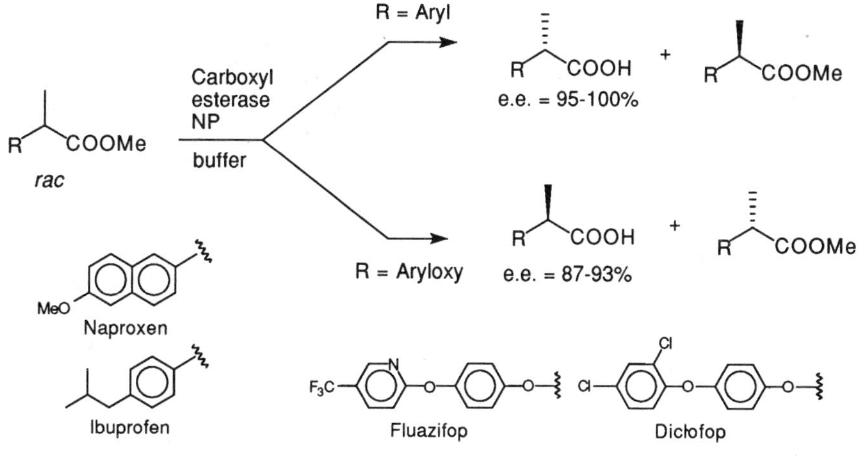

Esterase-Activity of Proteases

Numerous highly selective hydrolyses of non-natural esters catalysed by α-chymotrypsin [163] and papain have featured in excellent reviews [164] and the examples shown above should illustrate the synthetic potential, already fully established in 1976. The major requirements for substrates of type I (see Scheme 2.16) to be selectively hydrolysed by α-chymotrypsin are the presence of a polar and a hydrophobic group on the α-centre (R^1 and R^2, respectively) as are, indeed, present in the natural amino acid substrates.

An interesting example from the recent literature is shown in Scheme 2.34. Hydrolysis of the D-α-nitro esters leads to the formation of the corresponding labile α-nitro-acids, which readily decarboxylate to yield secondary nitro compounds. The L-configured enantiomers of the unhydrolysed esters could be obtained in high optical purity and these were further transformed into α-methyl L-amino acid derivatives [165].

Scheme 2.34. Resolution of α-nitro-α-methyl carboxylates by α-chymotrypsin.

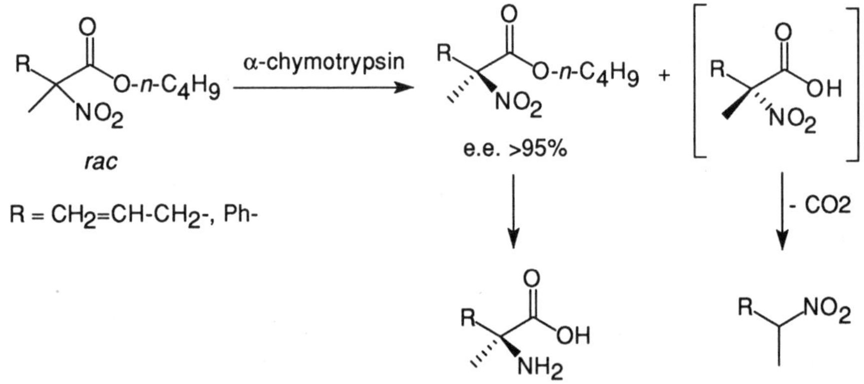

R = $CH_2=CH-CH_2-$, Ph-

Scheme 2.35. Regioselective ester hydrolysis by proteases.

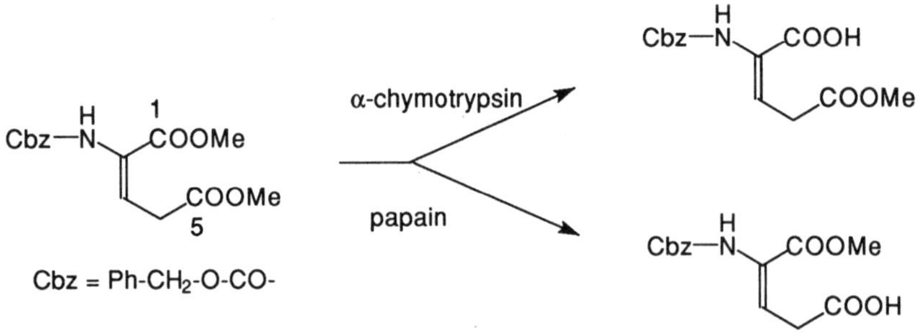

Cbz = $Ph-CH_2-O-CO-$

Proteases such as α-chymotrypsin, papain and subtilisin are also useful for regioselective hydrolytic transformations (see Scheme 2.35). For example, whereas regioselective hydrolysis of a dehydroglutamate diester at the 1-position can be achieved using α-chymotrypsin, the 5-ester is attacked by the

2.1 Hydrolytic Reactions

protease papain [166]. It is noteworthy that papain is one of the few enzymes used for organic synthetic transformations which originate from plant sources (the papaya in this case).

The use of subtilisin as a biocatalyst for the stereospecific hydrolysis of esters has increased over the past years [167-169] and some examples are given below.

Dibenzyl esters of aspartic and glutamic acid can be selectively deprotected at the 1-position by subtilisin-catalysed hydrolysis [170]. 1,2,3-Propantricarboxylic esters - 'retro-fats' - and citrates were hydrolysed in a highly regioselective manner by subtilisin, other possible regioisomeric esters were not detected [171].

Scheme 2.36. Regioselective ester-hydrolyses catalysed by subtilisin.

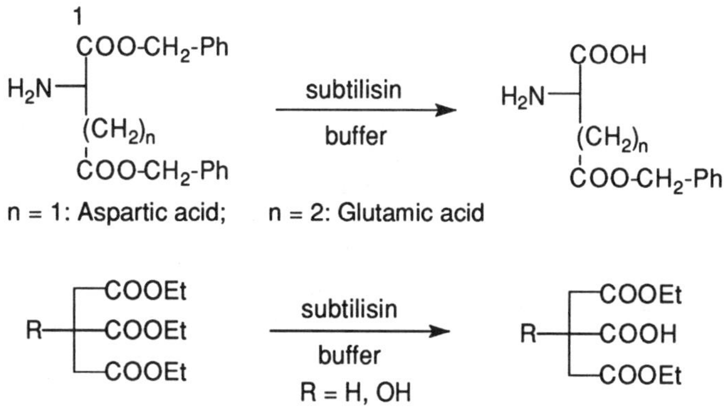

         1
         COO-CH₂-Ph   COOH
H₂N─┤   subtilisin   H₂N─┤
         (CH₂)ₙ   buffer   (CH₂)ₙ
         COO-CH₂-Ph   COO-CH₂-Ph

n = 1: Aspartic acid;   n = 2: Glutamic acid

         ┌─COOEt   ┌─COOEt
R──┼─COOEt   subtilisin   R──┼─COOH
         └─COOEt   buffer   └─COOEt
                         R = H, OH

Quite recently, two proteases have emerged as highly selective biocatalysts for specific purposes. Penicillin acylase is highly selective for the cleavage of phenylacetate esters; this makes it very useful for the selective removal of phenacetyl protective groups (Scheme 2.37) [172]. Furthermore, it can be used for the resolution of esters of primary [173] and secondary alcohols [174] as long as the acid moiety consists of a phenacetyl group. Some structural similarity of the alcohol moiety with that of the natural substrate penicillin G has been stated as being of an advantage.

A protease derived from *Aspergillus oryzae*, which has hitherto mainly been used for cheese-processing, has been shown to be particularly useful for the resolution of sterically hindered substrates such as α,α,α-trisubstituted carboxylates (Scheme 2.38). Whereas 'traditional' proteases such as subtilisin were plagued by slow reaction rates and low selectivities, the α-trifluoromethyl mandelic ester (which constitutes a precursor of a widely used chiral derivatisation agent 'Mosher´s acid' [175]) was successfully resolved by the *Aspergillus oryzae* protease [176].

Scheme 2.37. Chemo- and enantioselective ester-hydrolyses catalysed by penicillin acylase.

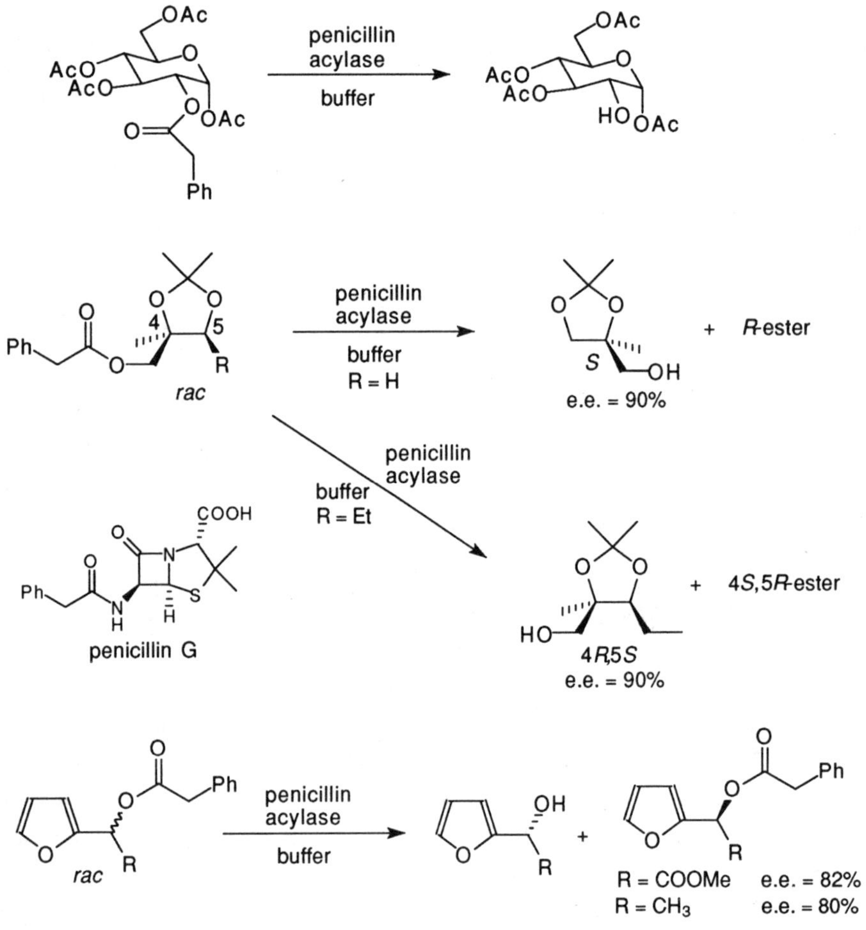

Another elegant example of a protease-catalysed hydrolysis of a carboxylic ester was demonstrated by the resolution of the anti-inflammatory agent 'Ketorolac' via hydrolysis of its ethyl ester by a protease derived from *Streptomyces griseus* (Scheme 2.39) [177]. When the hydrolysis was carried out at a pH of >9, spontaneous racemisation of the substrate ester provided more of the enzymatically hydrolysed *S*-enantiomer from its *R*-counterpart, thus raising the theoretical yield of the resolution of this racemate to 100%.

2.1 Hydrolytic Reactions

Scheme 2.38. Resolution of bulky esters by *Aspergillus oryzae* protease.

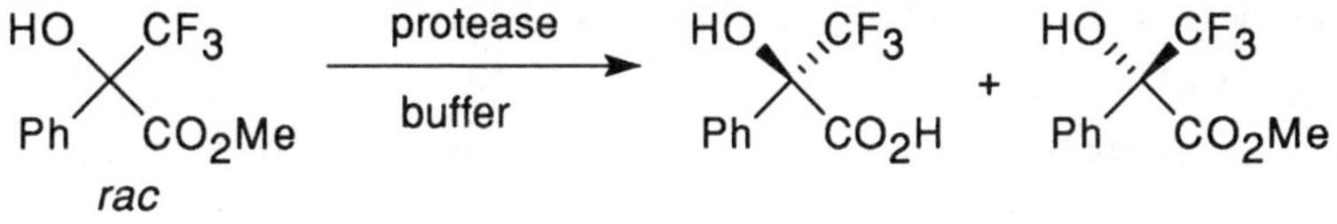

protease	e.e. [%]	e.e. [%]
subtilisin	25	25
Aspergillus oryzae prot.	88	88

Scheme 2.39. Resolution with in situ racemisation by protease from *Streptomyces griseus*.

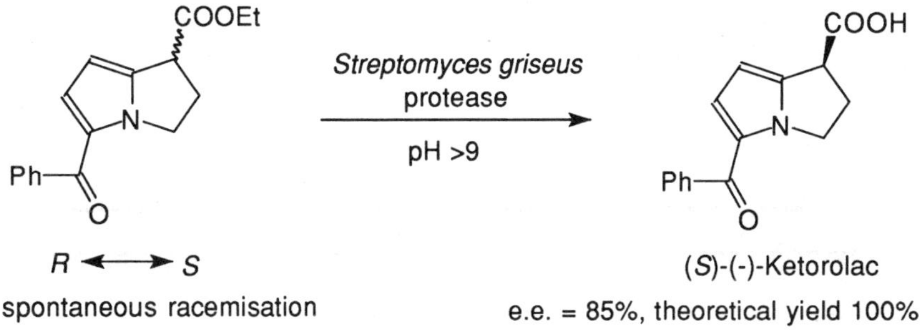

R ⇌ S
spontaneous racemisation

(S)-(-)-Ketorolac
e.e. = 85%, theoretical yield 100%

Optimisation of Selectivity

Many enantiodifferentiating enzymatic hydrolyses of non-natural esters do not show a perfect selectivity, but are often in the range of an e.e. of 50-90% which is considered as being 'moderate' to 'good'. In order to avoid tedious and material-consuming processes to enhance the optical purity of the product e.g. by crystallisation techniques, there are some methods to improve the selectivity of an enzymatic transformation itself [178, 179]. Many, but not all of them, can be valid for other types of enzymes.

Substrate-modification is one of the most promising techniques. This is applicable for all types of enzymatic transformations. As may be concluded from some of the foregoing examples, the ability of an enzyme to 'recognize' the chirality of a given substrate strongly depends on its steric shape. Thus by variation of the substrate structure, most easily performed by the addition or removal of protective groups of different size and/or polarity, a better fit of the substrate may be achieved which leads to an improved selectivity of the enzyme.

Scheme 2.40 shows the optimisation of a PLE-catalysed asymmetrisation of dimethyl 3-aminoglutarate esters using the 'substrate-modification' approach [180]. By varying the N-protective group in size and polarity, the optical purity of the hemiester could be significantly enhanced as compared to the unprotected derivative. In addition, a remarkable reversal in stereochemistry was achieved upon the stepwise increase of the size of the protective group X. This provides an elegant method to control the configuration of the product.

Scheme 2.40. Optimisation of PLE-catalysed hydrolysis by substrate-modification.

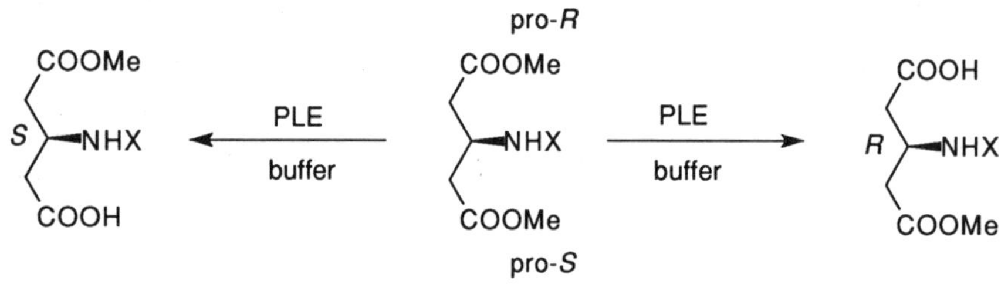

X	Configuration	e.e. [%]
H	R	41
CH_3-CO-	R	93
CH_2=CH-CO-	R	8
C_2H_5-CO-	R	6
n-C_4H_9-CO-	S	2
$(CH_3)_2$CH-CO-	S	54
C_6H_{11}-CO-	S	79
$(CH_3)_3$C-CO-	S	93
Ph-CH_2-O-CO-	S	93
(E)-CH_3-CH=CH-CO-	S	>97

The use of a second acid- or alcohol moiety with a predetermined chirality in the substrate - leading to diastereomeric esters - can be helpful in the case of substrate modification. This technique is called the 'bichiral-ester method' and has been shown to be particularly useful for lipase-catalysed resolutions. It is discussed in the next section.

2.1 Hydrolytic Reactions

Variation of the solvent system (also called 'medium engineering') by the addition of water-miscible organic cosolvents such as methanol, *tert*-butanol acetone, dioxane, acetonitrile, dimethyl formamide (DMF) and dimethyl sulphoxide (DMSO) is a promising and quite frequently used method to improve the selectivity of hydrolytic enzymes, in particular with esterases [181-183]. The concentration of cosolvent may vary from 10-50% of volume. At higher concentrations, however, enzyme-deactivation is unavoidable. Most of these studies have been performed with PLE, where often a significant selectivity enhancement has been obtained, especially by addition of dimethyl sulphoxide or lower alcohols such as methanol and *tert*-butanol. However, the price one usually has to pay on addition of water-miscible organic cosolvents to the aqueous reaction medium is a depletion in the reaction rate. The mechanistic action of such modified solvent systems on an enzyme is only poorly understood and predictions on the outcome of such a medium-engineering cannot be made. Research in this area would certainly be worth persuing.

Scheme 2.41. Selectivity enhancement of PLE by addition of organic cosolvents.

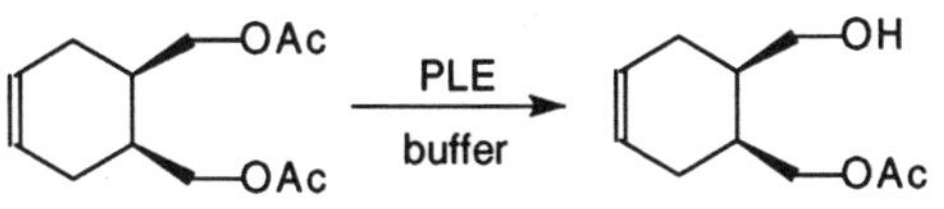

Medium:	e.e. [%]
H_2O	55
H_2O/DMSO 6:4	72
H_2O/DMF 8:2	84
H_2O/*t*-BuOH 9:1	96

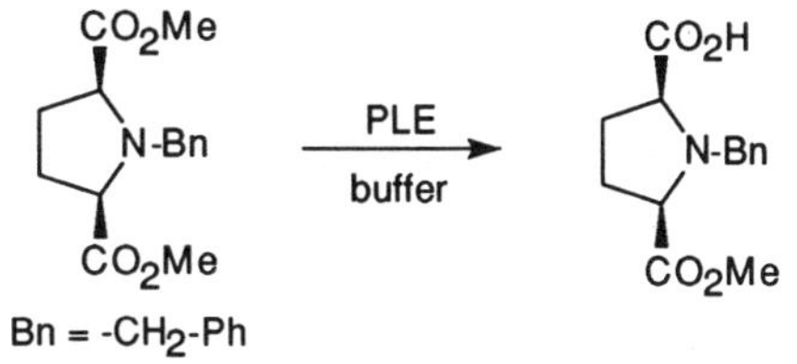

Medium:	e.e. [%]
H_2O	17
H_2O/MeOH 9:1	>97
H_2O/DMSO 3:1	>97

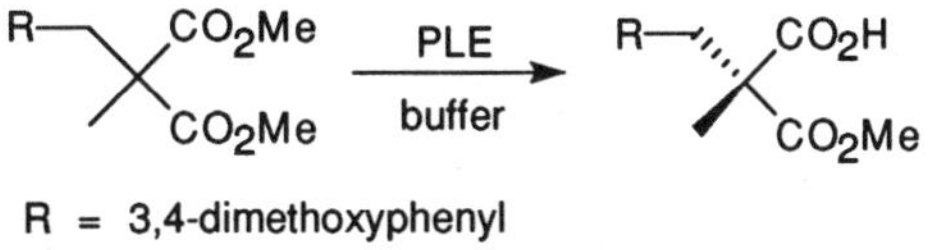

Medium:	e.e. [%]
H_2O	25
H_2O/DMSO 1:1	93

Bn = -CH$_2$-Ph

R = 3,4-dimethoxyphenyl

The selectivity enhancement of PLE-mediated hydrolyses upon the addition methanol, *tert*-butanol and dimethyl sulphoxide to the reaction medium is exemplified in Scheme 2.41. The optical purities of products were in a range of ~20-50% when a pure aqueous buffer system used, but the addition of methanol and/or DMSO led to a significant improvement [184].

The 'enantioselective inhibition' of lipases by the addition of chiral amines functioning as non-competitive inhibitors has been reported [185]. This phenomenon is therefore discussed in the next chapter.

Choice of a different biocatalyst which might possess a better selectivity towards a given substrate mainly depends on the number of available candidates of the same enzyme class. In the esterase-group it is therefore less promising than a substrate modification, but it may be a particularly viable option for lipase-catalysed reactions.

Optimisation of the reaction conditions such as adjustment of the pH or a variation of the temperature [186, 187] usually counts for only minor changes in selectivity. The choice of the right buffer system may have a positive effect in certain cases [188].

Adjustment of the reaction, bearing in mind the underlying kinetics, (Section 2.1.1) can be very helpful, but only if the reaction is of a type where the extent of conversion of the reaction affects the optical purity of substrate and/or product. Re-esterification of the enantiomerically enriched (but not yet optically pure) product and subjection of the material to a second enzymatic hydrolysis may be tedious process but it can certainly help.

Model Concepts

To avoid trial-and-error modification of the substrate structure and to provide suitable tools to predict the stereochemical outcome of enzymatic reactions on non-natural substrates, useful abstract 'models' for a couple of enzymes have been developed. By use of the models, one should be able to 'redesign' a substrate, if the initial results with respect to rate and/or selectivity of the reaction were not acceptable. Since the application of such 'models' holds a couple of potential pitfalls, the most important principles underlying their construction are discussed here.

A correct three-dimensional 'map' of the active site of an enzyme can be accurately determined by X-ray crystallography [189]. Since the tertiary structure of most enzymes is regarded as being closely related to the preferred form in a dissolved state [190], this method gives the most accureate description of the structure of the enzyme. Unfortunately, this can be done only with pure crystalline enzymes such as α-chymotrypsin [191], subtilisin [192],

2.1 Hydrolytic Reactions

and most recently with lipases from *Mucor* [193] and *Geotrichum candidum* [194], which are not the ones, which are commonly used for organic biotransformations.

If the amino acid sequence of an enzyme is known either wholly or even in part, computer-assisted calculations called 'molecular modelling' can provide estimated three-dimensional structures of enzymes [195]. This is accomplished by taking an analogy between known parts of the enzyme under investigation with other enzymes, whose amino acid sequence and three-dimensional structure is already known. Depending on the percentage of the similarity - the 'overlap' - of the amino acid sequences, the results are more or less secure. In general an overlap of about ~60% is regarded to give quite reliable results; below this threshold the results are less accurate.

If neither the amino acid sequence nor X-ray data are available for an enzyme (which is unfortunately the case for the majority of enzymes) one can proceed as follows. A number of artificial substrates having a broad variety of structures is subjected to an enzymatic reaction. The results (i.e. the rate of conversion and enantioselection) then allow one to create a general structure of an imagined 'ideal' substrate, which an actual substrate structure should simulate as closely as possible to ensure rapid acceptance by the enzyme and a high enantioselection. Of course this crude method gives rise to more reliable predictions, the larger the number of test-substrates used and the more rigid their structures. This idealized substrate structure is then called a 'substrate model'. Such models have been developed for PLE [196] and *Candida cylindracea* lipase [197, 198].

Figure 2.8. Substrate model for PLE.

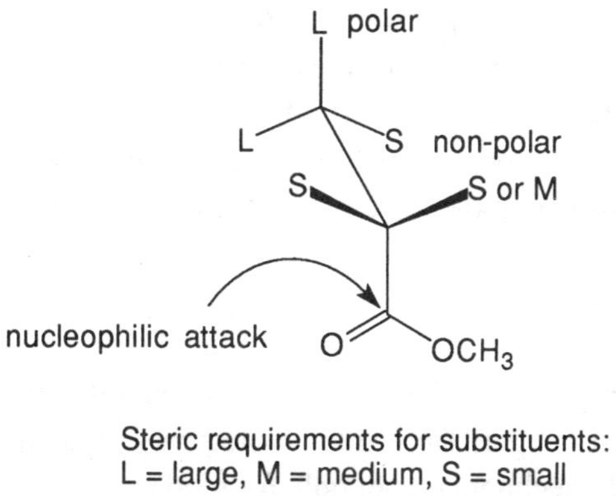

Steric requirements for substituents:
L = large, M = medium, S = small

To ensure an optimal selectivity by PLE, the α- and β-substituents of methyl carboxylates should be assigned according to their size (L = large, M = medium and S = small) with the preferably accepted enantiomer being shown in Figure 2.8.

Instead of developing an ideal *substrate structure* one also can try to picture the structure of the *active site* of the enzyme by the method described above. Certainly, the results from such studies are more reliable the larger the number of substrates used for the construction of the active site. Such *active site models* usually resemble an arrangement of assumed 'sites' or 'pockets' which are usually box- or cave-shaped. A relatively reliable active-site model for PLE [199] using cubic-space descriptors was based on the evaluation of the results obtained from over 100 substrates.

Figure 2.9. Active-site model for PLE.

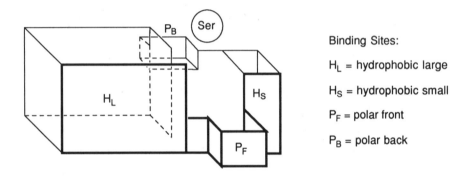

Binding Sites:

H_L = hydrophobic large

H_S = hydrophobic small

P_F = polar front

P_B = polar back

The boundaries of the model represent the space available for the accommodation of the substrate. The important binding regions which determine the specificity of the reaction are two hydrophobic pockets (H_L and H_S, with L = large and S = small) and two pockets of more polar character (P_F and P_B, with F = front and B = back). The best fit of a substrate is determined by locating the ester group to be hydrolysed within the locus of the serine residue and then arranging the remaining moieties in the H and P pockets.

2.1.3.2 Lipases

Lipases are enzymes which hydrolyse triglycerides into fatty acids and glycerol [200, 201]. Apart from their biological significance, they play an important rôle in biotechnology, not only for food and oil processing [202-204] but also for the preparation of chiral intermediates. In fact, about 15% of all

2.1 Hydrolytic Reactions

biotransformations reported up to date were performed with lipases. Although they can hydrolyse and form carboxylic ester bonds like proteases and esterases, their molecular mechanism is different and this gives rise to some unique properties [205] which will be discussed henceforth.

The most important difference between lipases and esterases is the physicochemical interaction with their substrates. In contrast to esterases, which show a 'normal' Michaelis-Menten activity depending on the substrate concentration [S] (i.e. a higher [S] leads to an increase in activity), lipases display almost no activity as long as they are in a dissolved monomeric state (see Figure 2.10). However, when the substrate concentration is gradually increased beyond its solubility limit, a sharp increase in lipase-activity takes place [206, 207]. The fact that lipases do not hydrolyse substrates under a critical concentration (the 'critical micellar concentration' CMC), but display a high activity beyond it, has been called the 'interfacial activation'.

The molecular rationale for this phenomenon has recently been shown to be a rearrangement process within the enzyme. A freely dissolved lipase in the absence of an aqueous/lipid interface resides in its inactive state [Enz], and a part of the enzyme molecule covers the active site. When the enzyme contacts the interface of a biphasic water-oil system, a short α-helix - the 'lid' - is folded back. Thus by opening its active site the lipase is rearranged into its active state [Enz]$^{\neq}$.

As a consequence, lipase-catalysed hydrolyses should be performed in a biphasic medium. It is sufficient to employ the substrate alone at elevated concentrations, such that it constitutes the second organic phase, or, alternatively, it may be dissolved in a water-immiscible organic solvent such as hexane, a dialkyl ether, or an aromatic liquid. Furthermore, physical parameters influencing the mass-transfer of substrate and product between the aqueous and organic phase such as stirring or shaking speed have a marked influence on the reaction rate of lipases. Triacylglycerols such as triolein or -butyrin are used as standard substrates for the determination of lipase-activity, whereas for esterases *p*-nitrophenyl acetate is the classic standard. For the above-mentioned reasons it is clear that the addition of water-immiscible organic solvents to lipase-catalysed reactions is a useful technique to improve activities. Water-soluble organic co-solvents are more often used in conjunction with esterases, which operate in a true solution.

Figure 2.10. Esterase- and lipase-kinetics.

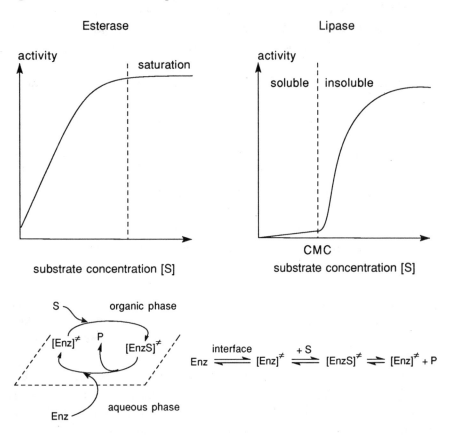

The fact, that many lipases have the ability to hydrolyse esters other than glycerides makes them particularly useful for organic synthesis. Furthermore, some lipases are also able to accept thioesters [208, 209]. In contrast to esterases, lipases have been used much more for the resolution of racemates than for effecting the '*meso*-trick'. Since the natural substrates are esters of a chiral alcohol - glycerol- with an achiral acid, it may be expected that lipases are most useful for hydrolysing esters of chiral alcohols rather than esters of chiral acids. Although this expectation is true for the majority of substrates (see substrate type I, Scheme 2.42), a minor fraction of lipases are also highly selective through recognizing the chirality of an acid moiety (substrate type II).

2.1 Hydrolytic Reactions

Scheme 2.42. Substrate-types for lipases.

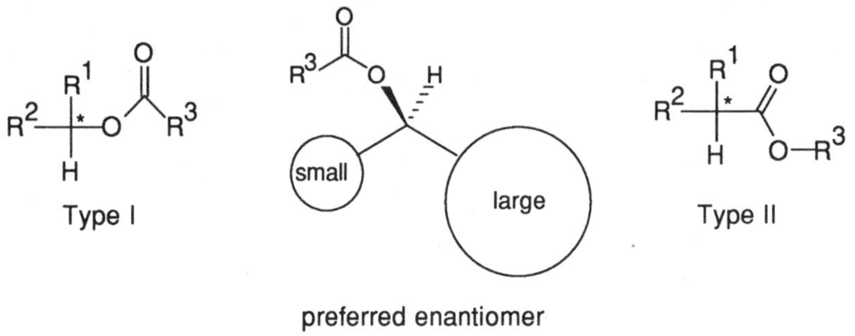

Type I — R^1, R^2, R^3 ester; preferred enantiomer (small/large); Type II

preferred enantiomer

Some of the general rules for substrate-construction are the same as those for esterase-substrates (see Scheme 2.16) such as the preferred close location of the chirality center and the necessity of having a hydrogen atom on the carbon atom bearing the chiral or prochiral centre. Other features are different. The acid moiety R^3 of lipase-substrate of type I should be of a straight-chain possessing at least three to four carbon units to ensure a high lipophilicity of the substrate. Whereas long-chain fatty acids such as oleates would be advantageous for a fast reaction rate, they cause operational problems such as a high boiling point of the substrate and a facilitated formation of foams and emulsions during extractive work-up. As a compromise between the two extremes - short chains for an ease of handling and long ones for a high reaction rate - n-butanoates are often the esters of choice. Furthermore, the majority of lipases show the same stereochemical preference for esters of secondary alcohols as shown in Scheme 2.42. Assuming that the Sequence Rule order of substituents R^1 and R^2 is large > small, the preferably accepted enantiomer lipase-substrate of type I possesses the (R)-configuration at the alcoholic center [210]. It should be mentioned that many proteases and in particular pig liver esterase exhibits a stereochemical preference opposite to that of lipases. Thus the stereochemical outcome of an asymmetric hydrolysis can be directed by choosing a hydrolase from a different class [211].

Substrate-type II represents the general structure of a smaller number of esters which were hydrolysed by lipases. When using lipases for type-II substrates, the alcohol moiety R^3 should preferentially consist of a long straight-chain alcohol such as n-butanol.

A large variety of different lipases is produced by bacteria or fungi and excreted as extracellular enzymes, which makes their production particularly easy. In contrast to esterases, only a minor fraction of lipases are isolated from mammalian sources. Since some lipases from the same genus (for instance from *Candida* or *Pseudomonas*) are supplied by different commercial sources, it should be pointed out that there are certainly some differences in selectivity and activity among the different preparations, but they are usually not in the the range of orders of magnitude. The actual enzyme content of a commercial lipase-preparation may vary significantly, in a range of less than 1% up to about 30%, and it should be noted that the selectivity of a lipase-preparation from the same microbial source does not necessarily increase with the price! From the ever increasing number of commercially available lipases (at present about 50) only those which have shown to be of a general applicability are discussed below.

Figure 2.11. Steric requirements of lipases.

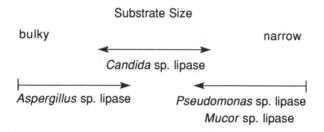

As a rule of thumb, the most widely used lipases may be characterized according to the steric requirements of their preferred substrate esters. Whereas *Aspergillus* sp. lipases are capable of accepting relatively bulky substrates and therefore exhibit low selectivities on 'narrow' ones, *Candida* sp. lipases are more liberal in this regard. Both the *Pseudomonas* and *Mucor* sp. lipases have been found to be often highly selective on substrates with limited steric requirements and hence are often unable to accept bulky compounds. As a consequence, substrates which are for instance recognized with moderate selectivities by a *Candida* lipase, are usually more selectively hydrolysed by a *Pseudomonas* type. Porcine pancreatic lipase (PPL) is not included in Figure 2.11 since it represents a crude mixture of different hydrolytic enzymes and predictions are difficult to make. However, pure PPL should be located with the *Pseudomonas* and *Mucor* lipases.

Porcine Pancreatic Lipase

The cheapest and hence most widely used lipase is isolated from porcine pancreas (PPL) [212-214]. The crudest preparations contain a significant number of other hydrolases besides the 'true PPL'; the latter is available at a high price in purified form. The crude preparation mostly used for biotransformations is called 'pancreatin' or 'steapsin' and it contains only about 6% of protein. The main hydrolase-impurities are α-chymotrypsin, cholesterol esterase, phospho-lipases and other unknown hydrolases. Phospholipases can usually be neglected as undesired hydrolase-impurities, because they are known to act only on negatively charged substrate esters which mimic their natural substrates - phospholipids. On the other hand, α-chymotrypsin and cholesterol esterase can be serious competitors in ester hydrolysis. Both of the latter proteins - and also other unknown hydrolases - can impair the selectivity of a desired PPL-catalysed ester hydrolysis by exhibiting a reaction of lower selectivity (or even of opposite stereochemistry). Thus, any models for PPL should be applied with great caution [215, 216]. Indeed, it has been shown in some cases that these hydrolase impurities were responsible for the highly selective transformation of substrates, which were not, in fact, accepted by purified 'true PPL'. Cholesterol esterase and α-chymotrypsin are likely to act on esters of primary and secondary alcohols whereas 'true PPL' is a highly selective catalyst for esters of primary alcohols. Despite the possible interference of different competing hydrolytic enzymes, numerous highly selective applications have been reported with crude PPL [217-219]. Unless otherwise stated, all of the examples shown below have been performed with steapsin.

Regioselective reactions are particularly important in the synthesis of biologically interesting carbohydrates, where selective protection and deprotection of hydroxyl groups is a central problem. Selective removal of acyl groups of peracylated carbohydrates from the anomeric center [220] or from primary hydroxyl groups [221, 222], leaving the secondary acyl groups intact, can be achieved with hydrolytic enzymes or chemical methods, but the regioselective discrimination between secondary acyl groups is a complicated task. PPL can selectively hydrolyse the butanoate ester on position 2 of the 1,6-anhydro-2,3,4-tri-O-butanoyl-galactopyranose derivative shown in Scheme 2.43 [223]. Only a minor fraction of the 2,4-deacylated product was formed.

Scheme 2.43. Regioselective hydrolysis of carbohydrate esters by PPL.

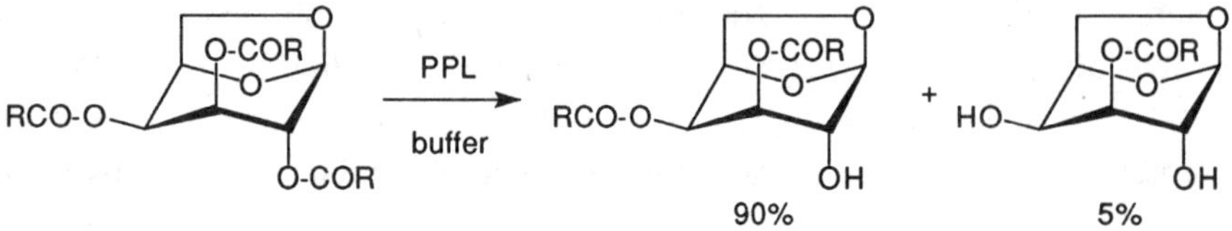

A simultaneous regio- and enantioselective hydrolysis of dimethyl 2-methyl-succinate has been reported with PPL [224] with a preference for the (S)-ester with the hydrolysis taking place on position 4. The residual unhydrolysed ester was obtained with >95% e.e. but the hemi-acid formed (73% e.e.) had to be recycled and subjected to a second hydrolytic step in order for it to be obtained in an optically pure form. It is interesting to note that α-chymotrypsin exhibited the same enantio- but the opposite regio-selectivity on this substrate preferably hydrolysing the ester on position 1 [225].

Scheme 2.44. Regio- and enantioselective hydrolysis of dimethyl α-methylsuccinate.

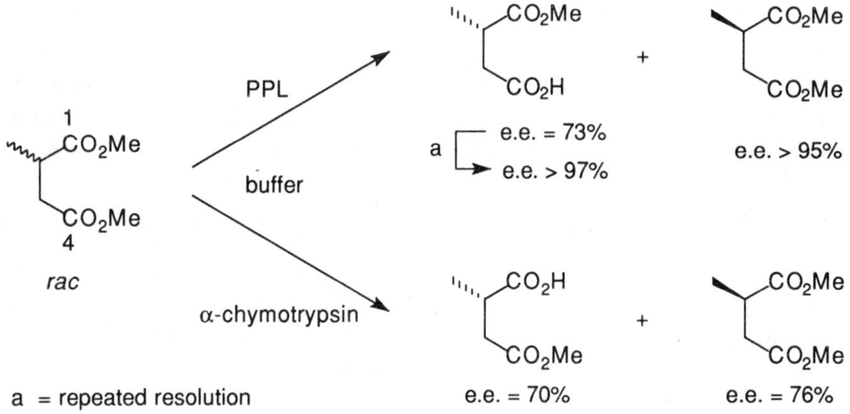

The asymmetric hydrolysis of cyclic *meso*-diacetates by PPL proved to be complementary to the PLE-catalysed hydrolysis of the corresponding *meso*-1,2-dicarboxylates (see Schemes 2.23 and 2.45). The cyclopentane derivative, which gave low e.e.'s using the PLE-method, was now obtained with 86% e.e. [226, 227]. This selectivity was later improved by substrate-modification of the cyclopentane-moiety [228], which gave access to a number of chiral cyclopentanoid building blocks which were used for the synthesis of

2.1 Hydrolytic Reactions

carbacyclic prostaglandin I_2 derivatives - potential therapeutic agents for the treatment of thrombotic deseases.

Scheme 2.45. Asymmetric hydrolysis of cyclic meso-diacetates by PPL.

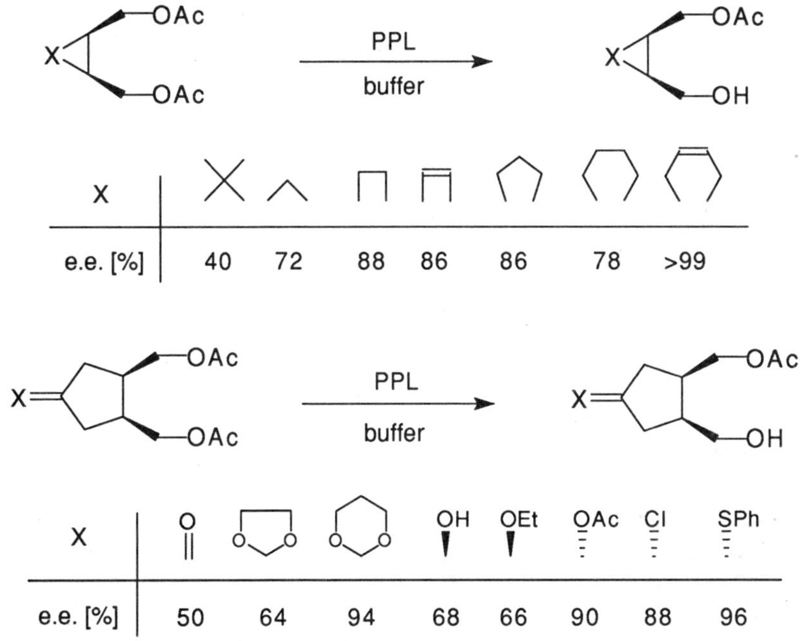

X	✕	∧	⊓	⊓	⟨⟩	⟨⟩	⟨⟩
e.e. [%]	40	72	88	86	86	78	>99

X	O=∥	⟨O O⟩	⟨O O⟩	OH	OEt	OAc	Cl	SPh
e.e. [%]	50	64	94	68	66	90	88	96

Optically active C_3-synthons - chiral glycerols - were obtained by asymmetric hydrolysis of prochiral 1,3-propanediol diesters using PPL. A remarkable influence of a π-system located on substitutents on position 2 on the optical purity of the products makes it clear that the selectivity of an enzyme does not depend on steric factors alone, but also on electronic issues [229]. Note that an *E*-carbon-carbon double bond or an aromatic system [230] on the 2-substituent led to an enhanced selectivity of the enzyme as compared to the corresponding saturated analogues. When the configuration of the double bond was Z, a reversal in the stereochemical preference took place, hand-inhand with an overall loss of selectivity. Additionally, this study shows a positive influence of a biphasic system (using di-*iso*-propyl ether or toluene [231] as water-immiscible organic cosolvent) on the enantioselectivity of the enzyme.

The last two of the entries of Scheme 2.46 represent a 'chiral glycerol' derivative where a positive influence of a long-chain fatty acid moiety on the selectivity of the enzyme was demonstrated [232, 233].

With the 2-phenyl derivative it was proved that the enzyme responsible for the asymmetric hydrolysis was not 'true PPL', which was totally inactive on this substrate, but a novel ester hydrolase which was isolated from crude pancreatin and which gave the corresponding S-monoacetate in >95% e.e. [234].

Scheme 2.46. Asymmetrisation of prochiral 1,3-propanediol diesters by PPL.

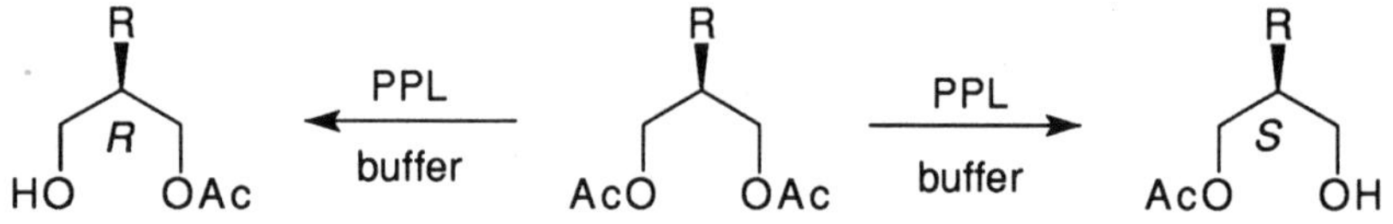

R	Cosolvent	Configuration	e.e. [%]
n-C_7H_{15}-	i-Pr_2O	S	70
(E)-n-C_5H_{11}-CH=CH-	none	S	84
(E)-n-C_5H_{11}-CH=CH-	i-Pr_2O	S	95
$(CH_3)_2$CH-$(CH_2)_2$-	i-Pr_2O	S	72
(E)-$(CH_3)_2$CH-CH=CH-	none	S	90
(E)-$(CH_3)_2$CH-CH=CH-	i-Pr_2O	S	97
(Z)-n-C_5H_{11}-CH=CH-	i-Pr_2O	R	53
(Z)-$(CH_3)_2$CH-CH=CH-	i-Pr_2O	R	15
Ph-	none	S	85-92
Ph-	toluene	S	99
p-CH_3-C_6H_4-	none	S	96
2-naphtyl-	none	S	>96
Ph-CH_2-O-	i-Pr_2O	R	88-91
Ph-CH_2-O-	i-Pr_2O	R	>94a

a The dipropionate ester was used here.

Chiral epoxy alcohols which are not easily available via the Sharpless-procedure [235], were successfully resolved with PPL. It is noteworthy that the lipase is not deactivated by reaction with the epoxide-moiety and that it is able to accept a large variety of structures [236, 237]. The significant influence of the nature of the acyl moiety on the selectivity of the resolution - again, the long-chain fatty acid esters gave better results than the corresponding acetate - may be attributed to the presence of different hydrolytic enzymes present in the crude PPL-preparation. In particular α-chymotrypsin and cholesterol esterase are

2.1 Hydrolytic Reactions

known to hydrolyse acetates of primary alcohols but not their long-chain counterparts. Thus they are more likely to be competitors of PPL when acetates are the substrates rather than butanoates. Lower selectivities were obtained with the corresponding *trans*-derivatives or - in line with the general model shown in Scheme 2.42 - when the distance between the ester moiety and the center(s) of chirality was increased.

Scheme 2.47. Resolution of epoxy-esters by PPL.

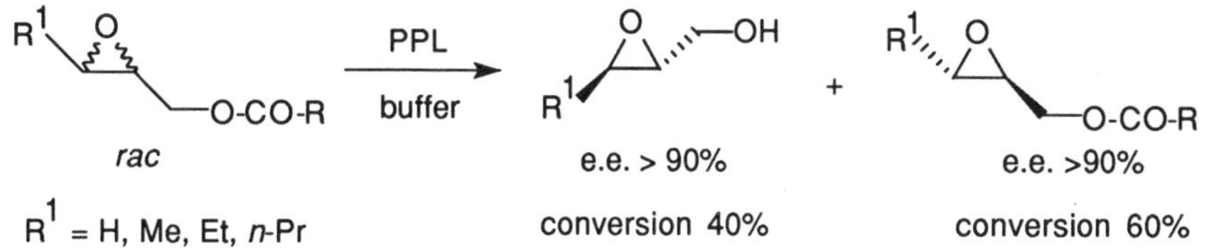

R^1 = H, Me, Et, *n*-Pr

rac → PPL, buffer → (e.e. > 90%, conversion 40%) + (e.e. >90%, conversion 60%)

R^1	R	e.e. [%]
H	CH_3	53
H	C_2H_5	88
H	n-C_3H_7	92
H	n-C_4H_9	96

During a study on the resolution of the sterically demanding bicyclic acetate shown in Scheme 2.48 [238], which represents an important chiral building block for the synthesis of leukotrienes [239], it was found that crude steapsin was a highly selective catalyst for its resolution. On the other hand, pure PPL and α-chymotrypsin were unable to hydrolyse the substrate. Cholesterol esterase, another known hydrolase-impurity in crude steapsin capable of accepting bulky substrates, was able to hydrolyse the ester but with low selectivity. Finally, a novel hydrolase which was isolated from crude PPL proved to be the enzyme responsible for the highly selective transformation.

PPL has been also used to open γ-substituted α-amino lactones in an enantioselective manner (Scheme 2.49) [240]. In this case the separation of the product from the remaining unhydrolysed substrate is particularly easy due to their difference in solubility at different pH-values. The unreacted lactone was extracted into ether from the aqueous phase at pH 7.5. Subsequent acidification led to the relactonisation of the hydroxy-acid.

Scheme 2.48. Resolution of bicyclic acetate by hydrolases present in crude PPL.

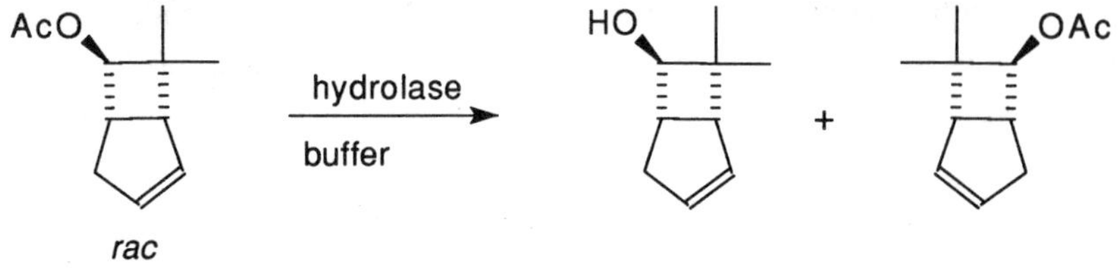

Enzyme	Reaction Rate	Selectivity (E)
crude PPL	good	>300
pure PPL	no reaction	---
α-chymotrypsin	no reaction	---
cholesterol esterase	fast	17
novel hydrolase	good	210

Scheme 2.49. Enantioselective hydrolysis of lactones by PPL.

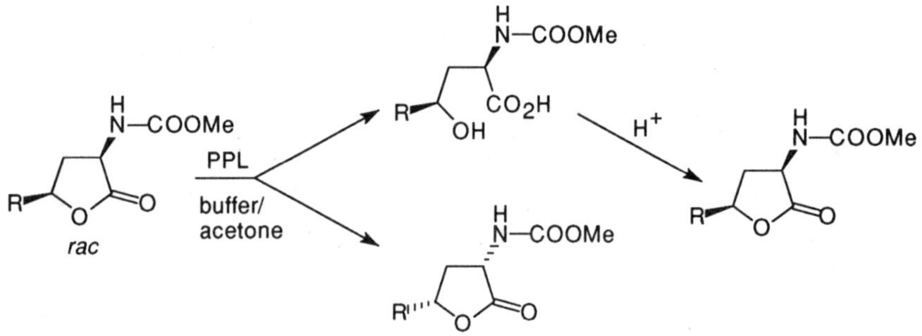

R	e.e. Acid [%]	e.e. Lactone [%]
H-	71	62
Ph-	86	32
$CH_2=CH-$	90	95

Candida sp. Lipases

Several crude lipase-preparations are available from the yeasts *Candida lipolytica*, *C. antarctica*, *C. rugosa* and *C. cylindracea* (CCL). The latter has been frequently used for the resolution of esters of secondary alcohols [241-246] and - to a lesser extent - for the resolution of α-substituted carboxylates [247, 248]. The CCL preparations from several commercial sources which

2.1 Hydrolytic Reactions

contain up to 16% of protein [249] differ to some extent in their activity but their selectivity is very similar [250]. As CCL is able to accommodate relatively bulky esters in its active site, it is the lipase of choice for the selective hydrolysis of esters of cyclic secondary alcohols. To illustrate this point, some representative examples are given below.

The racemic cyclohexyl enol ester shown in Scheme 2.50 was enzymatically resolved by CCL to give a ketoester with an S-stereo centre on the α-position (77% e.e.) coupled with a diastereoselective protonation of the liberated enol, which led to an R-configuration on the newly generated center on the γ-carbon atom. Only a trace of the corresponding S,S-diastereomer was formed; the remaining R-enol ester was obtained optically pure [251]. In accord with the concept that an acyloxy moiety is hydrolysed preferentially to a carboxylate by 'typical' lipases, no significant hydrolysis on the ethyl carboxylate was observed.

Scheme 2.50. Enzymatic resolution of a cyclic enol ester by CCL.

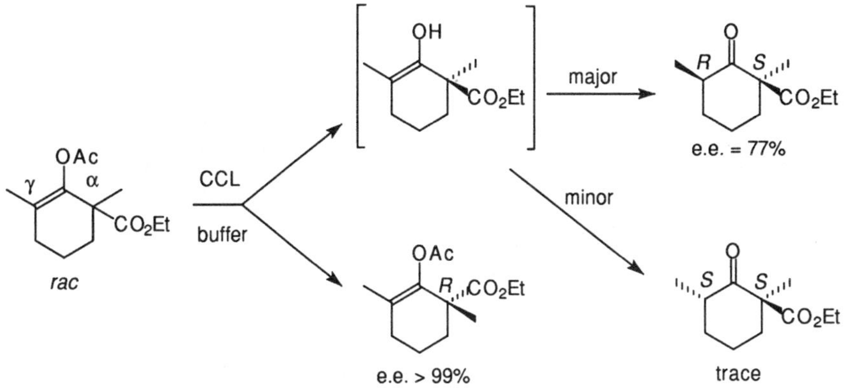

Racemic 2,3-dihydroxy carboxylates, protected as their respective acetonides, were resolved by CCL [252] by using their lipophilic n-butyl esters (Scheme 2.51). It is particularly noteworthy that the bulky α-methyl derivatives could be transformed, which are usually not accepted by hydrolases.

A number of cyclohexane 1,2,3-triols were obtained in optically active form via resolution of their esters using CCL and *Pseudomonas* sp. lipase (PSL) as shown in Scheme 2.52 [253]. To prevent acyl migration which would lead to racemisation of the product, two of the hydroxyl groups in the substrate molecule were protected as the corresponding acetal. In this case, a variation of the acyl chain from acetate to butanoate increased the reaction rate, but had no significant effect on the selectivity of the enzyme.

Scheme 2.51. Enzymatic resolution of 1,3-dioxolane-4-carboxylates by CCL.

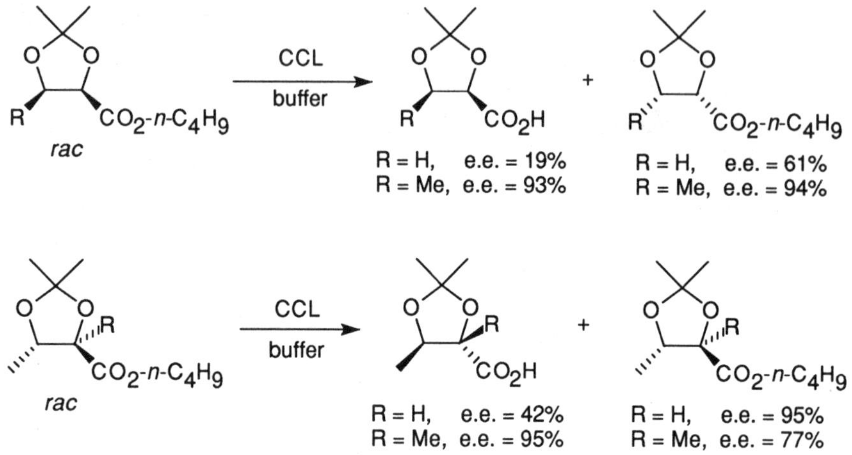
CCL
buffer
rac
R = H, e.e. = 19%
R = Me, e.e. = 93%
R = H, e.e. = 61%
R = Me, e.e. = 94%
CCL
buffer
rac
R = H, e.e. = 42%
R = Me, e.e. = 95%
R = H, e.e. = 95%
R = Me, e.e. = 77%

Scheme 2.52. Enzymatic resolution of cyclohexane-1,2,3-triol esters by CCL and PSL.

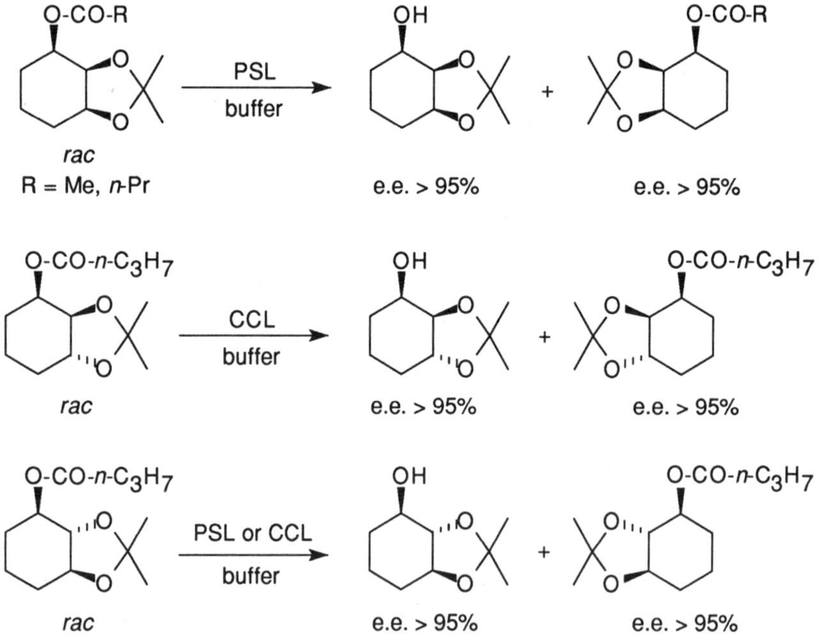
PSL
buffer
rac
R = Me, n-Pr
e.e. > 95%
e.e. > 95%
CCL
buffer
rac
e.e. > 95%
e.e. > 95%
PSL or CCL
buffer
rac
e.e. > 95%
e.e. > 95%

2.1 Hydrolytic Reactions

In order to provide a general tool which would allow to predict the sense and magnitude of the enantioselection of CCL involving esters of bicyclic secondary alcohols, a substrate model for CCL has been developed [254]. Evaluation of the results obtained from 25 structurally different esters possessing a rigid norbornane-type framework [255-257], through analysing the relationship between substrate structure and the selectivity of CCL, led a set of rules. This should allow the rational 'redesign' of substrate esters, which are transformed with insufficient selectivity and speed.

Figure 2.12. Substrate-model for CCL.

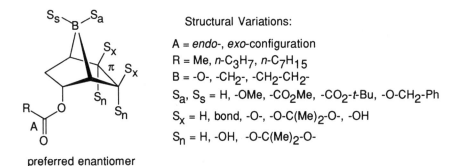

Structural Variations:

A = *endo*-, *exo*-configuration
R = Me, n-C_3H_7, n-C_7H_{15}
B = -O-, -CH_2-, -CH_2-CH_2-
S_a, S_s = H, -OMe, -CO_2Me, -CO_2-t-Bu, -O-CH_2-Ph
S_x = H, bond, -O-, -O-C(Me)$_2$-O-, -OH
S_n = H, -OH, -O-C(Me)$_2$-O-

preferred enantiomer

	Region	Requirements
A	Site of reaction	must be *endo*, R may be *n*-alkyl, preferably n-C_3H_7
B	Bridge	may contain hetero atoms, but must be small
S_a,S_s	*anti-syn* Substituents	methylene bridges may carry small ester, ether or acetal groups
S_x	*exo*-Substituents	'allowed region', substituents may be large
S_n	*endo*-Substituents	'forbidden region', substituents (if any) must be very small
π	π-Site	π-electrons in this site enhance the selectivity

The fact that crude *Candida cylindracea* lipase occasionally exhibits a moderate selectivity particularly on α-substituted carboxylic esters could be attributed to the presence of two isomeric forms of the enzyme present in the crude preparation [258]. Both forms (denoted as fraction A and B), could be separated by Sephadex chromatography and were shown to possess a qualitatively identical but quantitatively different stereoselectivity. Thus racemic α-phenyl propionate was resolved with low selectivity (E = 10) using crude CCL, whereas enzyme-fraction A was highly selective (E >100). The isomeric lipase fraction B showed almost the same moderate selectivity pattern as did the crude enzyme, although it possessed the same stereochemical preference. Treatment of form B with the surface-active agent deoxycholate and an organic solvent system (ethanol/ether) forced its transformation into the more stable conformer, form A, via an unfolding-refolding rearrangement. This non-covalent modification of the enzyme provided a method for the transformation of CCL of form B into its more stable and more selective isomer.

Scheme 2.53. Selectivity enhancement of CCL by non-covalent enzyme modification.

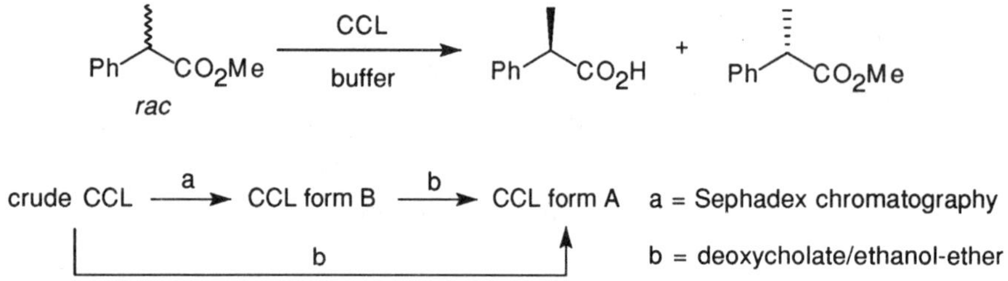

$$Ph\text{-CH(CO}_2Me) \text{ (rac)} \xrightarrow{\text{CCL, buffer}} Ph\text{-CH(CO}_2H) + Ph\text{-CH(CO}_2Me)$$

crude CCL $\xrightarrow{a}$ CCL form B $\xrightarrow{b}$ CCL form A   a = Sephadex chromatography

crude CCL $\xrightarrow{b}$ CCL form A   b = deoxycholate/ethanol-ether

Lipase	Selectivity (E)
crude CCL	10
CCL form A	>100
CCL form B	21
CCL form B[a]	>50

[a] After treatment with deoxycholate and ethanol/ether.

Pseudomonas sp. Lipases

Bacterial lipases isolated from *Pseudomonas fluorescens*, *P. aeruginosa* and *P. cepacia* are highly selective catalysts [259]. They seem to possess a more 'narrow' active site as compared to CCL, since they are often unable to accommodate bulky substrates, but they can be extremely selective on 'slim' counterparts [260-264]. Like the majority of the microbial lipases, the commercially available crude *Pseudomonas* sp. lipase-preparations (PSL) all possess a stereochemical preference for the hydrolysis of the *R*-esters of secondary alcohols, but the selectivity among the different preparations may differ to some extent [265]. recently, an active-site model for PSL has been proposed [266].

2.1 Hydrolytic Reactions

The exceptionally high selectivity of PSL on 'narrow' open-chain esters is demonstrated by the following examples.

The asymmetrisation of prochiral dithioacetal esters possessing up to five bonds between the prochiral centre and the ester carbonyl - the site of reaction - proceeded with high selectivity using a PSL [267]. This example of a highly selective chiral recognition of 'remote' chiral/prochiral centres is unusual amongst hydrolytic enzymes.

Scheme 2.54. Asymmetrisation of esters having remote prochiral centers by lipases.

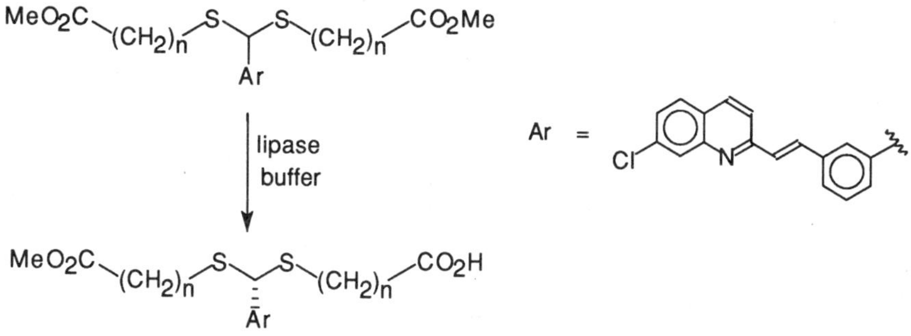

Lipase	n / Distance[a]	e.e. [%]
CCL	1 / 3	78
CCL	2 / 4	62
CCL	3 / 5	85
PSL	1 / 3	48
PSL	2 / 4	>98
PSL	3 / 5	79

[a] Number of bonds between the prochiral centre and the site of reaction.

Chirality need not reside on a quarternary carbon atom to be recognized by PSL but can be located on a sulphur atom. Thus optically pure aryl-sulphoxides were obtained by lipase-catalysed resolution of methyl sulphinyl acetates [268] in a biphasic medium containing toluene. The latter are important starting materials for the synthesis of chiral allylic alcohols via the 'SPAC'-reaction.

Scheme 2.55. Resolution of sulphoxide esters by PSL.

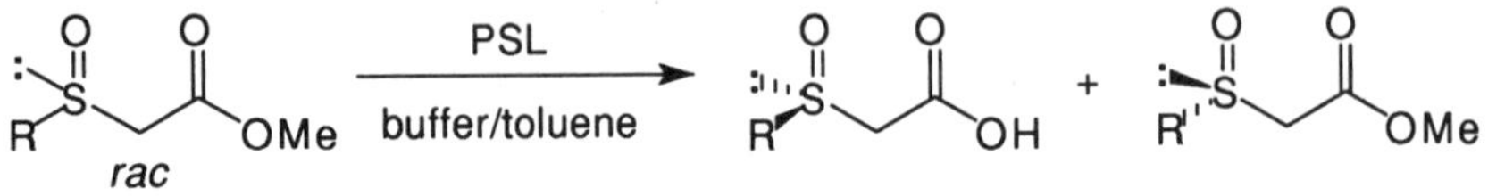

R	e.e. Acid [%]	e.e. Ester [%]
p-NO_2-C_6H_4-	97	>98
p-Cl-C_6H_4-	91	>98
Ph-	92	>98
C_6H_{11}-	>98	>98

Optically active α- and β-hydroxyaldehydes are useful chiral building blocks for the synthesis of bioactive natural products such as grahamimycin A_1 [269] and amino sugars [270]. These chiral synthons have been prepared from natural precursors or by microbial reduction of synthetic α- and β-ketoaldehyde derivatives [271], which allows the preparation of a single enantiomer. PSL-catalysed resolution of the corresponding hydroxy dithioacetal esters gave both enantiomers in excellent optical purity [272]. A significant selectivity enhancement caused by the sulphur atoms was demonstrated by the low optical purities obtained when the O-acetal esters were employed instead of the thioacetals (e.e.'s 11-27%).

Scheme 2.56. Resolution of dithioacetal esters by PSL.

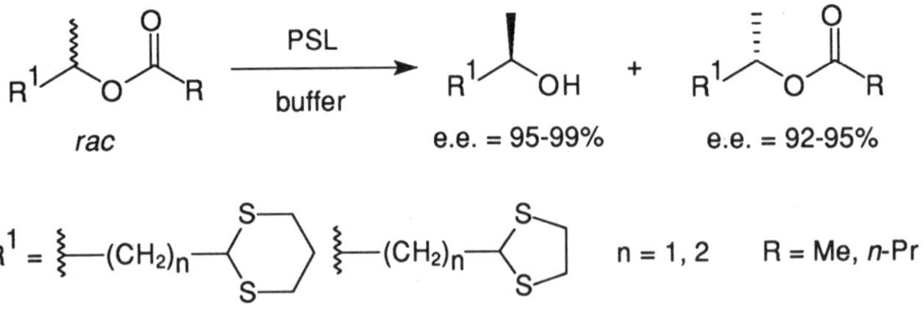

The selectivity of PSL-catalysed hydrolyses may be improved by substrate-modification through variation of the non-chiral acyl moiety [273]. Whereas alkyl- and chloroalkyl esters gave poor selectivities, the introduction of a sulphur atom to furnish the thioacetates proved to be advantageous. Thus, optically active β-hydroxynitriles, precursors of β-hydroxy acids and β-aminoalcohols, were conveniently resolved via the methyl- or phenyl-thiomethylcarbonyl derivatives.

Scheme 2.57. Resolution of β-acyloxynitriles by PSL.

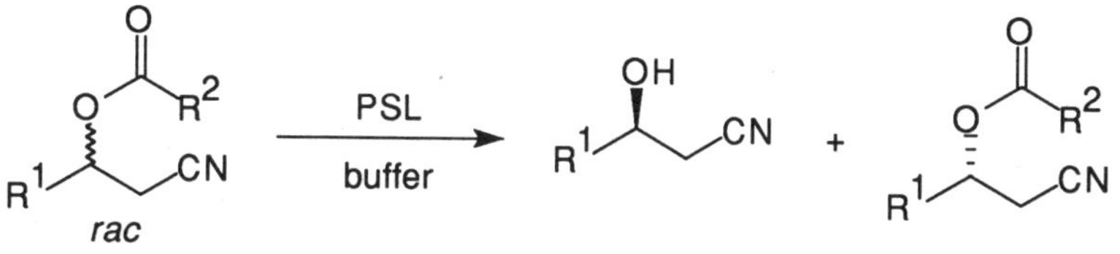

R^1	R^2	Selectivity (E)
Me-	Me	7
Me-	Cl-CH$_2$-	6
Me-	n-C$_3$H$_7$-	2
Me-	Me-O-CH$_2$-	14
Me-	Me-S-CH$_2$-	29
Me-	Ph-S-CH$_2$-	6
Ph-CH$_2$-CH$_2$-	Ph-S-CH$_2$-	36
PH-CH=CH-	Ph-S-CH$_2$-	55
Ph-	Ph-S-CH$_2$-	74

Scheme 2.58. Resolution of bicyclo[3.3.0]octane systems by CCL and PSL.

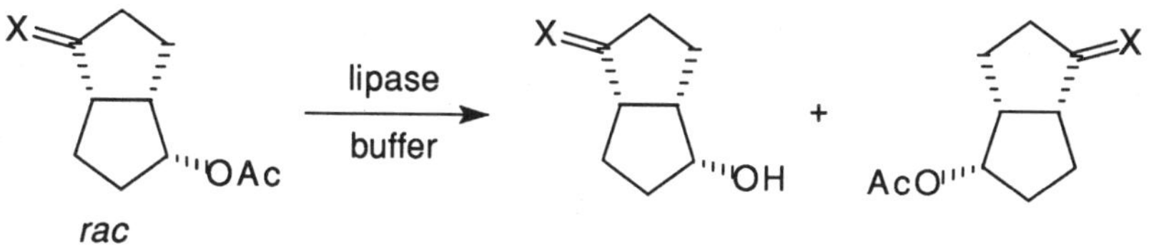

Lipase	X	e.e. Alcohol [%]	e.e. Ester [%]	Selectivity (E)
CCL	H	92	74	53
PSL	H	>99	>98	~1000
PSL	-O-(CH$_2$)$_2$-O-	99	99	~700

The trend that CCL often exhibits a lower stereoselectivity compared to PSL is demonstrated by the resolution of acetates derived from the rigid bicyclo[3.3.0]octane framework (Scheme 2.58) [274]. They were cleanly resolved into their enantiomers by PSL, whereas CCL exhibited a significantly lower selectivity.

Mucor sp. Lipases

Lipases from *Mucor* species (MSL) [275] such as *M. miehei* and *M. javanicus* have been used for biotransformations more recently [276, 277]. With respect to the steric requirements of substrates they seem to be related to the *Pseudomonas* sp. lipases. The different MSL-preparations are similar in their hydrolytic specificity [278]. The elucidation of the the three-dimensional structure of *Mucor miehei* lipase revealed that the chemical operator of this hydrolase is an Asp-His-Ser catalytic triad [279].

A case where only MSL showed good selectivity is shown in Scheme 2.59 [280]. The asymmetrisation of *meso*-dibutanoates of a tetrahydrofuran-2,5-dimethanol, which constitutes the central subunit of several naturally occurring polyether antibiotics [281] and platelet-activating-factor (PAF) antagonists, was investigated using different lipases. Whereas CCL showed low selectivity, PSL - as may be expected from the points mentioned above - was somewhat better. *Mucor* sp. lipase, however, was completely selective leading to optically pure hemiester products. This study also provided an interesting example of selectivity enhancement by medium engineering using PPL as the biocatalyst. Thus on performing the enzymatic hydrolysis in a biphasic system consisting of aqueous buffer and hexane (1:1), the selectivity was increased by a factor of more than four as compared to the results obtained with neat buffer. It should be noted that the analogous reaction of the acetate (R = H) with PLE at low temperature resulted in the formation of the opposite enantiomer [282].

The majority of lipase-catalysed transformations have been performed using PPL, CCL, PSL and MSL - the 'champion lipases' - and it may be assumed that most of the typical lipase-substrates may be resolved by choosing one of this group. The largely untapped potential of other lipases which are more rarely used, is illustrated by the following examples.

Sterically demanding α-substituted β-acyloxy esters were resolved using an *Aspergillus* sp. lipase (see Scheme 2.60). Again, introduction of a thioacetate as the acyl moiety improved the selectivity dramatically [283]. The diastereomeric *syn/anti*-conformation of the substrate was of critical importance due to the fact that, in contrast to the *syn*-substrates, only the *anti*-derivatives were resolved with high selectivities.

2.1 Hydrolytic Reactions

Scheme 2.59. Asymmetrisation of bis(acyloxymethyl)tetrahydrofurans by lipases.

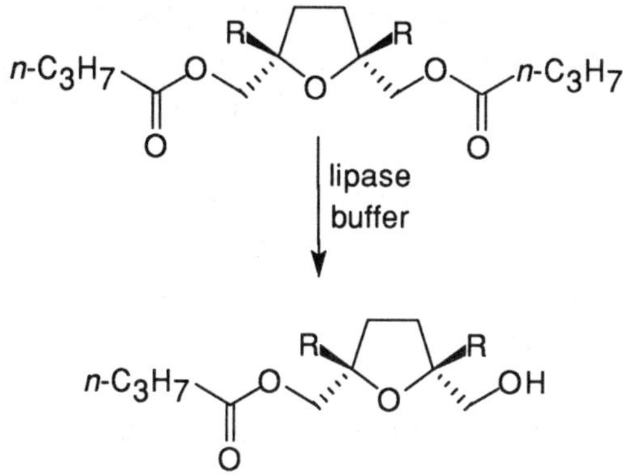

Lipase	R	Solvent system	e.e. [%]
CCL	H	buffer	12
PSL	H	buffer	81
MSL	H	buffer	>99
MSL	Me	buffer	>99
PPL	Me	buffer	20
PPL	Me	buffer/hexane	85

Scheme 2.60. Resolution of α-substituted β-acyloxy esters by *Aspergillus* sp. lipase.

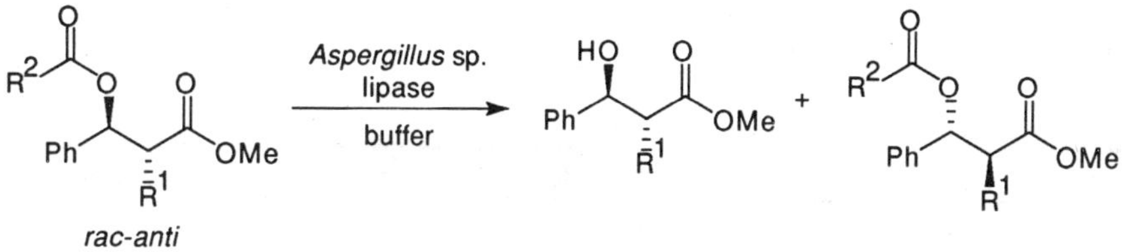

R^1	R^2	Selectivity (E)
Me-	Me-	80
Et-	Me-	20
Me-S-	Me-	30
Me-S-	Me-S-CH_2-	180

Optically pure cyanohydrins are required for the preparation of synthetic pyrethroids. The latter compounds are used as more environmentally acceptable agricultural pest control agents when compared to the classic highly chlorinated phenol derivatives. They also constitute important intermediates for the synthesis of chiral α-hydroxyacids, α-hydroxyaldehydes [284] and amino-alcohols [285]. They may be obtained via asymmetric hydrolysis of their respective acetates by microbial lipases [286]. By using this technique only the remaining non-accepted substrate enantiomer can be obtained in high optical purity. The cyanohydrin so-formed is spontaneously racemised since it is in equilibrium with the corresponding aldehyde, liberating hydrocyanic acid at pH values of above 4. However, it has recently been shown that the racemisation of the cyanohydrin can be avoided when the hydrolysis is carried out at pH 4.5 [287]. For the application of special solvent systems which can suppress this equilibrium see Section 3.1.

The resolution of the commercially important esters of (S)-α-cyano-3-phenoxybenzyl alcohol was not effected by lipases such as CCL and PSL. The best selectivities were obtained with lipases from *Chromobacterium* and *Arthrobacter* sp. [288], respectively.

Scheme 2.61. Hydrolysis of cyanohydrin esters using microbial lipases.

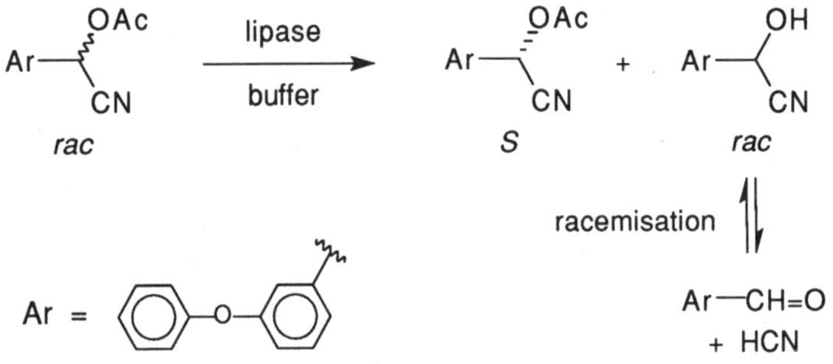

Lipase	e.e. Ester [%]	Configuration	Selectivity (E)
CCL	70	R	12
PSL	93	S	88
Alcaligenes sp.	93	S	88
Chromobacterium sp.	96	S	160
Arthrobacter sp.	>99	S	>1000

Another little known lipase is obtained from the mold *Geotrichum candidum* [289, 290]. The three-dimensional structure of this enzyme has recently been elucidated by X-ray crystallography [291] showing it to be a serine-hydrolase (like MSL), with a catalytic triad consisting of an *Glu*-His-Ser sequence, in contrast to the more usual *Asp*-His-Ser counterpart. It has a very high sequence homology to CCL (~40%) and shows a strong preference for oleic - i.e. (Z)-octadecenoic - and linoleic - i.e. (Z,Z)-9,12-octadecadienoic - acid moieties when acting on natural triacyl glycerols.

Amongst numerous lipases that have been tested, it was the only lipase which was able to resolve β-acyloxy fatty acid esters with chain lengths of C_{10} to C_{18} with acceptable selectivities. The corresponding β-hydroxy acids are structural components of several natural toxins of plants, microorganisms and higher animals [292].

Scheme 2.62. Hydrolysis of β-acyloxy esters using Geotrichum candidum lipase.

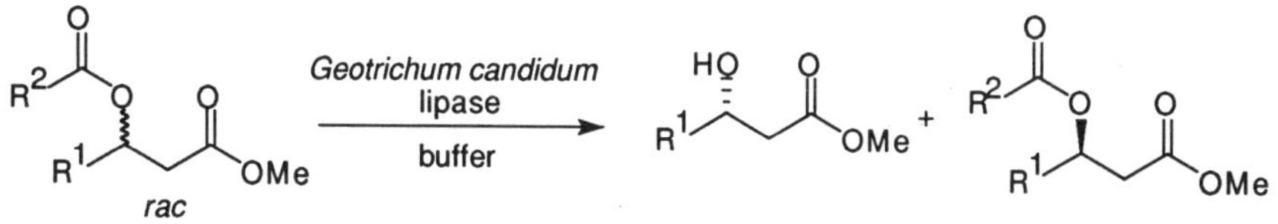

R^1	R^2	Selectivity (E)
$n\text{-}C_7H_{15}\text{-}$	$n\text{-}C_3H_7\text{-}$	11
$n\text{-}C_{11}H_{23}\text{-}$	$n\text{-}C_3H_7\text{-}$	20
$n\text{-}C_{15}H_{31}\text{-}$	$n\text{-}C_3H_7\text{-}$	20
$(n\text{-}C_4H_9)_2CH\text{-}(CH_2)_4\text{-}$	$n\text{-}C_3H_7\text{-}$	62
$n\text{-}C_{11}H_{23}\text{-}$	$Cl\text{-}CH_2\text{-}$	3
$n\text{-}C_{11}H_{23}\text{-}$	$n\text{-}C_7H_{15}\text{-}$	5

Another extracellular lipase, called 'Cutinase', is produced by the micoorganism *Fusarium solani pisi* for the hydrolysis of cutin - a wax ester which is excreted by plants in order to protect their leaves against microbial attack [293]. The enzyme has been purified to homogeineity and has been made readily available in sufficient quantity by genetic engineering. Up to now it has not yet been widely used for the biotransformation of non-natural esters, but it certainly has a potential for being a useful pure lipase to complement or even supersede one or more of the numerous crude preparations which are generally employed.

Optimisation of Selectivity

Many of the general techniques for an enzymatic selectivity-enhancement such as adjustment of temperature [294], pH, and the kinetic parameters of the reaction (product recycling or a sequential resolution) which were described for the hydrolysis of esters using esterases and proteases, are applicable to lipase-catalysed reactions as well. Due to the large number of available lipases the switch to another enzyme to obtain a better selectivity is relatively easy. Substrate-modification involving not only the chiral alcohol moiety of an ester but also its acyl group [295] as described above is a valuable technique for the improvement of lipase-catalysed transformations. Medium engineering with lipases has been shown to be more effective by applying biphasic systems (aqueous buffer plus water-immiscible organic solvent) instead of monophasic solvents (buffer plus water-miscible organic cosolvent).

Two special optimisation techniques which were developed with lipases in particular, are described below.

Bichiral esters. In general, the enzymatic resolution of an ester is carried out with substrates which have only one center of chirality. If it is located in the alcohol acid moiety (R/S), the acyl chain is usually non-chiral and vice versa. Since the selectivity of an enzyme depends on the whole structure of the substrate molecule, it is possible to improve the selectivity of the resolution of e.g. a racemic alcohol (RS) by using a chiral acid moiety (with a given predetermined configuration, R' or S', respectively, see Figure 2.13) instead of an achiral one. Thus a mixture of diastereomeric esters of type I is used as the substrate (R/R' and S/R' or R/S' and S/S'-esters, respectively). Of course, analogous considerations leading to an enhanced selectivity in the resolution of a racemic acid moiety by using a chiral alcohol with a predetermined centre are true (type II). To emphasize the presence of two different types of chiral centers this technique is called the 'bichiral ester-method' [296].

The asymmetric hydrolysis of the diastereomeric (S')-2-chloropropionate esters of (R/S)-1-phenylethanol catalysed by CCL proceeds with low selectivity by producing the (R)-alcohol and by leaving the (S'/S)-diastereomeric ester untouched (Scheme 2.63). On the other hand, when the acid moiety of opposite (R')-configuration was used instead, the selectivity was markedly increased [297].

During an extended study it was shown that the selectivity enhancement of an alcohol resolution may be significant (up to seven-fold) depending on the lipase used [298]. In acid-resolutions, however, the influence of a given chiral alcohol moiety on the selectivity was much smaller. This observation illustrates

2.1 Hydrolytic Reactions

the trend that lipases can recognize the chirality of an alcohol moiety rather than that of an acid residue.

Figure 2.13. Types of 'Bichiral esters'.

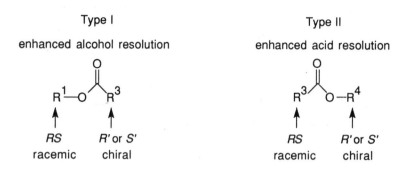

Scheme 2.63. Enhanced alcohol resolution by CCL using bichiral esters.

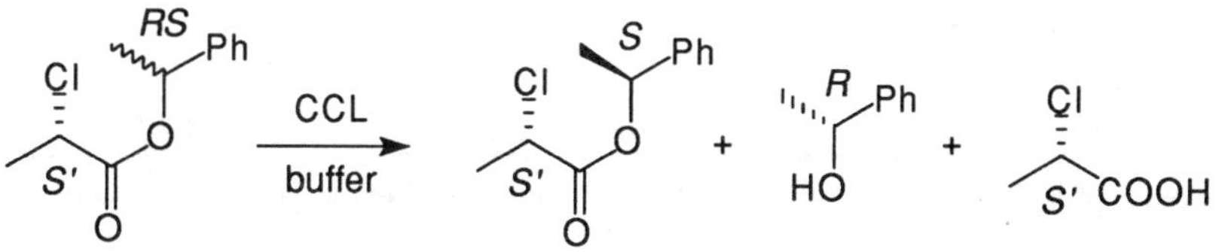

Reaction	Configuration of Acid	Selectivity (E)
S'/RS → S'/S + R + S'	S'	14
R'/RS → R'/S + R + R'	R'	32

Enantioselective inhibition of lipases. The addition of chiral bases such as amines and aminoalcohols has been found to have a strong influence on the selectivity of *Candida cylindracea* [299] and *Pseudomonas lipase* [300]. As shown in Scheme 2.64, the resolution of 2-aryloxypropionates by CCL proceeds with low to moderate selectivity in aqueous buffer alone. The addition of chiral bases of the morphinan-type to the medium led to a significant improvement, ca. one order of magnitude. Kinetic inhibition experiments revealed that the molecular action of the base on the lipase is a non-competitive inhibition - i.e. the base attaches itself to the lipase at a site other than the active site, causing the inhibition in the transformation of one enantiomer but not that of its mirror-image. Moreover, the chirality of the base has only a minor impact on the selectivity enhancement effect. Unfortunately, the general application of this technique is limited by the high cost of the morphinan alkaloids that are used.

Scheme 2.64. Selectivity enhancement of CCL by enantioselective inhibition.

Ar-O-CH(CO₂Me) (rac) →[CCL, buffer, inhibitor] Ar-O-CH(CO₂H) + Ar-O-CH(CO₂Me)

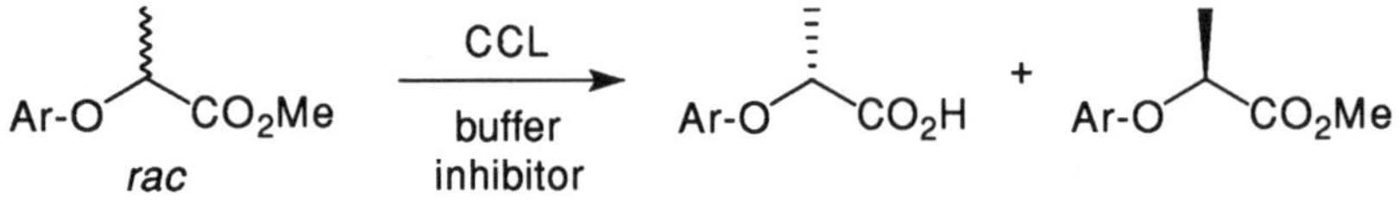

Ar-	Inhibitor	Selectivity (E)
2,4-dichlorophenyl-	none	1
2,4-dichlorophenyl-	morphine	2
2,4-dichlorophenyl-	quinine	4
2,4-dichlorophenyl-	dextromethorphan[a]	20
2,4-dichlorophenyl-	laevomethorphan	20
4-chlorophenyl-	none	17
4-chlorophenyl-	dextromethorphan	>100
4-chlorophenyl-	laevomethorphan	>100

[a] Dextro- or laevo-methorphan = D- or L-3-methoxy-N-methylmorphinane, respectively.

Scheme 2.65. Selectivity enhancement of PSL by enantioselective inhibition.

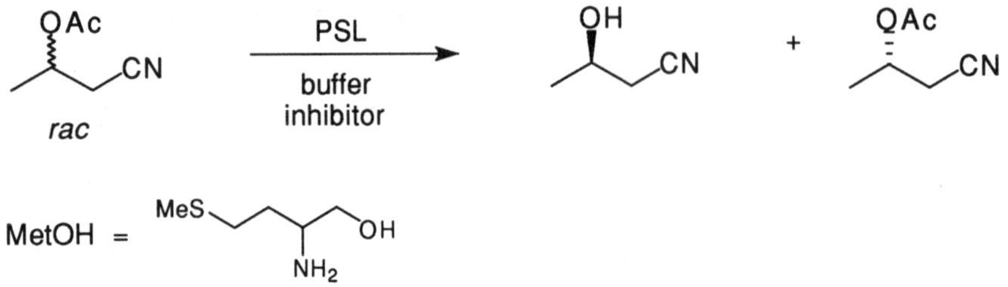

OAc-CH(CN) (rac) →[PSL, buffer, inhibitor] OH-CH(CN) + OAc-CH(CN)

MetOH = MeS-CH₂-CH(NH₂)-CH₂-OH

Inhibitor	Selectivity (E)
none	7
(RS)-MetOH[a]	7
(R)-MetOH	8
(S)-MetOH	14
dextromethorphan	15

[a] MetOH = methioninol = 2-amino-4-methylthio-1-butanol.

2.1 Hydrolytic Reactions

More recently, simple amino alcohols such as Methioninol (MetOH) were shown to have a similar effect on PSL catalysed resolutions. As depicted in Scheme 2.65, the chirality of the base had a significant influence on the selectivity of the reaction in this case.

2.1.4 Hydrolysis and Formation of Phosphate Esters

The hydrolysis of phosphate esters by phosphatases can often be achieved by chemical methods and the application of enzymes for this reaction is only advantageous if the substrate is susceptible to decomposition. Thus the enzymatic hydrolysis of phosphates has found only a limited number of applications. The same is true concerning the enantiospecific hydrolyses of racemic phosphates which afford a kinetic resolution.

However, the formation of phosphate esters is of importance, particularly when a regio- or enantioselective phosphorylation is required. Numerous bioactive agents display their highest activity only when they are transformed into phosphorylated analogues. Furthermore, a number of essential cofactors or cosubstrates for other enzyme-catalysed reactions of significant synthetic importance involve phosphate esters. For instance, nicotinamide adenine dinucleotide (NAD^+) or glucose-6-phosphate is an essential cofactor or cosubstrate, respectively, for dehydrogenase-catalysed reactions. Dihydroxyacetone phosphate is needed for enzymatic aldol-reactions, and adenosine triphosphate (ATP) represents the energy-rich phosphate donor for most biological phosphorylation reactions. Carbohydrate triphosphates are essential for glycosyl transfer reactions. Phosphatases have gained widespread use for the synthesis of such phosphate esters.

Hydrolysis of Phosphate Esters

Chemical hydrolysis of polyprenyl pyrophosphates is hampered by side reactions due to the lability of the molecule. Hydrolysis catalysed by acid phosphatase - an enzyme named because it displays its pH-optimum in the acidic range - readily afforded the corresponding dephosphorylated products in acceptable yields [301]. Similarly, the product from an aldolase-catalysed reaction is a labile 2-oxo-1,3,4-triol, which is phosphorylated in position 1. Dephos-phorylation under mild conditions without isolation of the intermediate phosphate species by using acid phosphatase is a method frequently used to obtain the chiral polyol products [302-305]. As shown in Scheme 2.66, enzymatic dephosphorylation of the aldol product obtained from 5-substituted hexanal derivatives gave the sensitive chiral keto-triol in good yield. In this case the latter product could be transformed into the sex pheromone of the bark beetle (+)-*exo*-brevicomin.

Scheme 2.66. Mild enzymatic hydrolysis of phosphate esters.

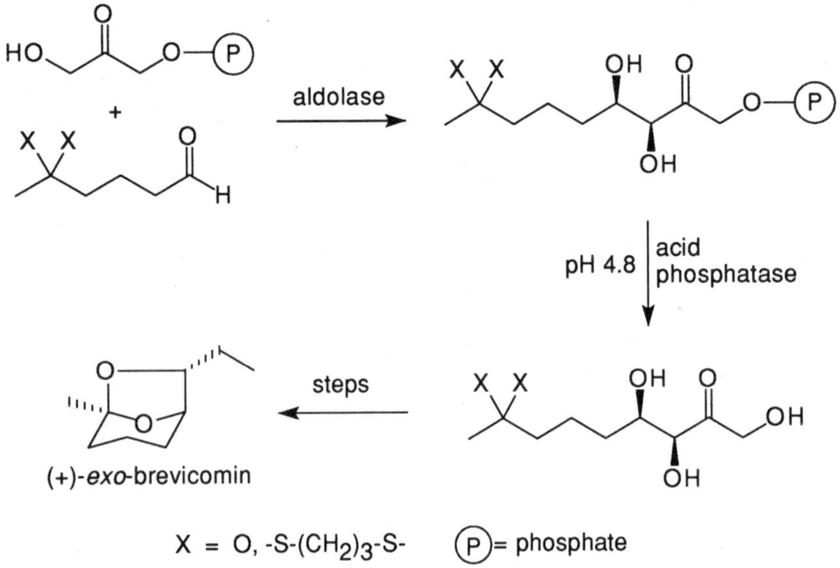

X = O, -S-(CH$_2$)$_3$-S-   (P) = phosphate

In comparison with the hydrolysis of carboxyl esters, enantioselective hydrolyses of phosphate esters have been scarcely reported. However, *rac*-threonine was resolved into its enantiomers via hydrolysis of its *O*-phosphate using acid phosphatase (see Scheme 2.67) [306]. As for the resolutions of amino acid derivatives using proteases, the natural L-enantiomer was hydrolysed leaving the D-counterpart behind. After separation of the D-phosphate from L-threonine, the D-enantiomer could be dephosphorylated using an unspecific alkaline phosphatase - an enzyme with the name derived from having its pH-optimum in the alkaline region.

Carbocyclic nucleoside analogues with potential antiviral activity such as aristeromycin [307] and fluorinated analogues of guanosine [308] were resolved via their 5'-phosphates using a 5'-ribonucleotide phosphohydrolase from snake venom (see Scheme 2.68). Again, the non-accepted enantiomer, possessing a configuration opposite to that of the natural ribose moiety, was dephos-phorylated by alkaline phosphatase.

2.1 Hydrolytic Reactions

Scheme 2.67. Resolution of threonine *O*-phosphate using acid phosphatase.

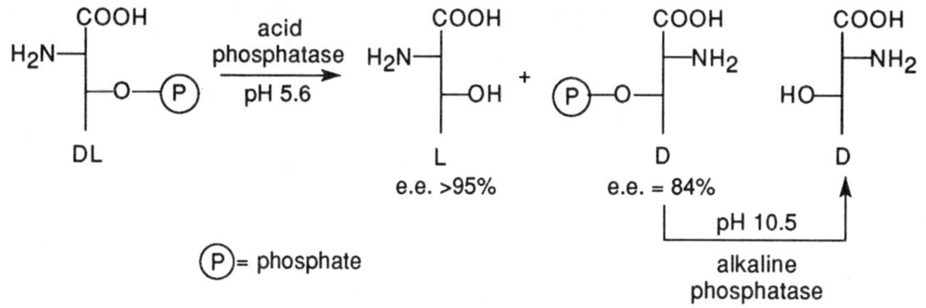

Ⓟ = phosphate

Scheme 2.68. Resolution of carbocyclic nucleoside analogues.

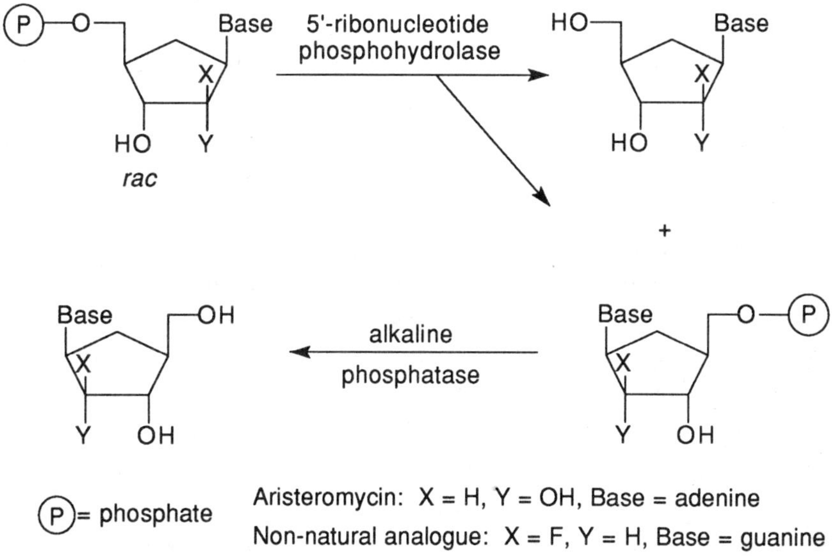

Ⓟ = phosphate

Aristeromycin: X = H, Y = OH, Base = adenine
Non-natural analogue: X = F, Y = H, Base = guanine

Formation of Phosphate Esters

The introduction of a phosphate moiety into a polyhydroxy compound by classic chemical methods usually requires a number of protection and deprotection steps. Employing enzymes for the regioslective formation of phosphate esters can eliminate many of these steps thus making these syntheses more efficient. Additionally, enantioselective transformations are also possible involving the asymmetrisation of prochiral or *meso*-diols or the resolution of racemates.

The formation of phosphate esters usually makes use of a sub-class of phosphatases - kinases -, which catalyse the transfer of a phosphate moiety (or a pyro- or triphosphate moiety in certain cases) from an energy-rich phosphate-donor, such as ATP which is used in the majority of cases. Due to the high price of these phosphate donors, they cannot be employed in stoicheometric amounts. Thus, efficient in situ-regeneration - recycling - is a precondition which has to be met to make phosphorylations economically feasible. ATP-recycling has been developed, especially by the group of G. M. Whitesides, to a stage where it has been made feasible on a molar scale [309, 310]. On the other hand, reversal of phosphate ester hydrolysis, i.e. the equivalent condensation reaction, has been performed in solvent systems with a reduced water content. Such systems would eliminate the use of expensive phosphate-donors but it is questionable if they will be of general use [311].

ATP-Recycling. All phosphorylating enzymes (kinases) require nucleoside triphosphates (in the majority ATP) as a cofactor. In living organisms, these energy-rich phosphates are regenerated by metabolic processes, but for biocatalytic processes, which are performed in vitro using purified enzymes, this does not occur. The (hypothetical) addition of stoicheometric amounts of these cofactors would not only be undesirable from a commercial standpoint but also for thermodynamic reasons. Quite often the accumulation of the inactive form of the consumed cofactor can tip the equilibrium of the reaction in the reverse direction. Thus, cofactors such as ATP are used in catalytic amounts and are continuously regenerated during the course of the reaction by an auxiliary system which usually consists of a second enzyme and a stoicheiometric quantity of an ultimate (cheap) phosphate donor. As the nucleoside triphosphates are intrinsically stable in solution, the XTP triphosphate species used in phosphorylation reactions is typically recycled in situ about 100 times, a value which is mainly limited by the need to obtain a convenient reaction rate. The total turnover numbers obtained in these reactions are in the range of about 10^6 to 10^8 mol of product per mol of enzyme.

Carbamate kinase catalyses the transfer of phosphate from carbamoyl phosphate to ADP [312]. The former can be conveniently prepared in situ from cyanate and phosphate, but it is spontaneously hydrolysed in the aqueous reaction medium. Furthermore, the unstable co-product carbamic acid spontaneously decarboxylates to form ammonia and carbon dioxide. Although this latter reaction would drive the phosphorylation to completion, it has a major drawback, that is ammonium ions displace essential Mg^{2+}-ions from the kinase. Thus this method has not gained widespread use.

2.1 Hydrolytic Reactions

The use of the phosphoenol pyruvate (PEP) / pyruvate kinase system is probably the most useful method for the regeneration of triphosphates [313]. PEP is not only very stable towards hydrolysis but it is also a stronger phosphorylating agent than ATP. Furthermore, nucleosides other than adenosine phosphates are also accepted by pyruvate kinase. The drawbacks of this system are the more complex synthesis of PEP [314, 315] and the fact that pyruvate kinase is inhibited by pyruvate at higher concentrations.

Scheme 2.69. Enzymatic recycling of ATP from ADP.

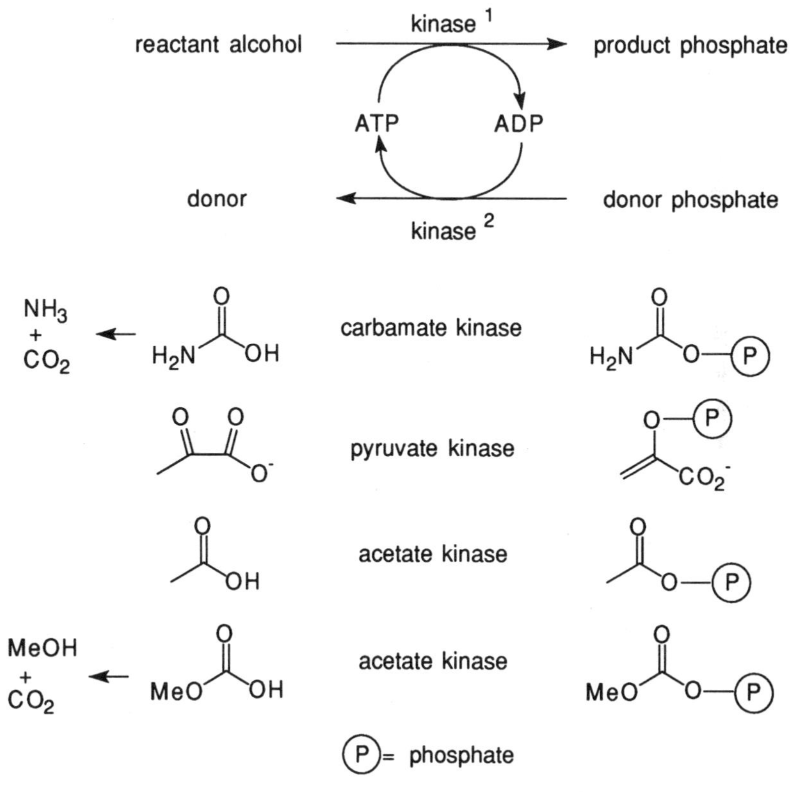

Acetyl phosphate, which can be made from acetic anhydride and phosphoric acid [316], is a commonly used regeneration system when used together with acetate kinase [317]. It is modestly stable in solution and while its phosphoryl donor potential is lower than that of PEP, it is considerably cheaper. As for pyruvate kinase, acetate kinase also can accept nucleoside phosphates other than adenosine, and it is inhibited by acetate ions.

As a cheap non-natural phosphate donor for the recycling of ATP, methoxycarbonyl phosphate has been proposed for acetate and carbamoyl kinase (but not for pyruvate kinase) [318]. This strong phosphorylating agent has the advantage that the liberated byproduct - methyl carbonate - readily decomposes to form relatively harmless materials (methanol and carbon dioxide), which can be physically removed from the reaction medium thus driving the reaction towards completion. Unfortunately, methoxycarbonyl phosphate has a relatively short half life time (20 min) for hydrolysis at pH 7.

A number of reactions which consume ATP generate AMP rather than ADP as a product. Still fewer produce adenosine [319]. A simple modification of the above- mentioned recycling systems for ADP makes possible the recycling of ATP from AMP. The addition of the enzyme adenosine kinase catalyses the phosphorylation of adenosine to give AMP, which in turn is further transformed to ADP. Both steps proceed with the consumption of ATP [320].

Scheme 2.70. Enzymatic recycling of ATP from adenosine and/or AMP.

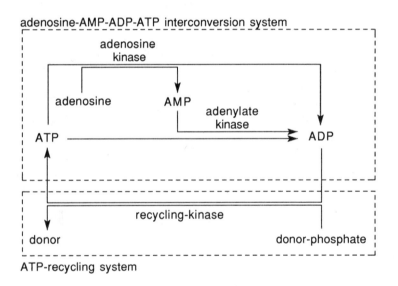

Regeneration of other nucleoside triphosphates (GTP, UTP and CTP) or the corresponding 2'-deoxynucleoside triphosphates (which are important

2.1 Hydrolytic Reactions

substrates for enzyme-catalysed glycosyl transfer reactions [321-323]) can be accomplished in the same manner using the acetate or pyruvate kinase systems.

Regioselective phosphorylation. The selective phosphorylation of hexoses (and their thia- or aza-analogues) on the primary alcohol moiety on position 6 can be achieved by a hexokinase [324]. The other secondary hydroxyl groups can be either removed or they can be exchanged for a fluorine atom. Such modified hexose-analogues represent potent enzyme-inhibitors and are therefore of interest as potential pharmaceuticals or pharmacological probes. In these studies ADP was recycled to ATP using pyruvate kinase and phosphoenol pyruvate as the ultimate phosphate donor.

Scheme 2.71. Phosphorylation of hexose derivatives by hexokinase.

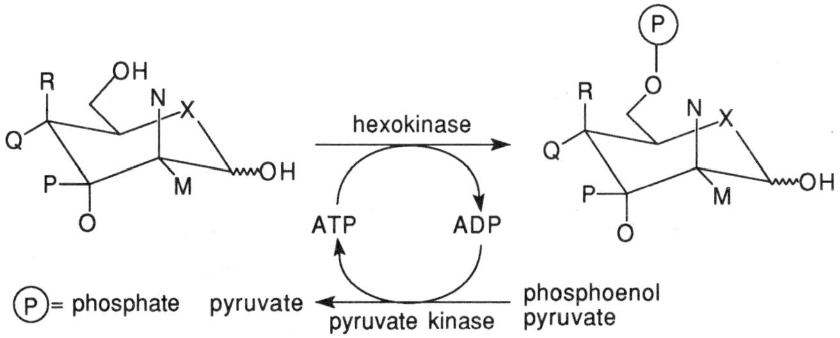

Substituents	Variations tolerated
M, P, Q	H, F, OH
N, O, R	H, F
X	O, S, NH

The most important compound in Scheme 2.71 is glucose-6-phosphate, which serves as a hydride source during the recycling of NADH when using glucose-6-phosphate dehydrogenase [325, 326] (see Section 2.2.1). Similarly, nicotinamide adenine dinucleotide (NAD$^+$) was converted into its more expensive phosphate NADP$^+$ using NAD kinase and acetyl phosphate as phosphate donor [327].

Scheme 2.72. Phosphorylation of dihydroxyacetone by glycerol kinase.

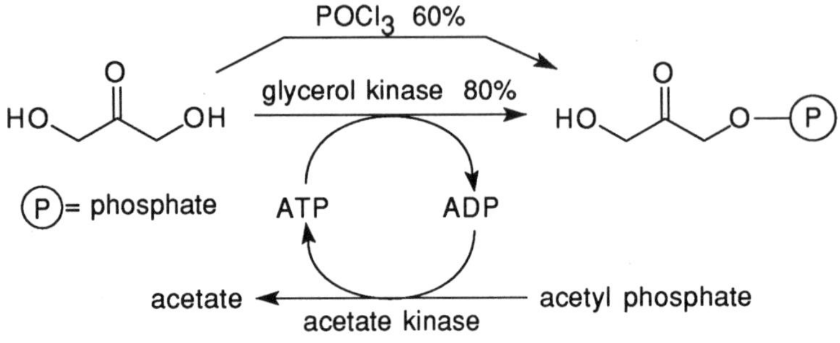

Scheme 2.73. Phosphorylation of D-ribose and enzymatic synthesis of UMP.

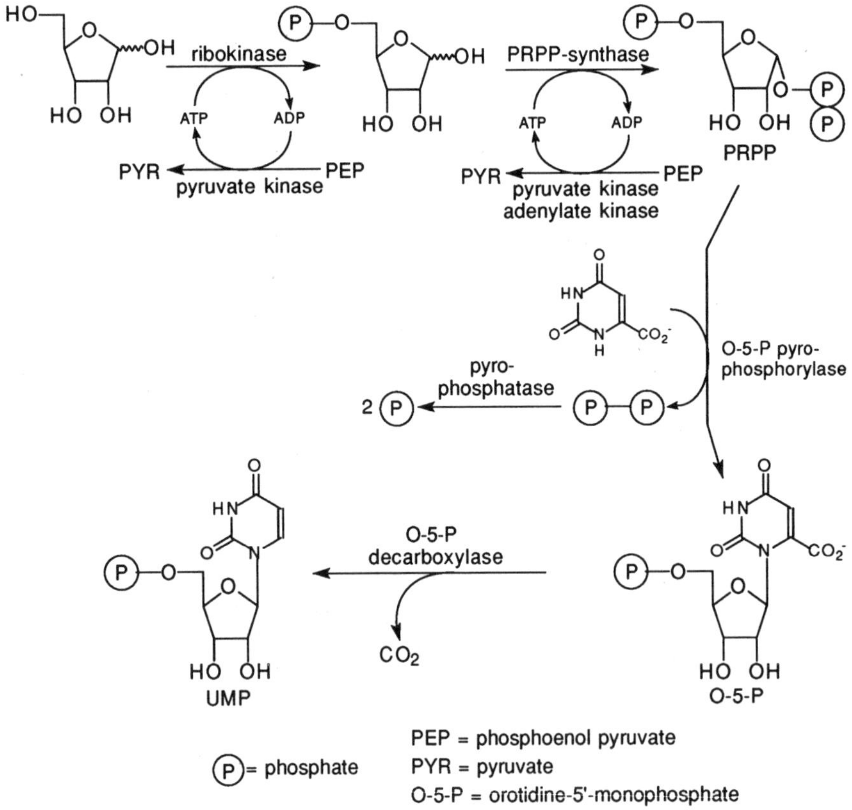

PEP = phosphoenol pyruvate
PYR = pyruvate
O-5-P = orotidine-5'-monophosphate

2.1 Hydrolytic Reactions

Another labile phosphate species, which is needed as cosubstrate for aldolase reactions, is dihydroxyacetone phosphate [328] (see Section 2.4.2). Its chemical synthesis using phosphorus oxychloride is hampered by moderate yields (60%) and by the occurrence of impurities remaining from the chemical steps which complicate the work-up procedure. Enzymatic phosphorylation, however, gives an 80% yield of a product, which is sufficiently pure so that it can be used directly in solution without isolation (Scheme 2.72).

5-Phospho-D-ribosyl-α-1-pyrophosphate (PRPP) serves as a key intermediate in the biosynthesis of purine, pyrimidine and pyridine nucleotides, such as nucleotide cofactors (ATP, UTP, GTP, CTP and NAD(P)H, Scheme 2.73). It was synthesized on a large scale from D-ribose using two consecutive phosphorylating steps [329]. Firstly, ribose-5-phosphate was obtained using ribokinase, and the pyrophosphate moiety was transferred from ATP onto the anomeric centre in the α-position by PRPP synthase. In this latter step, AMP rather than ADP was generated from ATP, so that an additional enzyme, adenylate kinase, was required for the recycling of ATP with phosphoenol pyruvate (PEP) as the phosphate donor. The PRPP obtained was subsequently transformed into uridine monophosphate (UMP) via two further enzymatic steps. The linkage of the base was accomplished by orotidine-5´-pyrophosphorylase (a transferase) and finally, decarboxylation of orotidine-5´-phosphate by orotidine-5´-phosphate decarboxylase (a lyase) led to UMP in 73% overall yield.

Enantioselective phosphorylation. Glycerol kinase [330] is not only able to accept its natural substrate - glycerol - to form *sn*-glycerol-3-phosphate [331], or close analogues of it such as dihydroxyacetone (see Scheme 2.72), but it is also able to transform a large variety of prochiral or racemic primary alcohols into chiral phosphates [332, 333]. The latter compounds represent synthetic precursors to phospholipids [334] and their analogues [335]. As depicted in Scheme 2.74, the glycerol backbone of the substrates may be structurally varied quite widely without affecting the high specificity of glycerol kinase. In resolutions of racemic substrates, both the phosphorylated species produced and the remaining substrate alcohols were obtained with excellent optical purities (88 to >94%). Interestingly, the phosphorylation of the amino alcohol shown in the last entry, occurred in an enantio- and chemoselective manner on the more nucleophilic nitrogen atom.

Scheme 2.74. Enantioselective phosphorylation of glycerol derivatives.

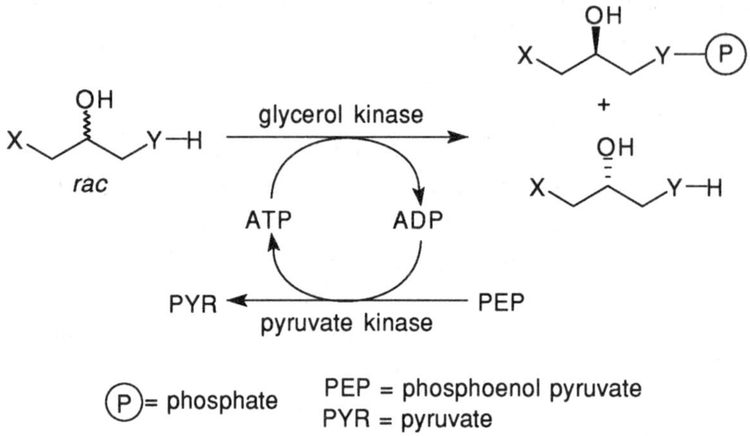

X	Y	e.e. Phosphate [%]	e.e. Alcohol [%]
Cl	O	>94	88
SH	O	94	n.d.
CH_3O	O	90	n.d.
Br	O	90	n.d.
OH	NH	>94	94

The evaluation of the data obtained from more than 50 substrates permitted the construction of a general model of a substrate that would be accepted by glycerol kinase.

Figure 2.14. Substrate model for glycerol kinase.

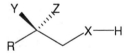

Position	Requirements
X	O, NH
Y	preferably OH, also H or F, but not NH_2
Z	H, OH (as hydrated ketone), small alkyl groups[a]
R	small groups, preferably polar, e.g. $-CH_2-OH$, $-CH_2-Cl$.

[a] Depending on enzyme source.

2.1.5 Hydrolysis of Epoxides

Epoxides are very useful compounds and are extensively employed as intermediates for the synthesis of enantiomerically pure bioactive compounds. Although several chemical methods are known for preparing them from optically active precursors, no efficient asymmetric syntheses involving asymmetrisation or resolution methods are known [336]. The only notably exception is the Sharpless-epoxidation, which is limited to allylic alcohols [337]. Microbial epoxidation of alkenes could provide an easy access to optically pure epoxides, but this technique has not yet been studied thoroughly enough to be of general use [338].

Enzymes catalysing the regio- and enantiospecific hydrolysis of epoxides - epoxide hydrolases (EH) [339] - play a key rôle in the metabolism of xenobiotics. In living cells, aromatics and olefins can be epoxidized by monooxygenases to form epoxides. Due to their electrophilic character the latter represent strong alkylating agents which makes them cancerogenic and teratogenic agents. In order to eliminate them from the cell, epoxide hydrolases catalyse their conversion into biologically harmless *trans*-1,2-diols, which can be further metabolized or excreted due to their higher water solubility. As a consequence, most of the epoxide-hydrolase activity found in higher organisms is located in organs such as the liver which are responsible for the detoxification of xenobiotics [340]. On the contrary, much less is known about corresponding enzymes from microbial sources.

Hepatic Epoxide Hydrolases

To date two main types of epoxide hydrolases from liver tissue have been characterized, i.e. a microsomal (MEH) and a cytosolic enzyme (CEH), which are different in their substrate specificities. As a rule of thumb, with non-natural epoxides, MEH has been shown to possess higher activities and selectivities as compared to its cytosolic counterpart.

Although pure MEH can be isolated from the liver of pigs, rabbits, mice or rats [341], a crude preparation of liver microsomes or even the 9000 g supernatant of homogenized liver can be a valuable source for EH-activity with little difference from that of the purified enzyme being observed. The enzyme is stable to storage at low temperature for several months [342]. Other enzyme-catalysed side-reactions such as ester-hydrolysis may occur with crude preparations. It must be noted that most of the transformations have been performed only on a millimolar scale without a specific preparative purpose, but scaling up should be possible.

The mechanism of enzymatic epoxide-hydrolysis is now firmly established to be of a base-catalysed S_N2-type [343, 344]. The imidazole unit of a histidine residue in the active site [345] provides nucleophilic assistance to a water molecule, which attacks the less hindered oxirane carbon atom with inversion of configuration [346]. The simultaneous transfer of a proton onto the anionic oxygen atom emanating from the ring-opening is postulated but not yet confirmed.

Figure 2.15. Mechanism of microsomal epoxide hydrolase.

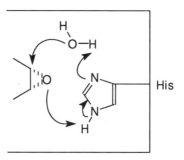

From the relatively large amount of data available for microsomal epoxide hydrolases, the following general rules can be deduced:

- Lipophilic aryl- or alkyl chains located close to the oxirane ring enhance the reaction rate, whereas polar groups, such as a hydroxy group, have an inhibitory effect. This latter behaviour may be explained by the fact that most xenobiotic substances are apolar compounds that need to be converted into more hydrophilic metabolites in order to facilitate their elimination [347].
- Depending on the substitution pattern of the substrates, terminal 1-substituted epoxides are hydrolysed more rapidly than their *cis*-1,2-disubstituted counterparts (see Scheme 2.75). The corresponding *trans*-1,2-disubstituted epoxides are not usually accepted by MEH (while they are by CEH [348]) and the sterically more demanding tri- and tetra-substituted derivatives are usually not substrates [349, 350].

One general limitation of enzymatic epoxide hydrolysis results from the limited stability of some oxiranes in the aqueous reaction medium at a pH value of around 9. Thus, unstable substrates are excluded a priori from this method since any undesired spontaneous hydrolysis would occur in a non-selective fashion and would deplete the optical purity of the product(s).

Scheme 2.75. Substrate types for MEH.

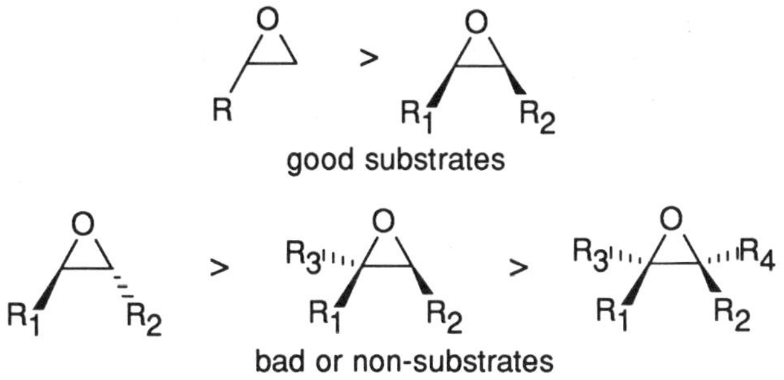

good substrates

bad or non-substrates

Racemic monosubstituted aryl- or alkyl epoxides may be resolved by MEH in accordance with the above mentioned mechanism, where in the more accessible ω-carbon atom of the (R)-epoxide is preferentially attacked to give the (R)-diol leaving the (S)-epoxide behind. As shown in Scheme 2.76, the nature of the substituent R is of significant importance in determining the selectivity. Whereas phenyl- [351], and straight-chain alkyl-epoxides were inefficiently resolved, branched alkyl chains increased the selectivity dramatically [352]. Furthermore, higher selectivities were observed at elevated substrate concentrations.

Scheme 2.76. Resolution of monosubstituted epoxides by MEH.

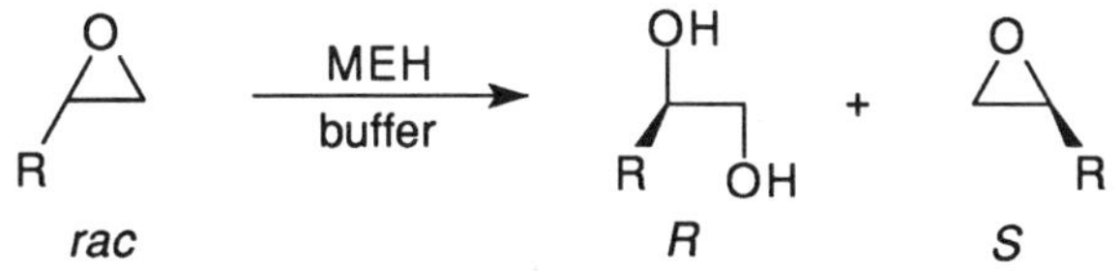

R	Conversion [%]	e.e. Diol [%]	Selectivity (E)
Ph-	18	<5	1.5
n-C$_4$H$_9$-	16	14	2
n-C$_8$H$_{17}$-	14	16	3
(CH$_3$)$_3$C-CH$_2$-	28	72	8
(CH$_3$)$_3$C-	38	92	42

1,2-Disubstituted aliphatic epoxides are also good substrates as long as they possess the *cis*-configuration [353]. Racemic *cis*-2-methyl-3-ethyloxirane was resolved by MEH leading to (R,R)-*threo*-2,3-pentanediol and residual (2R,3S)-

epoxide with excellent optical purity. The reaction proceeds in a complete regio- and enantioselective manner on the least hindered C-2 atom with a preference for the (2S)-enantiomer. Thus, the (R,R)-diol is produced by the S_N2-type of reaction via a Walden-inversion at carbon 2. According to the above-mentioned rules relating to the substrate structures preferred by MEH (Scheme 2.75), the corresponding *trans*-isomer was accepted at a much reduced reaction rate and with low enantioselectivity.

Scheme 2.77. Resolution of an aliphatic *cis*-disubstituted epoxide by MEH.

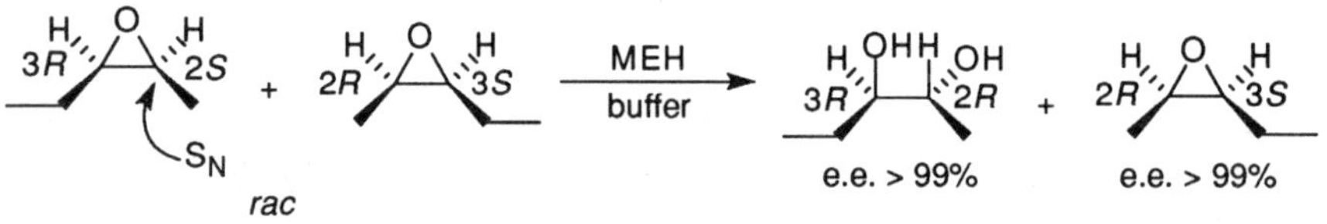

Cyclic *cis-meso*-epoxides can be asymmetrically hydrolysed to give *trans*-diols. In this case, once again the (S)-configurated oxirane carbon atom is attacked and inverted to yield an (R,R)-diol [354, 355]. In comparison to the microsomal epoxide hydrolase, cytosolic EH exhibited a lower specificity.

Scheme 2.78. Asymmetrisation of cyclic *cis-meso*-epoxides by EH.

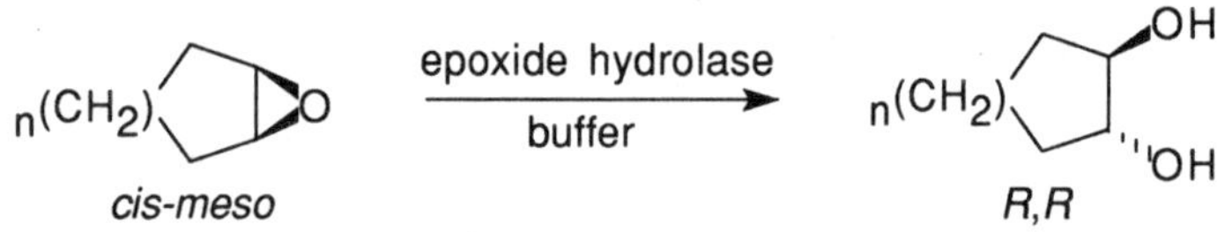

n	Optical Purity of Diol [%]	
	microsomal EH	cytosolic EH
1	90	60
2	94	22
3	40	30
4	70	no reaction

During a study on the topology of the active site of MEH using substituted cyclohexene oxides [356, 357], it has been shown that the preferentially hydrolysed enantiomer is the one in which the cyclohexane ring has 3,4-M helicity (Scheme 2.79) [358].

2.1 Hydrolytic Reactions

Scheme 2.79. Resolution of substituted cyclohexene oxides by MEH.

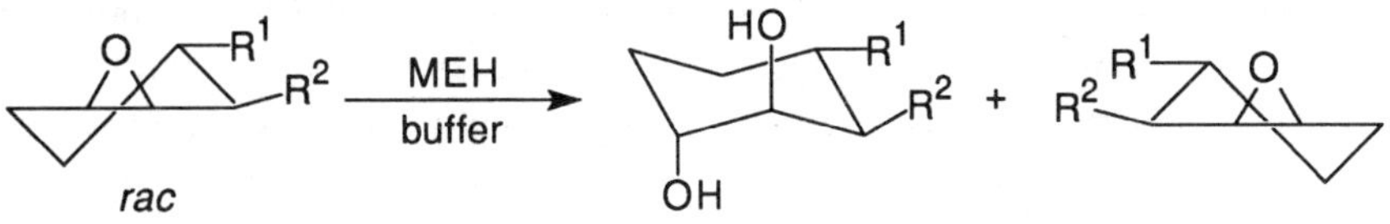

R^1	R^2	Conversion [%]	e.e. Diol [%]	Selectivity (E)
t-Bu	H	47	88	37
H	t-Bu	53	74	17

An interesting example of an enantioconvergent epoxide-hydrolysis was observed with racemic 3,4-epoxytetrahydropyran (Scheme 2.80) [359]. Both enantiomers of the substrate react at a similar rate leading to the (3R,4R)-diol as the sole product with an enantiomeric excess of >96%. This unusual stereochemical outcome of an attempted kinetic resolution was rationalized by postulating a diaxial opening of the epoxide with attack on the (S)-carbon atom in each enantiomer. Hence, one enantiomer is attacked at the C-3 position and the other at C-4.

Scheme 2.80. Enantioconvergent hydrolysis of (±)-3,4-epoxytetrahydropyran by MEH.

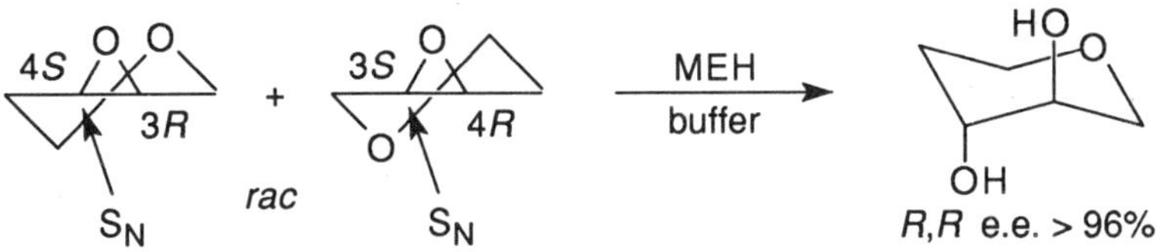

Scheme 2.81. Enzymatic hydrolysis of steroid epoxides and aziridines by MEH.

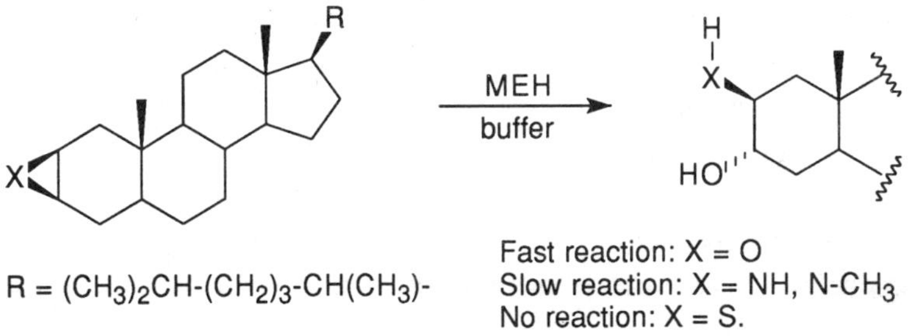

$R = (CH_3)_2CH-(CH_2)_3-CH(CH_3)-$

Fast reaction: X = O
Slow reaction: X = NH, N-CH$_3$
No reaction: X = S.

On steroid substrates, MEH has been shown to be able to hydrolyse epoxides and also (albeit at slower rates) the corresponding aziridines to form *trans*-1,2-aminoalcohols (Scheme 2.81) [360]. The fact that the thiirane was inert towards enzymatic hydrolysis was interpreted to be a reflection of the respective ring-strains. The source of enzyme activity was assumed to be the same microsomal epoxide hydrolase, but this assumption is still an open question.

Microbial Epoxide Hydrolases

Although microorganisms are known to possess epoxide hydrolases, the number of applications for preparative organic transformations is comparatively limited [361-363]. For example, a racemic allylic terpene alcohol containing a *cis*-trisubstituted epoxide moiety was hydrolysed by *Helminthosporium sativum* to yield the (*R,R*)-diol with concomitant oxidation of the alcoholic group: other minor metabolic products were observed. The mirror image (*S,S*)-epoxide was not transformed. Both optically pure enantiomers were then chemically converted into a juvenile hormone [364].

Scheme 2.82. Microbial resolution of a terpene epoxide.

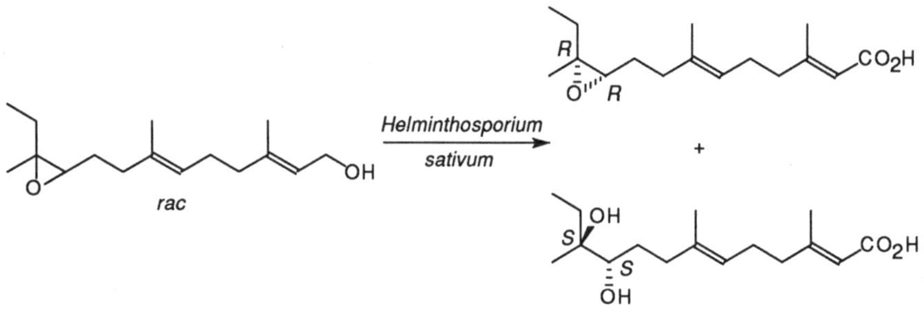

2.1.6 Hydrolysis of Nitriles

Organic compounds containing nitrile groups are found in the environment not only as natural products but also as a result of activities by man [365]. Naturally occurring nitriles are synthesized by several biological sources such as plants, fungi, bacteria, algae, sponges and insects, but not by mammals. Cyanide is highly toxic to living cells and interferes with biochemical pathways by three major mechanisms:
- Tight chelation to di- and tri-valent metal atoms in metallo-enzymes such as cytochromes,

2.1 Hydrolytic Reactions

- addition onto aldehydes or ketones to form cyanohydrin derivatives of enzyme substrates and
- reaction with Schiff's-base intermediates (e.g. in transamination reactions) to form stable nitrile derivatives [366].

As shown in Scheme 2.83, natural nitriles include cyanogenic glucosides which are produced by a wide range of plants including major crops such as cassava [367] and sorghum (millet). Plants and microorganisms are also able of producing aliphatic or aromatic nitriles (such as cyanolipids, ricinine and phenylacetonitrile [368]) often to protect themselves against attack by higher organisms.

Scheme 2.83. Naturally occurring organic nitriles.

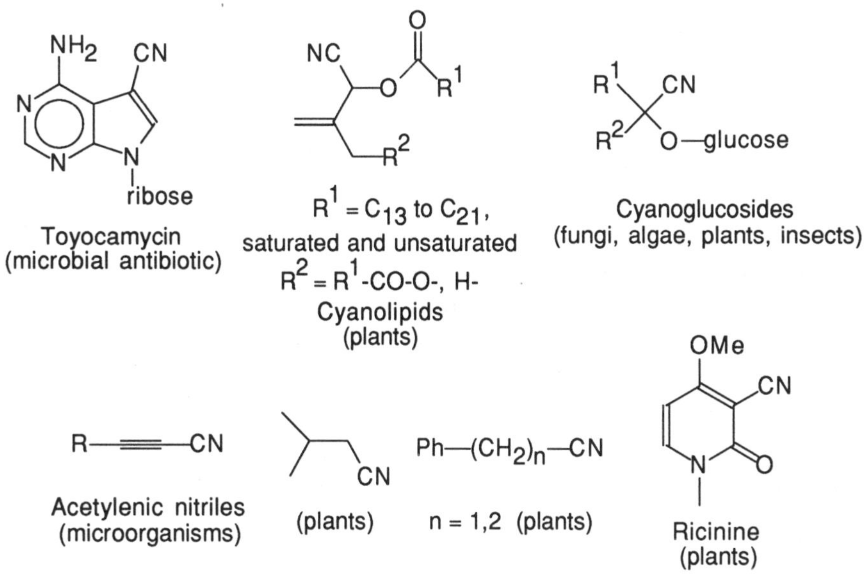

Toyocamycin (microbial antibiotic)

$R^1 = C_{13}$ to C_{21}, saturated and unsaturated
$R^2 = R^1\text{-CO-O-, H-}$
Cyanolipids (plants)

Cyanoglucosides (fungi, algae, plants, insects)

Acetylenic nitriles (microorganisms)

(plants)

$Ph\text{-}(CH_2)_n\text{-}CN$
$n = 1, 2$ (plants)

Ricinine (plants)

As a consequence, it is not unexpected that there are several biochemical pathways for nitrile-degradation, such as oxidation and - more important - by hydrolysis. Enzyme-catalysed hydrolysis may occur via two different pathways depending on the type of substrate [369, 370].
- Aliphatic nitriles are generally metabolized in two stages. First they are converted to the corresponding carboxamide by a nitrile hydratase and then to the carboxylic acid by an amidase enzyme (a protease) [371].
- Aromatic, heterocyclic and certain unsaturated aliphatic nitriles are directly hydrolysed to the correponding acids without formation of the intermediate

free amide by a so-called nitrilase enzyme. The nitrile hydratase and nitrilase enzyme use distinctively different mechanisms of action.

In view of more recent findings, however, this distinction should be regarded as less well defined [372].

Scheme 2.84. General pathways of the enzymatic hydrolysis of nitriles.

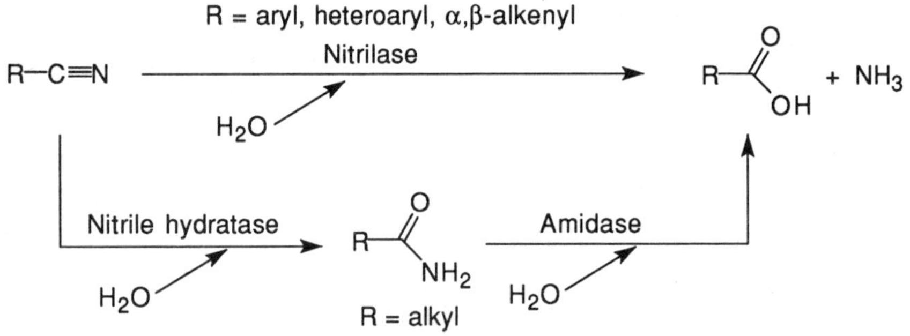

```
                    R = aryl, heteroaryl, α,β-alkenyl              O
                              Nitrilase                            ‖
    R—C≡N     ─────────────────────────────────────▶     R—C—OH   + NH₃
                        ↗
                      H₂O
      │                                                             ↑
      │                              O
      │ Nitrile hydratase            ‖            Amidase
      └──────────────────▶       R—C—NH₂    ──────────────────┘
                        ↗                          ↗
                      H₂O                        H₂O
                                   R = alkyl
```

Nitrile hydratases are known to possess a pyrroloquinoline quinone (PQQ) as a prostetic group; a tightly bound metal atom (iron or cobalt [373]) is necessary for catalytic action [374-377]. Interestingly, PQQ is also known to be a cofactor for redox enzymes such as amine oxidases. The mechanism which has been proposed for nitrile hydratases is shown in Scheme 2.85. The central ferric ion is equatorially coordinated, possibly by the nitrogen atoms of four imidazole units. One of the axial ligands is a thiolate donor, the other is most probably water. Upon approach of the substrate, the latter can be replaced by the nitrile. The 1,2-diketone PQQ, which is present in its stable monohydrate form, then delivers a water molecule onto the nitrile group through a network of hydrogen bonds, thus forming a carboxamide moiety in a single step.

On the contrary, nitrilases operate by a completely different mechanism. They possess neither coordinated metal atoms, nor cofactors such as PQQ, but do have an essential sulfhydryl residue of a cysteine [378]. Their mechanism of action is similar to general base-catalysed nitrile hydrolysis. Nucleophilic attack by the sulfhydryl residue within the enzyme on the nitrile carbon atom forms an enzyme-bound imine, which is hydrated to give a tetrahedral intermediate. An acyl-enzyme intermediate is then formed after the expulsion of ammonia. The former intermediate is hydrolysed to yield a carboxylic acid, liberating the enzyme.

2.1 Hydrolytic Reactions

Scheme 2.85. Proposed mechanism of nitrile hydratases.

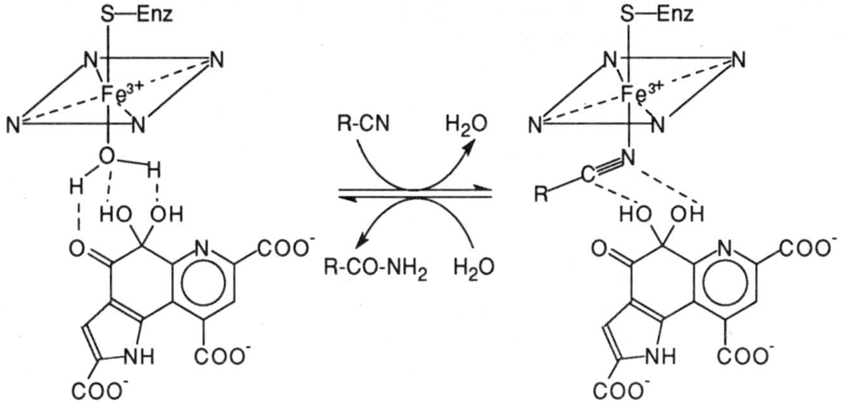

Scheme 2.86. Proposed mechanism of nitrilases.

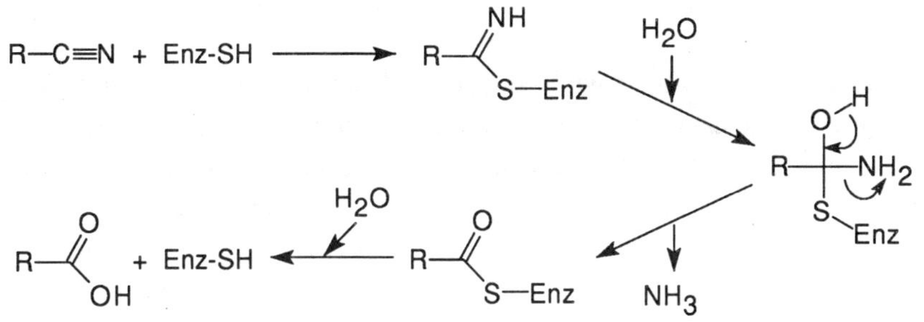

Enzymatic hydrolysis of nitriles is not only interesting from an academic standpoint, but also from a biotechnological point of view [379-382]. Cyanide represents a widely applicable C_1-synthon - a 'water-stable carbanion' - but the conditions usually required for the chemical hydrolysis of nitriles present several disadvantages. The reactions usually require either strongly acidic or basic media incompatible with other hydrolysable groups that may be present. Furthermore, energy consumption is high and unwanted side-products (such as cyanide itself or considerable amounts of other salts) are formed. Using enzymatic methods, conducted under mild conditions at or near physiological pH, most of these drawbacks can be avoided. Additionally, these transformations can often be achieved in a chemo-, regio- and (in certain cases) enantio-selective manner.

Another important aspect is the enzymatic hydrolysis of cyanide for the detoxification of industrial effluents [383-386].

Due to the fact that isolated enzymes involved in nitrile hydrolysis are generally very sensitive, the majority of transformations have been performed using whole cell systems.

Microbial Hydrolysis of Nitriles

The microorganisms used as sources of nitrile-hydrolysing enzymes usually belong to the genera *Bacillus*, *Brevibacterium*, *Micrococcus* and *Bacteridium* and they generally show metabolic diversity. Depending on the source of carbon and nitrogen - acting as 'inducer' - added to the culture medium, either nitrilases or nitrile hydratases are predominantly produced by the cell. Thus the desired hydrolytic pathway leading to an amide or a carboxylic acid can be biologically 'switched on' during the growth of the culture by using aliphatic or aromatic nitriles as inducers. In order to avoid substrate inhibition, which is a more common phenomenon with nitrile-hydrolysing enzymes than product inhibition [387], the substrates are fed continuously to the culture.

Acrylamide is one of the most important commodity chemicals for the synthesis of various polymers and is produced in an amount of about 200 000 tons per annum worldwide. In its conventional synthesis, the hydration of acrylonitrile is performed with copper catalysts. However, the preparative procedure for the catalyst, difficulties in its regeneration, problems associated with separation and purification of the formed acrylamide, and undesired polymerisation are serious drawbacks. Using whole cells of *Brevibacterium* sp. [388, 389], *Pseudomonas chlororapis* [390, 391] or *Rhodococcus rhodochrous* [392] acrylonitrile can be converted into acrylamide in yields of >99%; the formation of byproducts such as acrylic acid is circumvented by blocking of the amidase activity. The scale of this biotransformation has already exceeded 6000 tons per annum.

Scheme 2.87. Microbial hydrolysis of acrylonitrile.

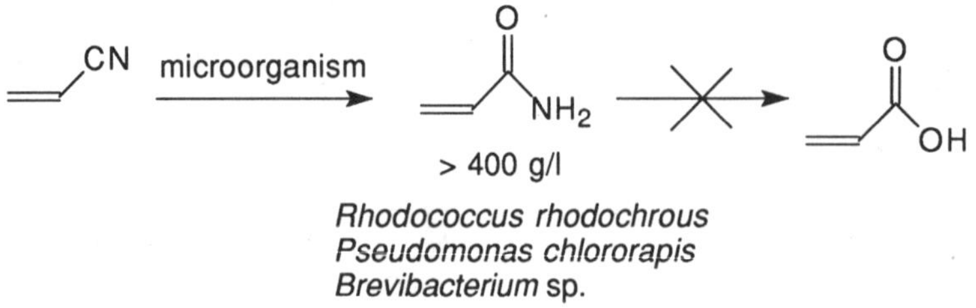

2.1 Hydrolytic Reactions

Aromatic and heteroaromatic nitriles were selectively transformed into the corresponding amides by *Rhodococcus rhodochrous* strain [393]; the products accumulated in the culture medium in significant amounts. In contrast to the hydrolyses performed by chemical means, the biochemical transformations were highly selective without the diminuating formation of the corresponding carboxylic acids.

Scheme 2.88. Microbial production of aromatic and heteroaromatic carboxamides.

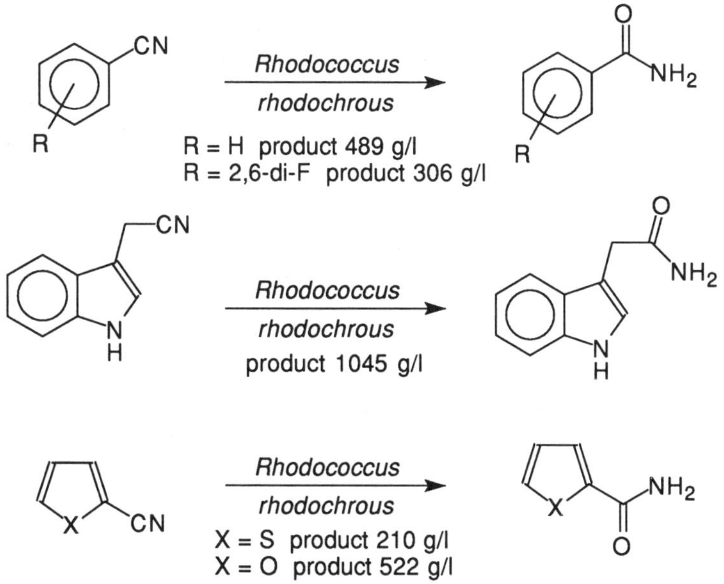

[Reaction 1: Aryl-CN with R substituent → Aryl-C(O)NH₂ via *Rhodococcus rhodochrous*; R = H product 489 g/l; R = 2,6-di-F product 306 g/l]

[Reaction 2: Indole-3-CH₂-CN → Indole-3-CH₂-C(O)NH₂ via *Rhodococcus rhodochrous*; product 1045 g/l]

[Reaction 3: 2-Furyl/Thienyl-CN → 2-Furyl/Thienyl-C(O)NH₂ via *Rhodococcus rhodochrous*; X = S product 210 g/l; X = O product 522 g/l]

Even more important from a commercial standpoint was that *o*-, *m*-, and *p*-substituted cyanopyridines were accepted as substrates [394, 395] to give picolinamide (a pharmaceutical), nicotinamide (a vitamin) and isonicotinamide (a precursor for isonicotinic acid hydrazide, a tuberculostatic). Extremely high productivities were obtained due to the fact that the less soluble carboxamide product readily crystallized from the reaction medium in greater than 99% purity. Nicotinamide is an important nutritional factor and therefore it is widely used as a vitamin additive for food and feed supplies. Pyrazinamide is also used as a tuberculostatic.

Scheme 2.89. Microbial production of aromatic and heteroaromatic carboxamides.

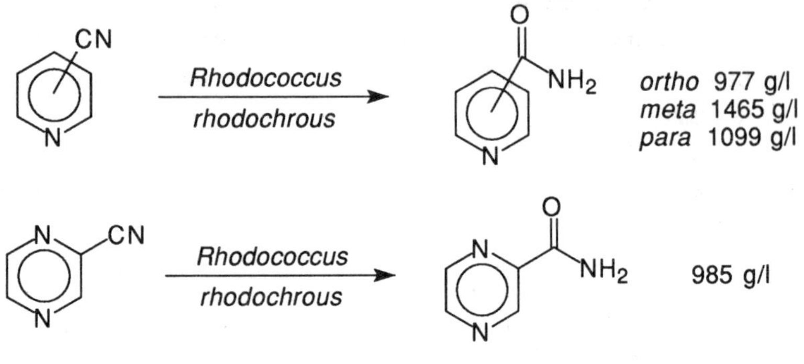

pyridine-CN → (Rhodococcus rhodochrous) → pyridine-C(O)NH$_2$ ortho 977 g/l
meta 1465 g/l
para 1099 g/l

pyrazine-CN → (Rhodococcus rhodochrous) → pyrazine-C(O)NH$_2$ 985 g/l

Scheme 2.90. Microbial production of aromatic and heteroaromatic carboxylic acids.

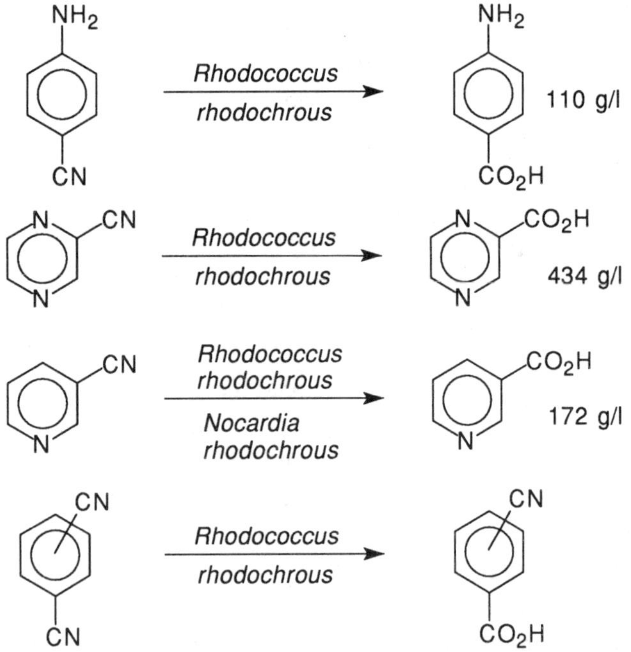

4-NH$_2$-C$_6$H$_4$-CN → (Rhodococcus rhodochrous) → 4-NH$_2$-C$_6$H$_4$-CO$_2$H   110 g/l

pyrazine-CN → (Rhodococcus rhodochrous) → pyrazine-CO$_2$H   434 g/l

3-cyanopyridine → (Rhodococcus rhodochrous / Nocardia rhodochrous) → nicotinic acid   172 g/l

dicyanobenzene → (Rhodococcus rhodochrous) → cyanobenzoic acid

By changing the biochemical pathway through using modified culture conditions, the corresponding carboxylic acids can be obtained as well (see Scheme 2.90). For instance, *p*-aminobenzoic acid, a member of the vitamin B

2.1 Hydrolytic Reactions

group, was obtained from *p*-aminobenzonitrile using whole cells of *Rhodococcus rhodochrous* [396]. Similarly, the antimycobacterial agent pyrazinoic acid was prepared in excellent purity from cyanopyrazine [397]. Like nicotinamide, nicotinic acid is a vitamin used in animal feed supplementation, in medicine and also as a biostimulator for the formation of activated sludge. Microbial hydrolysis of 3-cyanopyridine using *Rhodococcus rhodochrous* [398] or *Nocardia rhodochrous* [399] proceeds quantitatively whereas chemical hydrolysis is hampered by moderate yields. 1,3- and 1,4-dicyanobenzenes were selectively hydrolysed by *Rhodococcus rhodochrous* to give the corresponding monoacids [400, 401].

Tranexamic acid (*trans*-4-aminomethyl-cyclohexane-1-carboxylic acid, Scheme 2.91) is a hemostatic agent, which is synthesized from *trans*-1,4-dicyanocyclohexane. Complete regioselective hydrolysis was achieved by using an *Acremonium* sp. [402].

Scheme 2.91. Regioselective microbial hydrolysis of an alicyclic dinitrile.

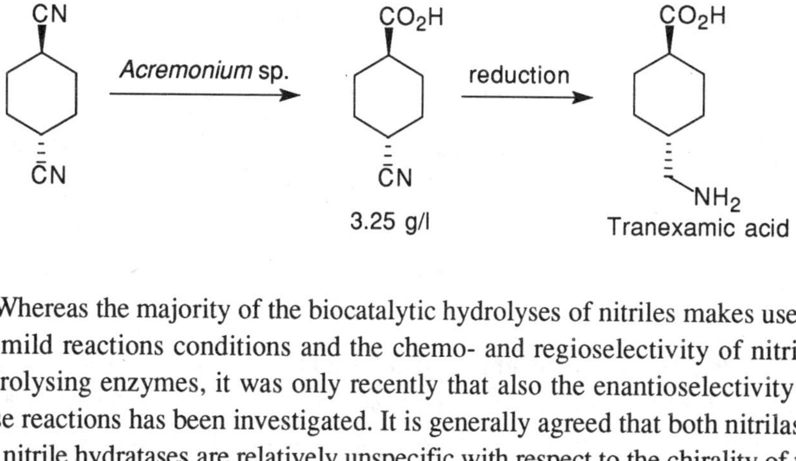

Whereas the majority of the biocatalytic hydrolyses of nitriles makes use of the mild reactions conditions and the chemo- and regioselectivity of nitrile-hydrolysing enzymes, it was only recently that also the enantioselectivity of these reactions has been investigated. It is generally agreed that both nitrilases and nitrile hydratases are relatively unspecific with respect to the chirality of the substrate, and that any enantiodiscrimination occurs during the hydrolysis of an intermediate carboxamide in the nitrile hydratase pathway by the amidase [403]. As a rule, the 'natural' L-configurated enantiomer is usually converted into the acid leaving the D-counterpart behind. This is not unexpected, if one remembers the high specificities of proteases on α-substituted carboxamides (see chapter 2.1.2).

Thus, α-hydroxy- and α-aminoacids can be obtained from the corresponding α-hydroxynitriles (cyanohydrins) and α-aminonitriles, which are easily synthesized in racemic form from the corresponding aldehyde precursors by addition of hydrogen cyanide or a Strecker synthesis, respectively.

Scheme 2.92. Enantioselective hydrolysis of α-hydroxynitriles.

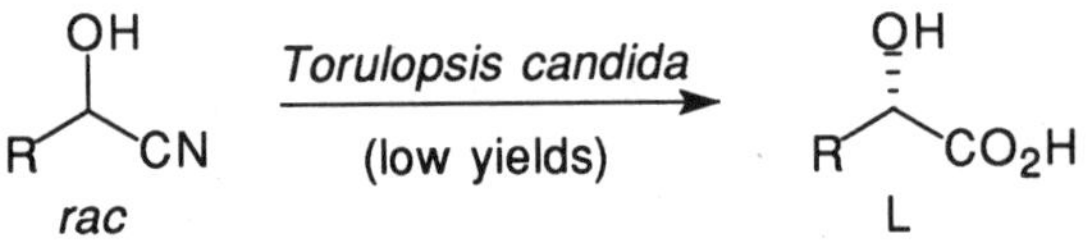

$$\underset{rac}{R\underset{CN}{\overset{OH}{\diagup}}} \xrightarrow[\text{(low yields)}]{\textit{Torulopsis candida}} \underset{L}{R\underset{CO_2H}{\overset{OH}{\diagup}}}$$

R	e.e. α-Hydroxyacid [%]
(CH3)2CH-	>90
(CH3)2CH-CH2-	>95

Whereas this method has been used only scarcely for the resolution of α-hydroxyacids [404], it has been applied to the hydrolysis of a large range of substrates of the α-aminonitrile-type [405-408].

Scheme 2.93. Enantioselective hydrolysis of α-aminonitriles.

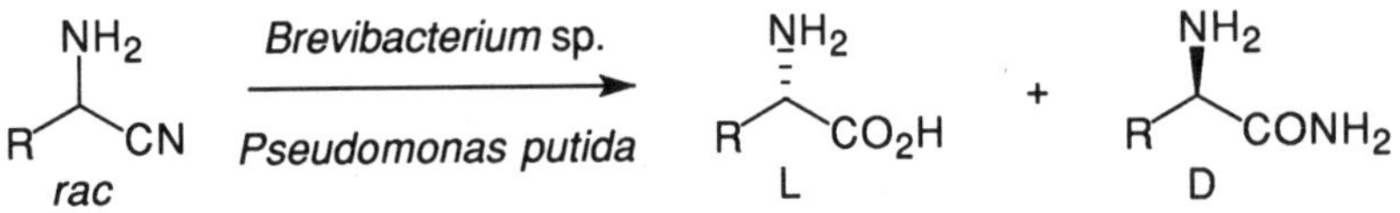

$$\underset{rac}{R\underset{CN}{\overset{NH_2}{\diagup}}} \xrightarrow[\textit{Pseudomonas putida}]{\textit{Brevibacterium sp.}} \underset{L}{R\underset{CO_2H}{\overset{NH_2}{\diagup}}} + \underset{D}{R\underset{CONH_2}{\overset{NH_2}{\diagup}}}$$

R = -CH3, -C2H5, (CH3)2CH-, (CH3)2CH-CH2-, -CH2-Ph, -CH2-S-CH3, -(CH2)2-CO2H, -(CH2)2-CONH2

An interesting example for the resolution of a secondary nitrile was recently reported [409]. Using a *Rhodococcus butanica* strain, several α-aryl substituted propionitriles were resolved with high selectivities. As expected, the 'natural' L-acid was formed and the D-amide accumulated in the medium. The former compounds constitute an important class of non-steroidal antiinflammatory agents. A minor amount of the (*S*)-nitrile, having an optical purity of 73%, was detected in one case. This latter observation may indicate that the enantiodiscrimination may not take place only during the second hydrolytic step of the reaction (i.e. amide-hydrolysis), as widely assumed, but also during the hydration of the nitrile-unit catalysed by the nitrile hydratase.

2.1 Hydrolytic Reactions

Scheme 2.94. Enantioselective hydrolysis of α-aryl propionitriles.

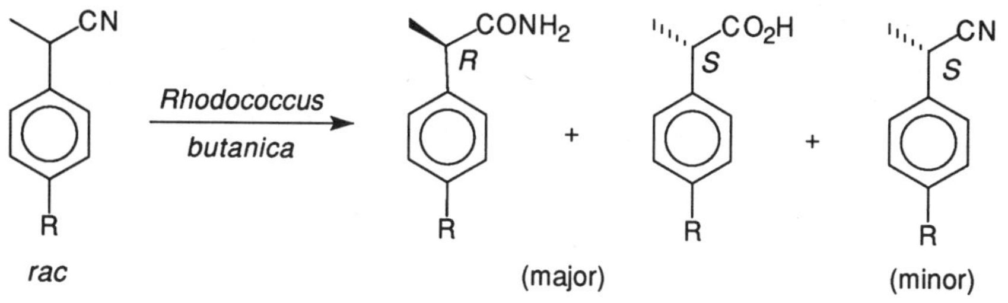

R	e.e. Amide [%]	e.e. Acid [%]	e.e. Nitrile [%]
CH$_3$-CH$_2$-CH(CH$_3$)-	99	87	73
Cl	76	>99	-
OCH$_3$	99	99	-

Enzymatic Hydrolysis of Nitriles

Enzymatic hydrolysis of nitriles using whole cell systems is certainly a powerful biocatalytic tool but it requires the experience of a microbiologist for the induction of the desired enzyme system(s) and elimination of undesired pathways by blocking or by using mutant strains. These procedures can be laborious and cannot be done in the average organic laboratory. To avoid this drawback, a ready-to-use crude enzyme preparation - denoted as SP 409 - derived from a *Rhodococcus* sp. containing a nitrile hydratase/amidase activity recently became available in immobilized form [410]. The enzyme system is able to accept both aliphatic and aromatic nitriles and converts them into carboxamides or acids, respectively [411-414]. Although preliminary studies indicate an influence of the substrate structure on the outcome of the reaction (i.e. whether amide or acid is produced), general rules cannot be proposed at present.

Aromatic dinitriles such as *m*- and *p*-dicyanobenzene were selectively hydrolysed to the corresponding cyano-acids in good yield. In case of the heterocyclic nitriles, the nature of the substrate determined the outcome of the product. Thus whereas pyrimidine-2-carbonitrile predominantly led to the formation of the carboxamide, the pyridazine-nitrile was converted into the acid.

Scheme 2.95. Regioselective enzymatic hydrolysis of aromatic dinitriles.

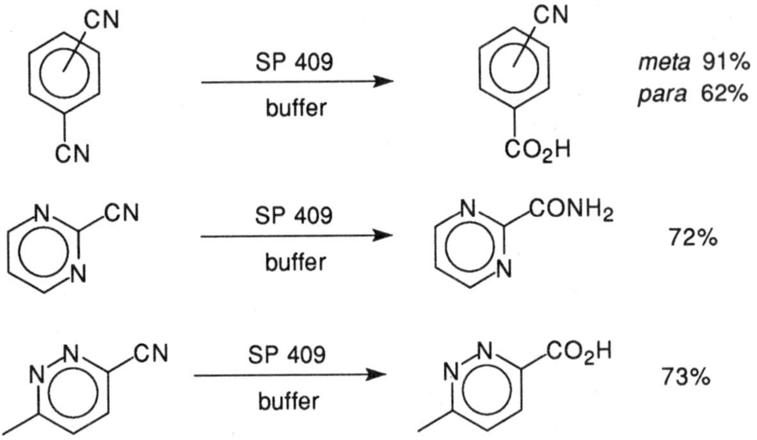

benzene-1,3-dicarbonitrile / benzene-1,4-dicarbonitrile →(SP 409, buffer)→ cyanobenzoic acid — meta 91%, para 62%

pyrazine-2-carbonitrile →(SP 409, buffer)→ pyrazine-2-carboxamide — 72%

6-methylpyridazine-3-carbonitrile →(SP 409, buffer)→ 6-methylpyridazine-3-carboxylic acid — 73%

The synthetic potential of the nitrile hydratase preparation SP 409 was demonstrated by the transformation of aliphatic nitriles bearing acid- or base-sensitive functional groups. As a rule, the corresponding carboxylic acids were obtained in good yields without the appearance of the side-products that were often seen in the corresponding chemical hydrolysis. For instance, base-labile 5-oxo-hexanenitrile was converted into the corresponding acid in 68% yield. Caution should be exercised when ester groups are present in the substrate. Methyl carboxylates may be hydrolysed by unspecific esterases present in the crude preparation, as can acetates of primary and secondary alcohols. In such cases, the use of the corresponding ethyl esters or 2,2-dimethylpropanoates (pivalates), which are not hydrolysed, is recommended. Surprisingly, epoxides may be hydrolytically cleaved as well by an enzymatic action. A selective mono-hydrolysis of aliphatic α,ω-dinitriles is also possible provided that the reaction is terminated at the monoacid stage, since prolonged reaction leads to the slow formation of the diacid. The selectivity is usually good with the short-chain derivatives and more complex product mixtures are only obtained when long-chain α,ω-dinitriles are employed.

2.1 Hydrolytic Reactions

Scheme 2.96. Enzymatic hydrolysis of primary aliphatic nitriles.

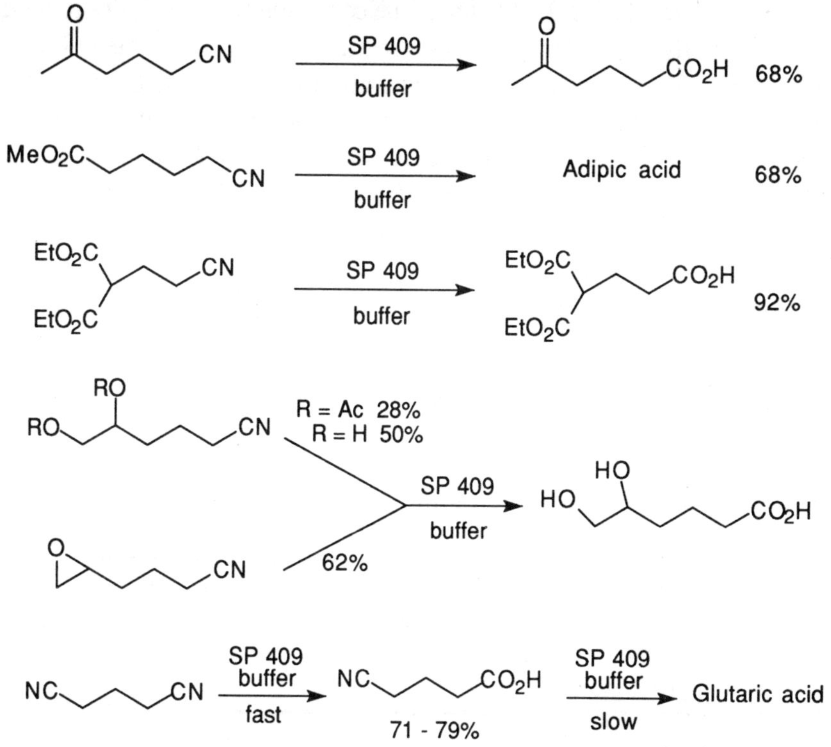

Secondary aliphatic nitriles are also accepted as substrates. For example, base-labile cyclopropyl carbonitrile was converted into the acid in excellent yield. Additional polar groups in the substrate, such as hydroxy moiety, do not seem to impede the hydrolysis. β-Hydroxynitriles, which are prone to elimination reactions [415], were transformed into the corresponding β-hydroxyacids without the formation of undesired elimination or epimerisation products.

Scheme 2.97. Enzymatic hydrolysis of secondary aliphatic nitriles.

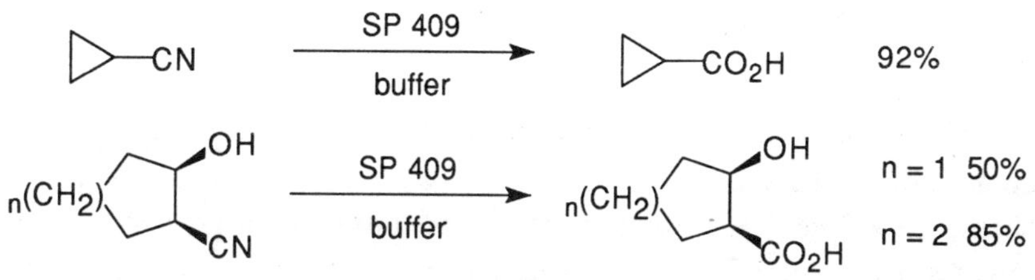

The solubility of the substrate in water may be a critical parameter for its successful conversion. As highly lipophilic compounds are generally converted at a substantially reduced reaction rate, substrates should be soluble in water, at least to some extent. The addition of water-miscible or -immiscible organic co-solvents such as methanol or toluene at various concentrations to effect solubilisation of the substrate led to instantaneous deactivation of the enzyme. As shown in Scheme 2.98, the more hydrophilic short-chain β-hydroxynitriles were quickly converted into the β-hydroxyacids with negligible formation of carboxamides. When the chain-lenth was extended, the overall reaction rate decreased and a considerable amount of amide was formed.

Scheme 2.98. Enzymatic hydrolysis of β-hydroxynitriles.

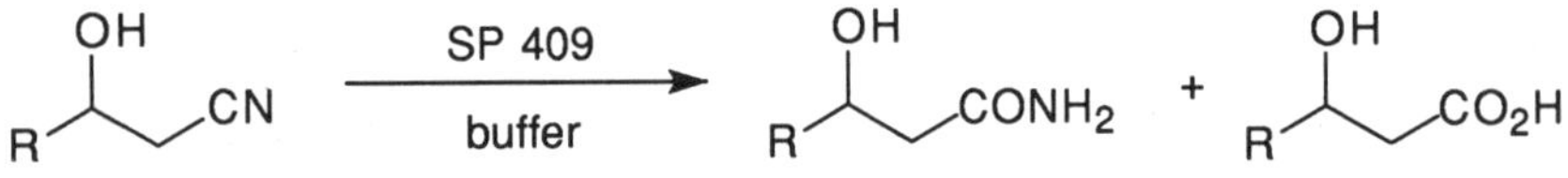

R	amide [%]	acid [%]
H	trace	63
n-C$_4$H$_9$-	trace	62
n-C$_6$H$_{13}$-	26	39

Scheme 2.99. Enzymatic hydrolysis of anomeric cyanoglucosides.

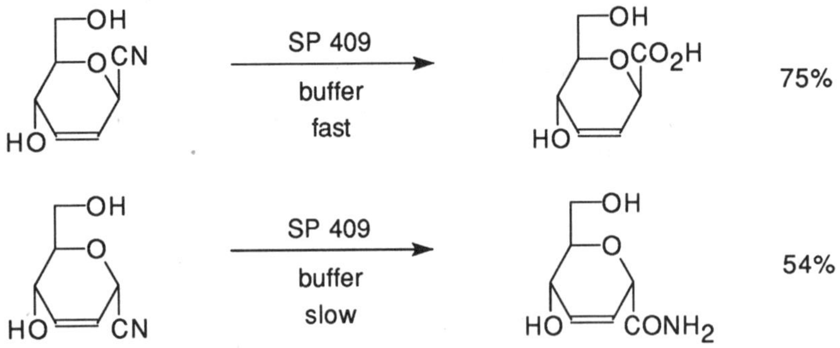

The fact that the nature of the product does not only depend on the lipophilicity of the substrate but also on its structure as a whole is shown in Scheme 2.99. The diastereomeric (anomeric) unsaturated glycosyl cyanides showed a divergent behaviour. Whereas the more reactive β-anomer gave the carboxylic acid in good yield after 48 hours, its α-anomeric counterpart was slowly transformed into the corresponding amide.

References

1. Fersht A (1985) Enzyme Structure and Mechanism, 2nd edn., Freeman, New York, pp 405
2. Jones JB, Beck JF (1976) Asymmetric Syntheses and resolutions using Enzymes. In: Jones JB, Sih CJ, Perlman D (eds) Applications of Biochemical Systems in Organic Chemistry, part I, p 107
3. Brady L, Brzozowski AM, Derewenda ZS, Dodson E, Dodson G, Tolley S, Turkenburg JP, Christiansen L, Huge-Jensen B, Norskov L, Thim L, Menge U (1990) Nature 343: 767
4. Kirchner G, Scollar MP, Klibanov AM (1986) J. Am. Chem. Soc. 107: 7072
5. Gotor V, Brieva R, Rebolledo F (1988) Tetrahedron Lett. 6973
6. Kitaguchi H, Fitzpatrick PA, Huber JE, Klibanov AM (1989) J. Am. Chem. Soc. 111: 3094
7. Björkling F, Godfredsen SE, O. Kirk O (1990) J. Chem. Soc., Chem. Commun. 1301
8. Gotor V, Astorga C, Rebolledo F (1990) Synlett. 387
9. Fastrez J, Fersht AR (1973) Biochemistry 12: 2025
10. Silver MS (1966) J. Am. Chem. Soc. 88: 4247
11. Chen C-S, Sih CJ (1989) Angew. Chem., Int. Ed. Engl. 28: 695
12. Matsumoto K, Ohta H (1989) Chem. Lett. 1589
13. Matsumoto K, Tsutsumi S, Ihori T, Ohta H (1990) J. Am. Chem. Soc. 112: 9614
14. Björkling F, Boutelje J, Gatenbeck S, Hult K, Norin T, Szmulik P (1985) Tetrahedron 41: 1347
15. Krisch K (1971) Carboxyl Ester Hydrolases. In: Boyer PD (ed) The Enzymes, 3rd edn., Academic Press, New York, vol. 5, pp 43-69
16. Ramos-Tombo GM, Schär H-P, Fernandez i Busquets X, Ghisalba O (1986) Tetrahedron Lett. 5707
17. Gais H-J, Lukas KL (1984) Angew. Chem., Int. Ed. Engl. 23: 142
18. Kasel W, Hultin PG, Jones JB (1985) J. Chem. Soc., Chem. Commun. 1563
19. Wang Y-F, Chen C-S, Girdaukas G, Sih CJ (1984) J. Am. Chem. Soc. 106: 3695
20. Wang Y-F, Chen C-S, Girdaukas G, Sih CJ (1985) Extending the Applicability of Esterases of Low Enantioselectivity in Asymmetric Synthesis. In: Porter R, Clark S (eds) Enzymes in Organic Synthesis, Ciba Foundation Symposium 111, Pitman, London, pp 128-145
21. Sih CJ, Wu S-H (1989) Resolution of Enantiomers via Biocatalysis; Eliel EL, Wilen SH (eds) Topics Stereochem. vol. 19, p 63, Wiley, New York
22. Dakin HD (1903) J. Physiol. 30: 253
23. Fülling G, Sih CJ (1987) J. Am. Chem. Soc. 109: 2845
24. Yamada H, Shimizu S, Shimada H, Tani Y, Takahashi S, Ohashi T (1980) Biochimie 62: 395
25. Klempier N, Faber K, Griengl H (1989) Synthesis 933
26. Chen C-S, Fujimoto Y, Girdaukas G, Sih CJ (1982) J. Am. Chem. Soc. 104: 7294
27. Martin VS, Woodard SS, Katsuki T, Yamada Y, Ikeda M, Sharpless KB (1981) J. Am. Chem. Soc. 103: 6237
28. Bredig G, Fajans K (1908) Ber. dtsch. chem. Ges. 41: 752
29. Jongejan JA, van Tol JBA, Geerlof A, Duine JA (1991) Rec. Trav. Chim. Pays-Bas 110: 247
30. van Tol JBA, Jongejan JA, Geerlof A, Duine JA (1991) Rec. Trav. Chim. Pays-Bas 110: 255
31. Oberhauser Th, Bodenteich M, Faber K, Penn G, Griengl H (1987) Tetrahedron 43: 3931
32. Chen C-S, Wu, S-H, Girdaukas G, Sih CJ (1987) J. Am. Chem. Soc. 109: 2812

33. For an alternative model see: Langrand G, Baratti J, Buono G, Tryantaphylides C (1988) Biocatalysis 1: 231
34. Guo Z-W, Wu S-H, Chen C-S, Girdaukas G, Sih CJ (1990) J. Am. Chem. Soc. 112: 4942
35. Kazlauskas RJ (1989) J. Am. Chem. Soc. 111: 4953
36. Wu S-H, Zhang L-Q, Chen C-S, Girdaukas G, Sih CJ (1985) Tetrahedron Lett. 4323
37. Macfarlane ELA, Roberts SM, Turner NJ (1990) J. Chem. Soc., Chem. Commun. 569
38. Chen C-S, Liu Y-C (1991) J. Org. Chem. 56: 1966
39. Williams RM (1989) Synthesis of Optically Active α-Amino Acids. Pergamon Press, Oxford
40. Soda K, Tanaka H, Esaki N (1983) Amino Acids. In: Rehm H-J, Reed G (eds) Biotechnology, Verlag Chemie, Weinheim, volume 3, pp 479
41. Schmidt-Kastner G, Egerer P (1984) Amino Acids and Peptides. In: Rehm H-J, Reed G (eds) Biotechnology, Verlag Chemie, Weinheim, volume 6a, pp 387
42. Meijer EM, Boesten WHJ, Schoemaker HE, van Balken JAM (1985) Use of Biocatalysts in the Industrial Production of Specialty Chemicals. In: Tramper J, van der Plas HC, Linko P (eds) Biocatalysts in Organic Synthesis, Elsevier, Amsterdam, pp 135
43. Leuenberger HGW, Kieslich K (1982) Biotransformationen. In: Präve P, Faust U, Sittig W, Sukatsch DA (eds) Handbuch der Biotechnologie, Akademische Verlagsges., Wiesbaden, pp 453
44. Enei H, Shibai H, Hirose Y (1982) Amino Acids and Related Compounds. In: Tsao GT (ed) Annual Reports on Fermentation Processes, Academic Press, New York, volume 5, pp 79
45. Abbott BJ (1976) Adv. Appl. Microbiol. 20: 203
46. Yonaha K, Soda K (1986) Adv. Biochem. Eng. Biotechnol. 33: 95
47. Soda K, Tanaka H, Esaki N (1983) Amino Acids; Rehm H-J, Reed G (eds) Biotechnology, vol. 3, p 479, Verlag Chemie, Weinheim
48. Wagner I, Musso H (1983) Angew. Chem., Int. Ed. Engl. 22: 816
49. Keller JW, Hamilton BJ (1986) Tetrahedron Lett. 1249
50. Yamada S, Hongo C, Yoshioka R, Chibata I (1983) J. Org. Chem. 48: 843
51. Esaki N, Shimoi H, Tanaka H, Soda K (1989) Biotechnol. Bioeng. 34: 1231
52. Warburg O (1906) Hoppe-Seyler's Z. Physiol Chem.48: 205
53. Jones M, Page MI (1991) J. Chem. Soc. 316
54. Miyazawa T, Takitani T, Ueji S, Yamada T, Kuwata S (1988) J. Chem. Soc., Chem. Commun. 1214
55. Miyazawa T, Iwanaga H, Ueji S, Yamada T, Kuwata S (1989) Chem. Lett. 2219
56. Jones JB, Beck JF (1976) Applications of Chymotrypsin in Resolutions and Asymmetric Synthesis. In: Jones JB, Sih CJ, Perlman D (eds) Applications of Biochemical Systems in Organic Chemistry, Wiley, New York, part I, pp 137
57. Dirlam NC, Moore BS, Urban FJ (1987) J. Org. Chem. 52: 3287
58. Berger A, Smolarsky M, Kurn N, Bosshard HR (1973) J. Org. Chem. 38: 457
59. Cohen SG (1969) Trans. N. Y. Acad. Sci. 31: 705
60. Hess PD (1971) Chymotrypsin-Chemical Properties and Catalysis; Boyer GP (ed) The Enzymes, vol. 3, p 213, Academic Press, New York
61. Chenevert R, Letourneau M, Thiboutot S (1990) Can. J. Chem. 68: 960
62. Roper JM, Bauer DP (1983) Synthesis 1041
63. Izquierdo MC, Stein RL (1990) J. Am. Chem. Soc., 112: 6054.
64. Chen S-T, Wang K-T, Wong C-H (1986) J. Chem. Soc., Chem. Commun. 1514
65. Glänzer BI, Faber K, Griengl H (1987) Tetrahedron 43: 771
66. Jones JB, Kunitake T, Niemann C, Hein GE (1965) J. Am. Chem. Soc. 87: 1777
67. Pattabiraman TN, Lawson WB (1972) Biochem. J. 126: 645 and 659
68. Greenstein JP, Winitz M (1961) Chemistry of the Amino Acids. Wiley, New York, p 715
69. Boesten WHJ, Dassen BHN, Kerkhoffs PL, Roberts MJA, Cals MJH, Peters PJH, van Balken JAM, Meijer EM, Schoemaker HE (1986) Efficient Enzymic Production of

Enantiomerically Pure Amino Acids. In: Schneider MP (ed) Enzymes as Catalysts in Organic Synthesis, Reidel, Dordrecht, p 355
70. Sambale C, Kula M-R (1987) Biotechnol. Appl. Biochem. 9: 251
71. Chibata I, Tosa T, Sato T, Mori T (1976) Methods Enzymol. 44: 746
72. Greenstein JP (1957) Methods Enzymol. 3: 554
73. Chibata I, Ishikawa T, Tosa T (1970) Methods Enzymol. 19: 756
74. Tosa T, Mori T, Fuse N, Chibata I (1967) Biotechnol. Bioeng. 9: 603
75. Chenault HK, Dahmer J, Whitesides GM (1989) J. Am. Chem. Soc. 111: 6354
76. Mori K, Otsuka T (1985) Tetrahedron 41: 547
77. Baldwin JE, Christie MA, Haber SB, Kruse LI (1976) J. Am. Chem. Soc. 98: 3045
78. Solodenko VA, Kasheva TN, Kukhar VP, Kozlova EV, Mironenko DA, Svedas VK (1991) Tetrahedron 47: 3989
79. Bücherer HT, Steiner W, (1934) J. Prakt. Chem. 140: 291
80. Yamada H, Takahashi S, Kii Y, Kumagai H (1978) J. Ferment. Technol. 56: 484
81. Wallach DP, Grisolia S (1957) J. Biol. Chem. 226: 277
82. Olivieri R, Fascetti E, Angelini L (1981) Biotechnol. Bioeng. 23: 2173
83. Yamada H, Shimizu S, Shimada H, Tani Y, Takahashi S, Ohashi T (1980) Biochimie 62: 395
84. Guivarch M, Gillonnier C, Brunie J-C (1980) Bull. Soc. Chim. Fr. 91
85. Yamada H, Takahashi S, Yoshiaki K, Kumagai H (1978) J. Ferment. Technol. 56: 484
86. Zhu L-M, Tedford MC (1990) Tetrahedron 46: 6587; It should be mentioned that some references and absolute configurations are wrong in this paper
87. Ohno M, Otsuka M (1989) Org. React. 37: 1
88. Pearson AJ, Bansal HS, Lai Y-S (1987) J. Chem. Soc., Chem. Commun. 519
89. Johnson CR, Penning TD (1986) J. Am. Chem. Soc. 108: 5655
90. Suemune H, Harabe T, Xie Z-F, Sakai K (1988) Chem. Pharm. Bull. 36: 4337
91. Dropsy EP, Klibanov AM (1984) Biotechnol. Bioeng. 26: 911
92. Kazlauskas RJ (1989) J. Am. Chem. Soc. 111: 4953
93. Kotani H, Kuze Y, Uchida S, Miyabe T, Limori T, Okano K, Kobayashi S, Ohno M (1983) Agric. Biol. Chem. 47: 1363
94. Jones JB (1980) In: Dunnill P, Wiseman A, Blakeborough N (eds) Enzymic and Non-enzymic Catalysis, Horwood/Wiley, New York, p 54
95. Schubert Wright C (1972) J. Mol. Biol. 67: 151
96. Philipp M, Bender ML (1983) Mol. Cell. Biochem. 51: 5
97. Fruton JS (1971) Pepsin. In: Boyer PD (ed) The Enzymes, volume 3, pp 119, Academic Press, London
98. Bianchi D, Cabri W, Cesti P, Francalanci F, Ricci M (1988) J. Org. Chem. 53: 104
99. Fancetic O, Deretic V, Marjanovic N, Glisin V (1988) Biotechnol. Forum 5: 90
100. Savidge TA, Cole M (1976) Methods Enzymol. 43: 705
101. Bender ML, Killheffer JV (1973) Crit. Rev. Biochem. 1: 149
102. Barnier J-P, Blanco L, Guibe-Jampel E, Rousseau G (1989) Tetrahedron 45: 5051
103. Heymann E, Junge W (1979) Eur. J. Biochem. 95: 509
104. Lam LKP, Brown CM, De Jeso B, Lym L, Toone EJ, Jones JB (1988) J. Am. Chem. Soc. 110: 4409
105. Öhrner N, Mattson A, Norin T, Hult K (1990) Biocatalysis 4: 81
106. Seebach D, Eberle M (1986) Chimia 40: 315
107. De Jeso B, Belair N, Deleuze H, Rascle M-C, Maillard B (1990) Tetrahedron Lett. 653
108. Jongejan JA, Duine JA (1987) Tetrahedron Lett. 2767
109. Burger U, Erne-Zellweger D, Mayerl CM (1987) Helv. Chem. Acta 70: 587
110. Hazato A, Tanaka T, Toru T, Okamura N, Bannai K, Sugiura S, Manabe K, Kurozumi S (1983) Nippon Kagaku Kaishi 9: 1390; Chem. Abstr. (1984) 100: 120720q
111. Papageorgiou C, Benezra C (1985) J. Org. Chem. 50: 1145
112. Sicsic S, Leroy J, Wakselman C (1987) Synthesis 155
113. Björkling F, Boutelje J, Gatenbeck S, Hult K, Norin T, Szmulik P (1985) Tetrahedron 41: 1347

114. Schneider M, Engel N, Boensmann H (1984) Angew. Chem., Int. Ed. Engl. 23: 66
115. Luyten M, Müller S, Herzog B, Keese R (1987) Helv. Chim. Acta 70: 1250
116. Huang F-C, Lee LFH, Mittal RSD, Ravikumar PR, Chan JA, Sih CJ, Capsi E, Eck CR (1975) J. Am. Chem. Soc. 97: 4144
117. Herold P, Mohr P, Tamm C (1983) Helv. Chim. Acta 76: 744
118. Adachi K, Kobayashi S, Ohno M (1986) Chimia 40: 311
119. Ohno M, Kobayashi S, Iimori T, Wang Y-F, Izawa T (1981) J. Am. Chem. Soc. 103: 2405
120. Mohr P, Waespe-Sarcevic, Tamm C, Gawronska K, Gawronski JK (1983) Helv. Chim. Acta 66: 2501
121. Cohen SG, Khedouri E (1961) J. Am. Chem. Soc. 83: 1093
122. Cohen SG, Khedouri E (1961) J. Am. Cherm. Soc. 83: 4228
123. Roy R, Rey AW (1987) Tetrahedron Lett. 4935
124. Santaniello E, Chiari M, Ferraboschi P, Trave S (1988) J. Org. Chem. 53: 1567
125. Gopalan AS, Sih CJ (1984) Tetrahedron Lett. 5235
126. Mohr P, Waespe-Sarcevic N, Tamm C, Gawronska K, Gawronski JK (1983) Helv. Chim. Acta 66: 2501
127. Schregenberger C, Seebach D (1986) Liebigs Ann. Chem. 2081
128. Sabbioni G, Jones JB (1987) J. Org. Chem. 52: 4565
129. Björkling F, Boutelje J, Gatenbeck S, Hult K, Norin T (1985) Appl. Microbiol. Biotechnol. 21: 16
130. Bloch R, Guibe-Jampel E, Girard G (1985) Tetrahedron Lett. 4087
131. Adachi K, Kobayashi S, Ohno M (1986) Chimia 40: 311
132. Laumen K, Schneider M (1984) Tetrahedron Lett. 5875
133. Harre M, Raddatz P, Walenta R, Winterfeldt (1982) Angew. Chem., Int. Ed. Engl. 21: 480
134. Deardorff DR, Mathews AJ, McMeekin DS, Craney CL (1986) Tetrahedron Lett. 1255
135. Wang Y-F, Sih CJ (1984) Tetrahedron Lett. 4999
136. Iriuchijima S, Hasegawa K, Tsuchihashi G (1982) Agric. Biol. Chem. 46: 1907
137. Eberle M, Missbach M, Seebach D (1990) Org. Synth. 69: 19
138. Mohr P, Rösslein L, Tamm C (1987) Helv. Chim. Acta 70: 142
139. Alcock NW, Crout DHG, Henderson CM, Thomas SE (1988) J. Chem. Soc., Chem. Commun. 746
140. Ramaswamy S, Hui RAHF, Jones JB (1986) J. Chem. Soc., Chem. Commun. 1545
141. Sicsic S, Ikbal M, Le Goffic F (1987) Tetrahedron Lett. 1887
142. Suemune H, Tanaka M, Ohaishi H, Sakai K (1988) Chem. Pharm. Bull. 36: 15
143. Klunder AJH, Huizinga WB, Hulshof AJM, Zwanenburg B (1986) Tetrahedron Lett. 2543
144. Crout DHG, Gaudet VSB, Laumen K, Schneider MP (1986) J. Chem. Soc., Chem. Commun. 808
145. Sugai T, Kuwahara S, Hishino C, Matsuo N, Mori K (1982) Agric. Biol. Chem. 46: 2579
146. Oritani T, Yamashita K (1980) Agric. Biol. Chem. 44: 2407
147. Ohta H, Miyamae Y, Kimura Y (1989) Chem. Lett. 379
148. Matsumoto K, Tsutsumi S, Ihori T, Ohta H (1990) J. Am. Chem. Soc. 112: 9614
149. Ziffer H, Kawai K, Kasai K, Imuta M, Froussios C (1983) J. Org. Chem. 48: 3017
150. Takaishi Y, Yang Y-L, DiTullio D, Sih CJ (1982) Tetrahedron Lett. 5489
151. Glänzer BI, Faber K, Griengl H (1987) Tetrahedron 43: 5791
152. Glänzer BI, Faber K, Griengl H (1988) Enzyme Microb. Technol. 10: 689
153. Stork G, Poirier JM (1983) J. Am. Chem. Soc. 105: 1073
154. Franck-Neumann M, Sedrati M, Vigneron J-P, Bloy V (1985) Angew. Chem. 97: 995
155. Fried J, Sih CJ (1973) Tetrahedron Lett. 3899
156. Kobayashi Y, Okamoto S, Shimazaki T, Ochiai Y, Sato F (1987) Tetrahedron Lett. 3959
157. Midland MM, Lee PE (1981) J. Org. Chem. 46: 3934
158. Chan K-K, Specian AC, Saucy G (1978) J. Org. Chem. 43: 3435
159. Buynak JD, Mathew J, Rao MN (1986) J. Chem. Soc., Chem. Commun. 941

2.1 Hydrolytic Reactions

160. Overman LE, Bell KL (1981) J. Am. Chem. Soc. 103: 1851
161. Mutsaers JHGM, Kooreman HJ (1991) Recl. Trav. Chim. Pays-Bas 110: 185
162. Fliche C, Braun J, Le Goffic F (1991) Synth. Commun. 21: 1429
163. Berger A, Smolarsky M, Kurn N, Bosshard HR (1973) J. Org. Chem. 38: 457
164. Jones JB, Beck JF (1976) Applications of Chymotrypsin in Resolution and Asymmetric Synthesis. In: Jones JB, Sih CJ, Perlman D (eds) Applications of Biochemical Systems in Organic Synthesis, pp 137, Wiley, New York.
165. Lalonde JJ, Bergbreiter DE, Wong C-H (1988) J. Org. Chem. 53: 2323
166. Shin C, Seki M, Takahashi N (1990) Chem. Lett. 2089
167. Uemura A, Nozaki K, Yamashita J, Yasumoto M (1989) Tetrahedron Lett. 3819
168. Kvittingen L, Partali V, Braenden JU, Anthonsen T (1991) Biotechnol. Lett. 13: 13
169. Pugniere M, San Juan C, Previero A (1990) Tetrahedron Lett. 4883-6
170. Chen S-T, Wang K-T (1987) Synthesis 581
171. Kvittingen L, Partali V, Braenden JU, Anthonsen T (1991) Biotechnol. Lett. 13: 13
172. Waldmann H (1988) Liebigs Ann. Chem. 1175
173. Fuganti C, Grasselli P, Servi S, Lazzarini A, Casati P (1988) Tetrahedron 44: 2575
174. Waldmann, H. (1989) Tetrahedron Lett. 3057
175. Dale JA, Dull DL, Mosher HS (1969) J. Org. Chem. 34: 2543
176. Feichter C, Faber K, Griengl H (1991) J. Chem. Soc., Perkin Trans. I, 653
177. Fülling G, Sih CJ (1987) J. Am. Chem. Soc. 109: 2845
178. Chen C-S, Sih CJ (1989) Angew. Chem., Int. Ed. Engl. 28: 695
179. Otto PPHL (1990) Chem. Biochem. Eng. Q, 4: 137
180. Adachi K, Kobayashi S, Ohno M (1986) Chimia 40: 311
181. Björkling F, Boutelje J, Gatenbeck S, Hult K, Norin T (1986) Bioorg. Chem. 14: 176
182. Guanti G, Banfi L, Narisano E, Riva R, Thea S (1986) Tetrahedron Lett. 4639
183. Santaniello E, Ferraboschi P, Grisenti P, Aragozzini F, Maconi E (1991) J. Chem. Soc., Perkin Trans. I, 601
184. Björkling F, Boutelje J, Gatenbeck, Hult K, Norin T (1985) Tetrahedron Lett. 4957
185. Guo Z-W, Sih CJ (1989) J. Am. Chem. Soc. 111: 6836
186. Lam LKP, Hui RAHF, Jones JB (1986) J. Org. Chem. 51: 2047
187. Holmberg E, Hult K (1991) Biotechnol. Lett. 13: 323
188. Boutelje J, Hjalmarsson M, Szmulik P, Norin T, Hult K (1987) Pig Liver Esterase in Asymmetric Synthesis. Steric Requirements and Control of the Reaction Conditions. In: Laane C, Tramper J, Lilly MD (eds) Biocatalysis in Organic Media, Elsevier, Amsterdam, p 361
189. Jansonius JN (1987) Enzyme Mechanism: What X-Ray Crystallography Can(not) Tell Us. In: Moras D, Drenth J, Strandlberg B, Suck D, Wilson K (eds) Crystallography in Molecular Biology, Plenum Press, New York, p 229
190. Fersht A (1977) Enzyme Structure and Mechanism, Freeman, San Francisco, p 15
191. Cohen SG (1969) Trans. N. Y. Acad. Sci. 31: 705
192. Schubert Wright C (1972) J. Mol. Biol. 67: 151
193. Brady L, Brzozowski AM, Derewenda ZS, Dodson E, Dodson G, Tolley S, Turkenburg JP, Christiansen L, Huge-Jensen B, Norskov L, Thim L, Menge U (1990) Nature 343: 767
194. Schrag JD, Li Y, Wu S, Cygler M (1991) Nature 351: 761
195. Burkert U, Allinger NL (1982) Molecular Mechanics; Caserio MC (ed) ACS Monograph, volume 177, Am. Chem. Soc., Washington
196. Mohr P. Waespe-Sarcevic N, Tamm C, Gawronska K, Gawronski JK (1983) Helv. Chim. Acta 66: 2501
197. Oberhauser Th, Faber K, Griengl H (1989) Tetrahedron 45: 1679
198. Kazlauskas RJ, Weissfloch ANE, Rappaport AT, Cuccia LA (1991) J. Org. Chem. 56: 2656
199. Toone EJ, Werth MJ, Jones JB (1990) J. Am. Chem. Soc. 112: 4946
200. Desnuelle P (1972) The Lipases. In: The Enzymes, Boyer PO (ed) volume 7, p 575, Academic Press, New York
201. Huang AHC (1987) Lipases. In: Biochemistry of Plants, volume 9, p 91, Academic Press, New York
202. Macrae (1983) J. Am. Oil Chem. Soc. 60: 291

203. Gunstone FD (1979) Compr. Org. Chem. 5: 641
204. Nielsen T (1985) Fette, Seifen Anstrichmittel 87: 15
205. Lipases: Structure, Mechanism and Genetic Engineering (1991) Alberghina L, Schmidt RD, Verger R (eds.) GBF Monographs, vol. 16, Verlag Chemie, Weinheim
206. Sarda L, Desnuelle P (1958) Biochim. Biophys. Acta 30: 513
207. Schonheyder F, Volqvartz K (1945) Acta Physiol. Scand. 9: 57
208. Bianchi D, Cesti P (1990) J. Org. Chem. 55: 5657
209. Iriuchijima S, Kojima N (1981) J. Chem. Soc., Chem. Commun. 185
210. Kazlauskas RJ, Weissfloch ANE, Rappaport AT, Cuccia LA (1991) J. Org. Chem. 56: 2656
211. Nagao Y, Kume M, Wakabayashi RC, Nakamura T, Ochiai M (1989) Chem. Lett. 239
212. Brockerhoff H (1968) Biochim. Biophys. Acta 159: 296
213. Brockmann HL (1981) Methods Enzymol. 71: 619
214. Desnuelle P (1961) Adv. Enzymol. 23: 129
215. Lutz D, Güldner A, Thums R, Schreier P (1990) Tetrahedron: Asymmetry 1: 783
216. Ehrler J, Seebach D (1990) Liebigs Ann. Chem. 379
217. Claßen A, Wershofen S, Yusufoglu A, Scharf H-D (1987) Liebigs Ann. Chem. 629
218. Jones JB, Hinks RS (1987) Can. J. Chem. 65: 704
219. Kloosterman H, Mosmuller EWJ, Schoemaker HE, Meijer EM (1987) Tetrahedron Lett. 2989
220. Shaw JF, Klibanov AM (1987) Biotechnol. Bioeng. 29: 648
221. Sweers HM, Wong C-H (1986) J. Am. Chem. Soc. 108: 6421
222. Hennen WJ, Sweers HM, Wang Y-F, Wong C-H (1988) J. Org. Chem. 53: 4939
223. Ballesteros A, Bernabé M, Cruzado C, Martin-Lomas M, Otero C (1989) Tetrahedron 45: 7077
224. Guibé-Jampel E, Rousseau G, Salaun J (1987) J. Chem. Soc., Chem. Commun. 1080
225. Cohen SG, Milovanovic A (1968) J. Am. Chem. Soc. 90: 3495
226. Kasel W, Hultin PG, Jones JB (1985) J. Chem. Soc., Chem. Commun. 1563
227. Laumen K, Schneider M (1985) Tetrahedron Lett. 2073
228. Hemmerle H, Gais H-J (1987) Tetrahedron Lett. 3471
229. Guanti G, Banfi L, Narisano E (1990) Tetrahedron: Asymmetry 1: 721
230. Guanti G, Narisano E, Podgorski T, Thea S, Williams A (1990) Tetrahedron 46: 7081
231. Patel RN, Robison RS, Szarka LJ (1990) Appl. Microbiol. Biotechnol. 34: 10
232. Kerscher V, Kreiser W (1987) Tetrahedron Lett. 531
233. Breitgoff D, Laumen K, Schneider MP (1986) J. Chem. Soc., Chem. Commun. 1523
234. Ramos-Tombo GM, Schär H-P, Fernandez i Busquets X, Ghisalba O (1986) Tetrahedron Lett. 5707
235. Gao Y, Hanson RM, Klunder JM, Ko SY, Masamune H, Sharpless KB (1987) J. Am. Chem. Soc. 109: 5765
236. Marples BA, Roger-Evans M (1989) Tetrahedron Lett. 261
237. Ladner WE, Whitesides GM (1984) J. Am. Chem. Soc. 106: 7250
238. Cotterill IC, Sutherland AG, Roberts SM, Grobbauer R, Spreitz J, Faber K (1991) J. Chem. Soc., Perkin Trans. I, 1365
239. Cotterill IC, Dorman G, Faber K, Jaouhari R, Roberts SM, Scheinmann F, Spreitz J, Sutherland AG, Winders JA, Wakefield BJ (1990) J. Chem. Soc., Chem. Commun. 1661
240. Gutman AL, Zuobi K, Guibe-Jampel E (1990) Tetrahedron Lett. 2037
241. Cotterill IC, Finch H, Reynolds DP, Roberts SM, Rzepa HS, Short KM, Slawin AMZ, Wallis CJ, Williams DJ (1988) J. Chem. Soc., Chem. Commun. 470
242. Naemura K, Matsumura T, Komatsu M, Hirose Y, Chikamatsu H (1988) J. Chem. Soc., Chem. Commun. 239
243. Pearson AJ, Lai Y-S (1988) J. Chem. Soc., Chem. Commun. 442
244. Pearson AJ, Lai Y-S, Lu W, Pinkerton AA (1989) J. Org. Chem. 54: 3882
245. Gautier A, Vial C, Morel C, Lander M, Näf F (1987) Helv. Chim. Acta 70: 2039
246. Pawlak JL, Berchtold GA (1987) J. Org. Chem. 52: 1765
247. Sugai T, Kakeya H, Ohta H (1990) J. Org. Chem. 55: 4643

248. Kitazume T, Sato T, Kobayashi T, Lin JT (1986) J. Org. Chem. 51: 1003
249. Abramowicz DA, Keese CR (1989) Biotechnol. Bioeng. 33: 149
250. Hoshino O, Itoh K, Umezawa B, Akita H, Oishi T (1988) Tetrahedron Lett. 567
251. Sugai T, Kakeya H, Ohta H, Morooka M, Ohba S (1989) Tetrahedron 45: 6135
252. Pottie M, Van der Eycken J, Vandevalle M, Dewanckele JM, Röper H (1989) Tetrahedron Lett. 5319
253. Dumortier, Van der Eycken J, Vandewalle M (1989) Tetrahedron Lett. 3201
254. Oberhauser Th, Faber K, Griengl H (1989) Tetrahedron 45: 1679
255. Oberhauser Th, Bodenteich M, Faber K, Penn G, Griengl H (1987) Tetrahedron 43: 3931
256. Saf R, Faber K, Penn G, Griengl H (1988) Tetrahedron 44: 389
257. Königsberger K, Faber K, Marschner C, Penn G, Baumgartner P, Griengl H (1989) Tetrahedron 45: 673
258. Wu S-H, Guo Z-W, Sih CJ (1990) J. Am. Chem. Soc. 112: 1990
259. Xie Z-F (1991) Tetrahedron: Asymmetry 2: 733
260. Xie Z-F, Nakamura I, Suemune H, Sakai K (1988) J. Chem. Soc., Chem. Commun. 966
261. Xie Z-F, Sakai K (1989) Chem. Pharm. Bull. 37: 1650
262. Seemayer R, Schneider MP (1990) J. Chem. Soc., Perkin Trans. I, 2359
263. Laumen K, Schneider MP (1988) J. Chem. Soc., Chem. Commun. 598
264. Kalaritis P, Regenye RW, Partridge JJ, Coffen DL (1990) J. Org. Chem. 55: 812
265. Klempier N, Geymayer P, Stadler P, Faber K, Griengl H (1990) Tetrahedron: Asymmetry 1: 111
266. Kloosterman M, Kierkels JGT, Guit RPM, Vleugels LFW, Gelade ETF, van den Tweel WJJ, Elferink VHM, Hulshof LA, Kamphuis J (1991) Lipases: Biotransformations, Active Site Models and Kinetics. In: Lipases: Structure, Mechanism and Genetic Engineering Alberghina L, Schmidt RD, Verger R (eds.) GBF Monographs, vol. 16, Verlag Chemie, Weinheim, page 187
267. Hughes DL, bergan JJ, Amato JS, Bhupathy M, Leazer JL, McNamara JM, Sidler DR, Reider PJ, Grabowski EJJ (1990) J. Org. Chem. 55: 6252
268. Burgess K, Henderson I (1989) Tetrahedron Lett. 3633
269. Ghiringhelli O (1983) Tetrahedron Lett. 287
270. Hanessian S, Kloss J (1985) Tetrahedron Lett. 1261
271. Fujisawa T, Kojima E, Itoh T, Sato T (1985) Chem. Lett. 1751
272. Bianchi D, Cesti P, Golini P (1989) Tetrahedron 45: 869
273. Itoh T, Takagi Y, Nishiyama S (1991) J. Org. Chem. 56: 1521
274. Xie Z-F, Suemune H, Sakai K (1990) Tetrahedron: Asymmetry 1: 395
275. Huge Jensen B, Galluzzo DR, Jensen RG (1987) Lipids 22: 559
276. Chan C, Cox PB, Roberts SM (1988) J. Chem. Soc., Chem. Commun. 971
277. Cotterill IC, Finch H, Reynolds DP, Roberts SM, Rzepa HS, Short KM, Slawin AMZ, Wallis CJ, Williams DJ (1988) J. Chem. Soc., Chem. Commun. 470
278. Klempier N, Faber K, Griengl H (1989) Synthesis 933
279. Brady L, Brzozowski AM, Derewenda ZS, Dodson E, Dodson G, Tolley S, Turkenburg JP, Christiansen L, Huge-Jensen B, Norskov L, Thim L, Menge U (1990) Nature 343: 767
280. Estermann H, Prasad K, Shapiro MJ (1990) Tetrahedron Lett. 445
281. Hilgenfeld R, Saenger W (1982) Topics in Current Chemistry 101: 1
282. Naemura K, Takahashi N, Chikamatsu H (1988) Chem. Lett. 1717
283. Itoh T, Kuroda K, Tomosada M, Takagi Y (1991) J. Org. Chem. 56: 797
284. Tinapp P (1971) Chem. Ber. 104: 2266
285. Satoh T, Suzuki S, Suzuki Y, Miyaji Y, Imai Z (1969) Tetrahedron Lett. 4555
286. Mitsuda S, Yamamoto H, Umemura T, Hirohara H, Nabeshima S (1990) Agric. Biol. Chem. 54: 2907
287. Effenberger F, Gutterer B, Ziegler T, Eckhardt E, Aichholz R (1991) Liebigs Ann. Chem. 47
288. Mitsuda S, Nabeshima S, Hirohara H (1989) Appl. Microbiol. Biotechnol. 31: 334
289. Okumura S, Iwai M, Tsujisaka Y (1979) Biochim. Biophys. Acta 575: 156
290. Jensen RG (1974) Lipids 9: 149

291. Schrag JD, Li Y, Wu S, Cygler M (1991) Nature 351: 761
292. Rietschel ET (1984) Chemistry of Endotoxin. In: Handbook of Endotoxin, Rietschel ET (ed), vol. 1, Elsevier, Amsterdam
293. Martinez C, Abergel C, Cambillau, de Geus P, Lauwereys M (1991) Crystallographic Study of a Recombinant Cutinase from *Fusarium solani pisi*. In: Lipases: Structure, Mechanism and Genetic Engineering Alberghina L, Schmidt RD, Verger R (eds.) GBF Monographs, vol. 16, Verlag Chemie, Weinheim , page 67
294. Holmberg E, Hult K (1991) Biotechnol. Lett. 323
295. Holmberg E, Szmulik P, Norin T, Hult K (1989) Biocatalysis 2: 217
296. Fowler PW, Macfarlane ELA, Roberts SM (1991) J. Chem. Soc., Chem. Commun. 453
297. Holmberg E, Dahlen E, Norin T, Hult K (1991) Biocatalysis 4: 305
298. Rabiller CG, Königsberger K, Faber K, Griengl H (1990) Tetrahedron 46: 4231
299. Guo Z-W, Sih CJ (1989) J. Am. Chem. Soc. 111: 6836
300. Itoh T, Ohira E, Takagi Y, Nishiyama S, Nakamura K (1991) Bull. Chem. Soc. Jpn. 64: 624
301. Fujii H, Koyama T, Ogura K (1982) Biochim. Biophys. Acta 712: 716
302. Durrwachter JR, Wong C-H (1988) J. Org. Chem. 53: 4175
303. Straub A, Effenberger F, Fischer P (1990) J. Org. Chem. 55: 3926
304. Bednarski MD, Simon ES, Bischofberger N, Fessner W-D, Kim M-J, Lees W, Saito T, Waldmann H, Whitesides GM (1989) J. Am. Chem. Soc. 111: 627
305. Schultz M, Waldmann H, Kunz H, Vogt W (1990) Liebigs Ann. Chem. 1019
306. Scollar MP, Sigal G, Klibanov AM (1985) Biotechnol. Bioeng. 27: 247
307 Herdewijn P, Balzarini J, De Clerq E, Vanderhaege H (1985) J. Med. Chem. 28: 1385
308. Borthwick AD, Butt S, Biggadike K, Exall AM, Roberts SM, Youds PM, Kirk BE, Booth BR, Cameron JM, Cox SW, Marr CLP, Shill MD (1988) J. Chem. Soc., Chem. Commun. 656
309. Langer RS, Hamilton BC, Gardner CR, Archer MD, Colton CC (1976) AIChE Journal 22: 1079
310. Chenault HK, Simon ES, Whitesides GM (1988) Biotechnol. Gen. Eng. Rev. 6: 221
311. Pradines A, Klaébé A, Périé J, Paul F, Monsan P (1988) Tetrahedron 44: 6373
312. Marshall DL (1973) Biotechnol. Bioeng. 15: 447
313. Wong C-H, Haynie SL, Whitesides GM (1983) J. Am. Chem. Soc. 105: 115
314. Hirschbein BL, Mazenod FP, Whitesides GM (1982) J. Org. Chem. 47: 3765
315. Simon ES, Grabowski S, Whitesides GM (1989) J. Am. Chem. Soc. 111: 8920
316. Crans DC, Whitesides GM (1983) J. Org. Chem. 48: 3130
317. Bolte J, Whitesides GM (1984) Bioorg. Chem. 12: 170
318. Kazlauskas RJ, Whitesides GM (1985) J. Org. Chem. 50: 1069
319. Cantoni GL (1975) Ann. Rev. Biochem. 44: 435
320. Baughn RL, Adalsteinsson O, Whitesides GM (1978) J. Am. Chem. Soc. 100: 304
321. Augé C, Mathieu C, Mérienne C (1986) Carbohydr. Res. 151: 147
322. Wong C-H, Haynie SL, Whitesides GM (1982) J. Org. Chem. 5416
323. Wong C-H, Haynie SL, Whitesides GM (1983) J. Am. Chem. Soc. 105: 115
324. Drueckhammer DG, Wong C-H (1985) J. Org. Chem. 50: 5912
325. Pollak A, Baughn RL, Whitesides GM (1977) J. Am. Chem. Soc. 77: 2366
326. Wong C-H, Whitesides GM (1981) J. Am. Chem. Soc. 103: 4890
327. Walt DR, Findeis MA, Rios-Mercadillo VM, Augé J, Whitesides GM (1984) J. Am. Chem. Soc. 106: 234
328. Wong C-H, Whitesides GM (1983) J. Org. Chem. 48: 3199
329. Gross A, Abril O, Lewis JM, Geresh S, Whitesides GM (1983) J. Am. Chem. Soc. 105: 7428
330. Thorner JW, Paulus H (1973) in: The Enzymes; Boyer PD (ed) Academic Press, New York, volume 8, p 487
331. Rios-Mercadillo VM, Whitesides GM (1979) J. Am. Chem. Soc. 101: 5829
332. Crans DC, Whitesides GM (1985) J. Am. Chem. Soc. 107: 7019
333. Crans DC, Whitesides GM (1985) J. Am. Chem. Soc. 107: 7008
334. Eibl H (1980) Chem. Phys. Lipids 26: 405
335. Vasilenko I, Dekruijff B, Verkleij A (1982) Biochim. Biophys. Acta 685: 144

336. Scott JW (1984) Chiral Carbon Fragments and their Use in Synthesis. In: Asymmetric Synthesis, Morrison JD, Scott JW (eds) vol 4, p 5, Academic Press, Orlando
337. Finn MG, Sharpless KB (1985) On the Mechanism of Asymmetric Epoxidation with Titanium-tartrate Catalysts. In: Asymmetric Synthesis, Scott JW (eds) vol 5, p 247, Academic Press, Orlando
338. Hartmans S (1989) FEMS Microbiol. Rev. 63: 235
339. Epoxide hydrolases are occasionally also called 'epoxide hydratases' or 'epoxide hydrases'
340. Seidegard J, Pierre JW (1983) Biochim. Biophys. Acta 695: 251
341. Lu AYH, Levin W (1978) Methods Enzymol. 52: 193
342. Berti G (1986) Enantio- and diastereoselectivity of microsomal epoxide hydrolase: Potential Applications to the Preparation of Non-racemic Epoxides and Diols. In: Enzymes as Catalysts in Organic Synthesis, Schneider MP (ed) NATO ASI Series, vol 178, p 349, Reidel, Dordrecht
343. Bellucci G, Berti G, Ferretti M, Marioni F, Re F (1981) Biochem. Biophys. Res. Commun. 102: 838
344. Armstrong RN, Levin W, Jerina DM (1980) J. Biol. Chem. 255: 4698
345. DuBois GC, Appella E, Levin W, Lu AYH, Jerina DM (1978) J. Biol. Chem. 253: 2932
346. Hanzlik RP, Heidemann S, Smith D (1978) Biochem. Biophys. Res. Commun. 82: 310
347. Jerina DM, Daly JW (1974) Science 185: 573
348. Mumby SM, Hammock BD (1979) Pestic. Biochem. Physiol. 11: 275
349. Oesch F (1972) Xenobiotica 3: 305
350. Oesch F (1974) Biochem. J. 139: 77
351. Jerina DM, Ziffer H, Daly JW (1970) J. Am. Chem. Soc. 92: 1056
352. Bellucci G, Chiappe C, Conti L, Marioni F, Pierini G (1989) J. Org. Chem. 54: 5978
353. Wistuba D, Schurig V (1986) Angew. Chem., Int. Ed. Engl. 25: 1032
354. Bellucci G, Chiappe C, Marioni F (1989) J. Chem. Soc., Perkin Trans. I 2369
355. Bellucci G, Capitani I, Chiappe C, Marioni F (1989) J. Chem. Soc., Chem. Commun. 1170
356. Bellucci G, Berti G, Ingrosso G, Mastrorilli E (1980) J. Org. Chem. 45: 299
357. Bellucci G, Berti G, Bianchini R, Cetera P, Mastrorilli E (1982) J. Org. Chem. 47: 3105
358. Cahn RS, Ingold C, Prelog V (1966) Angew. Chem., Int. Ed. Engl. 5: 385
359. Bellucci G, Berti G, Catelani G, Mastrorilli E (1981) J. Org. Chem. 46: 5148
360. Watabe T, Suzuki S (1972) Biochem. Biophys. Res. Commun. 46: 1120
361. Weijers CAGM, De Haan A, De Bont JAM, (1988) Appl. Microbiol. Biotechnol. 27: 337
362. Kolattukudy PE, Brown L (1975) Arch. Biochem. Biophys. 166: 599
363. For an unusual *cis*-hydration see: Allen RH, Jakoby WB (1969) J. Biol. Chem. 244: 2078
364. Imai K, Marumo S, Mori K (1974) J. Am. Chem. Soc. 96: 5925
365. Legras JL, Chuzel G, Arnaud A, Galzy P (1990) World Microbiol. Biotechnol. 6: 83
366. Solomonson LP (1981) Cyanide as a Metabolic Inhibitor. In: Vennesland B et al. (eds) Cyanide in Biology, p 11. Academic Press, London
367. Legras JL, Jory M, Arnaud A, Galzy P (1990) Appl. Microbiol. Biotechnol. 33: 529
368. Jallageas J-C (1980) Adv. Biochem. Eng. 14: 1
369. Thompson LA, Knowles CJ, Linton EA, Wyatt JM (1988) Chem. Brit. 900
370. Nagasawa T, Yamada H (1989) Trends Biotechnol. 7: 153
371. Arnaud A, Galzy P, Jallageas JC (1976) Folia Microbiol. 21: 178
372. Hjort CM, Godtfredsen SE, Emborg C (1990) J. Chem. Technol. Biotechnol. 48: 217
373. Nagasawa T, Takeuchi K, Yamada H (1988) Biochem. Biophys. Res. Commun. 155: 1008
374. Nagasawa T, Nanba H, Ryuno K, Takeuchi K, Yamada H (1987) Eur. J. Biochem. 162: 691
375. Nagasawa T, Yamada H (1986) Biochem. Biophys. Res. Commun. 147: 701

376. Sugiura Y, Kuwahara J, Nagasawa T, Yamada H (1987) J. Am. Chem. Soc. 109: 5848
377. Nagume T (1991) J. Mol. Biol. 220: 221
378. Asano Y, Fujishiro K, Tani Y, Yamada H (1982) Agric. Biol. Chem. 46: 1165
379. Ingvorsen K, Yde B, Godtfredsen SE, Tsuchiya RT (1988) Microbial Hydrolysis of Organic Nitriles and Amides. In: Cyanide Compounds in Biology. Ciba Foundation Symposium 140, Wiley, Chichester, p 16
380. Wyatt JM, Linton EA (1988) The Industrial Potential of Microbial Nitrile Biochemistry. In: Cyanide Compounds in Biology. Ciba Foundation Symposium 140, Wiley, Chichester, p 32
381. Knowles CJ (1988) Cyanide Utilization and Degradation by Microorganisms. In: Cyanide Compounds in Biology. Ciba Foundation Symposium 140, Wiley, Chichester, p 3
382. Nagasawa T, Yamada H (1990) Large-Scale Bioconversion of Nitriles into Useful Amides and Acids. In: Biocatalysis, Abramowicz (ed) Van Nostrand Reinhold, New York, p 277
383. Nazly N, Knowles CJ, Beardsmore AJ, Naylor WT, Corcoran EG (1983) J. Chem. Technol. Biotechnol. 33: 119
384. Knowles CJ, Wyatt JM (1988) World Biotechnol. Rep. 1: 60
385. Wyatt JM (1988) Microbiol. Sci. 5: 186
386. Ingvorsen K, Hojer-Pedersen B, Godtfredsen SE (1991) Appl. Environ. Microbiol. 57: 1783
387. Maestracci M, Thiéry A, Arnaud A, Galzy P (1988) Indian J. Microbiol. 28: 34
388. Bui K, Arnaud A, Galzy P (1982) Enzyme Microb. Technol. 4: 195
389. Lee CY, Chang HN (1990) Biotechnol. Lett. 12: 23
390. Asano Y, Yasuda T, Tani Y, Yamada H (1982) Agric. Biol. Chem. 46: 1183
391. Ryuno K, Nagasawa T, Yamada H (1988) Agric. Biol. Chem. 52: 1813
392. Watanabe I (1987) Methods Enzymol. 136: 523
393. Mauger J, Nagasawa T, Yamada H (1989) Tetrahedron 45: 1347
394. Mauger J, Nagasawa T, Yamada H (1988) J. Biotechnol. 8: 87
395. Nagasawa T, Mathew CD, Mauger J, Yamada H (1988) Appl. Environ. Microbiol. 54: 1766
396. Kobayashi M, Nagasawa T, Yanaka N, Yamada H (1989) Biotechnol. Lett. 11: 27
397. Kobayashi M, Yanaka N, Nagasawa T, Yamada H (1990) J. Antibiot. 43: 1316
398. Mathew CD, Nagasawa T, Kobayashi M, Yamada H (1988) Appl. Environ. Microbiol. 54: 1030
399. Vaughan PA, Cheetham PSJ, Knowles CJ (1988) J. Gen. Microbiol. 134: 1099
400. Bengis-Garber C, Gutman AL (1988) Tetrahedron Lett. 2589
401. Kobayashi M, Nagasawa T, Yamada H (1988) Appl. Microbiol. Biotechnol. 29: 231
402. Nishise H, Kurihara M, Tani Y (1987) Agric. Biol. Chem. 51: 2613
403. Kieny-L'Homme M-P, Arnaud A, Galzy P (1981) J. Gen. Appl. Microbiol. 27: 307
404. Fukuda Y, Harada T, Izumi Y (1973) J. Ferment. Technol. 51: 393
405. Macadam AM, Knowles CJ (1985) Biotechnol. Lett. 7: 865
406. Maestracci M, Bui K, Thiéry A, Arnaud A, Galzy P (1988) Adv. Biochem. Eng./Biotechnol. 36: 67
407. Mundy BP, Liu FHS, Strobel GA (1973) Can. J. Biochem. 51: 1440
408. Arnaud A, Galzy P, Jallageas J-C (1980) Bull. Soc. Chim. Fr. II: 87
409. Kakeya H, Sakai N, Sugai T, Ohta H (1991) Tetrahedron Lett. 1343
410. Preparation SP 409 from NOVO Industri, Denmark
411. Hönicke-Schmidt P, Schneider MP (1990) J. Chem. Soc., Chem. Commun. 648
412. Cohen MA, Sawden J, Turner NJ (1990) Tetrahedron Lett. 7223
413. Klempier N, de Raadt A, Faber K, Griengl H (1991) Tetrahedron Lett. 341
414. de Raadt A, Klempier N, Faber K, Griengl H (1992) J. Chem. Soc., Perkin Trans. I, 137
415. Huisgen R, Christl M (1973) Chem. Ber. 106: 3291

2.2 Reduction Reactions

The enzymes employed in redox reactions are classified into three categories: dehydrogenases, oxygenases and oxidases [1-3]. The former group of enzymes has been widely used for the reduction of carbonyl groups of aldehydes or ketones and of carbon-carbon double bonds. Depending on the substitutional pattern of the substrate, both of these reactions offer potential asymmetrisation of a prochiral substrate leading to a chiral product, and they are therefore of importance. The reverse process (e.g. alcohol oxidations or dehydrogenation reactions) usually go in hand with the destruction of a chiral center, and are therefore generally of limited use.

Scheme 2.100. Reduction reactions catalysed by dehydrogenases.

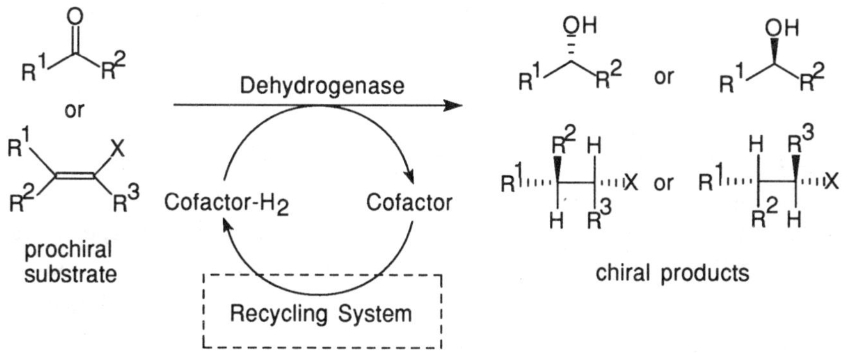

In contrast, oxygenases - named for using molecular oxygen as cosubstrate - have been shown to be particularly useful for oxidation reactions since they are able to catalyse the functionalisation of non-activated C-H or C=C bonds, affording a hydroxylation or epoxidation procedure, respectively (see Section 2.3). Oxidases, which are responsible for the transfer of electrons, play only a minor rôle in the biotransformation of non-natural organic compounds.

2.2.1 Recycling of Cofactors

The major and crucial distinction between redox enzymes and hydrolases described in the previous Chapter, is that they require redox cofactors, which donate or accept the chemical equivalents for reduction (or oxidation). For the majority of redox enzymes, nicotinamide adenine dinucleotide [NAD(H)] and its respective phosphate [NADP(H)] are required by about 80% and 10% of redox enzymes, respectively. Flavines (FMN, FAD) and pyrroloquinoline quinone (PQQ) are encountered more rarely. All of these cofactors have one feature in common, that is they are sensitive and prohibitively expensive if used in stoichiometric amounts [4]. Since it is only the oxidation state of the cofactor which changes during the reaction it may be regenerated in situ by using a second redox-reaction to allow it to enter the reaction cycle again. Thus, the expensive cofactor is needed only in catalytic amounts and this leads to a drastic reduction in cost. The efficiency of such a recycling process is measured by the number of cycles which can be achieved before a cofactor molecule is destroyed. It is expressed as the 'total turnover number' (TTN) - analogous to the 'productivity number' - which is the total number of moles of product formed per mole of cofactor during the course of a complete reaction. As a rule of thumb, a few thousand cycles (10^3-10^4) are sufficient for redox-reactions on a laboratory scale, whereas for technical purposes, total turnover numbers of at least 10^5 are highly desirable. The economic barrier to large-scale reactions posed by cofactor costs has been recognized for many years and a large part of the research concerning dehydrogenases has been expended to solve the problem of recycling [5-7].

Cofactor-recycling is no problem when whole microbial cells are used as biocatalysts for redox reactions. In this case, cheap sources of redox equivalents such as carbohydrates can be used since the microorganism possesses all the enzymes and cofactors which are required for metabolism. The advantages and disadvantages of using whole cell systems are discussed in Section 2.2.3.

Recycling of Reduced Nicotinamide Cofactors

The easiest but least efficient method of regenerating NADH from NAD+ is the non-enzymic reduction using a reducing agent such as sodium dithionite ($Na_2S_2O_4$) [8]. The corresponding turnover numbers of this process are low (TTN ≤ 100). Furthermore, enzyme deactivation caused by $Na_2S_2O_4$ may occur, presumably owing to modification of thiol groups in the protein [9]. Similarly, electrochemical [10-12] and photochemical regeneration methods [13-16], which would be a cheap and easy-to-use alternative, suffer from poor regioselectivity, occurrence of side-reactions and low to moderate turnover numbers (TTN <1000). On the other hand, enzymic methods for NADH- or NADPH-recycling have been shown to be much more efficient; nowadays these represent the methods of choice. They may be conveniently subdivided into coupled-substrate and coupled-enzyme types.

2.2 Reduction Reactions

In the coupled-substrate process the cofactor required for the transformation of the main substrate is constantly regenerated by addition of a second auxiliary substrate (donor) which is transformed by the same enzyme, but into the opposite direction [17-19]. To shift the equilibrium of the reaction in the desired direction, the donor is usually applied in excess leading to turnover numbers of up to 10^3 [20]. Although this approach is applicable in principle to both directions of redox reactions, it has been mainly used for reductions due to the fact that the equilibrium of dehydrogenase reactions lies heavily in favour of reduction [21]. Some of the disadvantages encountered in coupled-substrate cofactor recycling are as follows:
- The product has to be purified from large amounts of auxiliary substrate, and
- Enzyme deactivation may occur when highly reactive carbonyl species such as acetaldehyde or cyclohexenone are involved.
- Enzyme inhibition caused by the high concentrations of the auxiliary substrate are common.

Scheme 2.101. Cofactor-recycling by the coupled-substrate method.

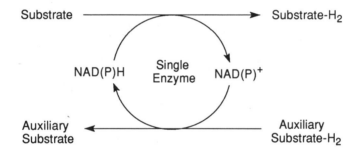

The coupled-enzyme approach is more advantageous. In this case the two parallel redox reactions - conversion of the main substrate and cofactor-recycling - are catalysed by two different enzymes [22]. To achieve optimal results, both of the enzymes should have sufficiently different specificities for their respective substrates whereupon the two enzymic reactions can proceed independently from each other and, as a consequence, both the substrate and the auxiliary substrate do not have to compete for the active site of a single enzyme, but are efficiently converted by the two biocatalysts independently. Several good and several excellent methods, each having its own particular pros and cons, have been developed to regenerate NADH. On the other hand, NADPH may be generated on a lab-scale but a really inexpensive and convenient method is still needed for a large scale work.

Scheme 2.102. Cofactor-recycling by the coupled-enzyme method.

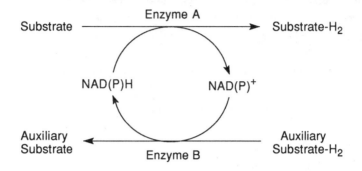

The best and most widely used method for recycling NADH uses formate dehydrogenase (FDH) to catalyse the oxidation of formate to CO_2 (Scheme 2.103) [23, 24]. This method has the advantage that both the auxiliary substrate and the co-product are innocous to enzymes and are easily removed from the reaction. FDH is commercially available, readily immobilized and stable, if protected from autooxidation. The only disadvantage of this system is the high cost of FDH and its low specific activity (3U/mg). However, both drawbacks can be readily circumvented by using an immobilized [25] or membrane-retained FDH-system [26]. Overall, the formate/FDH-system is the most convenient and most economical method for regenerating NADH, particularly for large-scale and repetitions applications, with TTN ranging from 10^3 to 10^5. However, it is not adaptable to allow the regeneration of NADPH.

Another useful method for recycling NAD(P)H makes use of the oxidation of glucose, catalysed by glucose dehydrogenase (GDH, Scheme 2.103) [27, 28]. The gluconolactone formed is spontaneously hydrolysed to give gluconic acid. The glucose dehydrogenase from *Bacillus cereus* is highly stable [29] and accepts either NAD^+ or $NADP^+$ with high specific activity. Like FDH, however, GDH is expensive and product isolation from gluconate may complicate the work-up. In the absence of purification problems, this method is attractive for laboratory use, and it is certainly the most convenient way to regenerate NADPH.

Similarly, glucose-6-phosphate dehydrogenase (G6PDH) catalyses the oxidation of glucose-6-phosphate (G6P) to 6-phosphoglucolactone, which spontaneously hydrolyses to the corresponding gluconate (Scheme 2.103). The enzyme from *Leuconostoc mesenteroides* is inexpensive, stable and accepts both NAD^+ and $NADP^+$ [30, 31], whereas yeast-G6PDH accepts only $NADP^+$. A major disadvantage of this system is the high cost of G6P. Thus, if used on a large scale, it must be enzymatically prepared from glucose using hexokinase

2.2 Reduction Reactions

and this involves the regeneration of ATP using kinases (see Section 2.1.4). To avoid problems arising from multi-enzyme systems, glucose-6-sulphate and G6PDH from *Saccharomyces cerevisiae* may be used to regenerate NADPH [32]. The sulphate does not act as an acid catalyst for the hydrolysis of NADPH and is more easily prepared than the corresponding phosphate [33]. The G6P/G6PDH system complements glucose/GDH as an excellent method for regenerating NADPH and is a good method for regenerating NADH.

Scheme 2.103. Enzymatic regeneration of reduced nicotinamide cofactors.

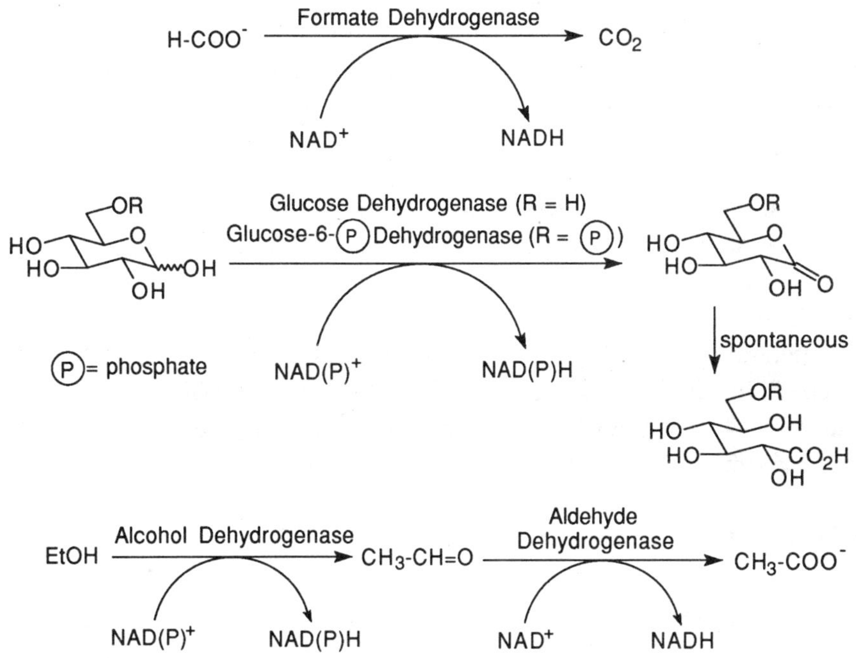

Ethanol and alcohol dehydrogenase (ADH) have been used extensively to regenerate NADH and NADPH [34, 35]. The low to moderate cost of ADH and the volatility of both ethanol and acetaldehyde make this system attractive for lab-scale reactions. An alcohol dehydrogenase from yeast reduces NAD+, while an ADH from *L. mesenteroides* is used to regenerate NADPH (Scheme 2.103). However, due to the low redox potential, only activated carbonyl-substrates such as aldehydes and cyclic ketones are reduced in good yields. With other substrates, the equilibrium must be driven by using ethanol in excess or by removing acetaldehyde. The latter result may be achieved by sweeping with nitrogen [36] or by further oxidizing acetaldehyde to acetate [37]

using aldehyde dehydrogenase concomitantly generating a second equivalent of reduced cofactor. All of these methods, however, give low TTN or involve complex and unstable multienzyme systems. Furthermore, even low concentrations of ethanol or acetaldehyde inhibit or deactivate enzymes.

A particularly attractive alternative for the regeneration of NADH makes use of hydrogenase enzymes, which are able to accept molecular hydrogen directly as hydrogen donor [38, 39]. The latter is strongly reducing, innocous to enzymes and nicotinamide cofactors, and its consumption leaves no by-product. For organic chemists, however, this method cannot be generally recommended because hydrogenase, which is extremely sensitive to oxidation, is not commercially available and thus requires undertaking complicated preparative fermentation procedures. Furthermore, some of the organic dyes, which serve as mediators for the transport of redox equivalents from the donor onto the cofactor are relatively toxic (see below).

Recycling of Oxidized Nicotinamide Cofactors

The best and most widely applied method for the regeneration of nicotinamide cofactors in their oxidized form involves the use of glutamate dehydrogenase (GluDH) which catalyses the reductive amination of α-ketoglutarate to give L-glutamate [40, 41]. Both NADH and NADPH are accepted. In addition, α-ketoadipate can be used instead of the corresponding glutarate [42], leading to the formation of a useful byproduct, L-α-aminoadipate.

Scheme 2.104. Enzymatic regeneration of oxidized nicotinamide cofactors.

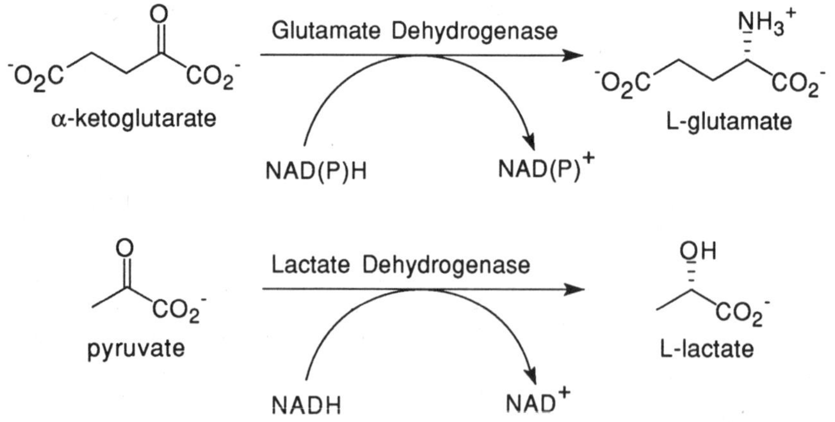

2.2 Reduction Reactions

Using pyruvate together with lactate dehydrogenase (LDH) to regenerate NAD+ offers the advantage, that LDH is less expensive and exhibits a higher specific activity than GluDH [43]. However, the redox potential is less favourable and LDH does not accept NADP+.

Acetaldehyde and yeast-ADH have also been used to regenerate NAD+ from NADH [44]. Although the total turnover numbers achieved were quite impressive (10^3-10^4), the above mentioned disadvantages of enzyme-deactivation and self-condensation of acetaldehyde out-weigh the merits of the low cost of yeast-ADH and the volatility of the reagents involved.

2.2.2 Reduction of Aldehydes and Ketones using Isolated Enzymes

A broad range of ketones can be reduced stereoselectively using dehydrogenases to give chiral secondary alcohols [45]. During the course of the reaction, the enzyme delivers the hydride either from the *si*- or the *re*-side of the ketone to give (*R*)- or (*S*)-alcohols, respectively. For most cases, the stereochemical course of the reaction, which is mainly dependent on the steric requirements of the substrate, may be predicted from a simple model which is generally referred to as 'Prelog´s rule' [46].

It is based on the stereochemistry of microbial reductions using *Curvularia falcata* cells and it states that the dehydrogenase delivers the hydride ion from the *re*-face of a prochiral ketone predominantly. The majority of the commercially available dehydrogenases used for the stereospecific reduction of ketones [such as yeast alcohol dehydrogenase (YADH), horse liver alcohol dehydrogenase (HLADH)] and the majority of microorganisms (for instance baker´s yeast) follow Prelog´s rule (Scheme 2.105). *Thermoanaerobium brockii* alcohol dehydrogenase (TBADH) also obeys this rule yielding (*S*)-alcohols when large ketones are used as substrates, but the behaviour is reversed with small substrates. Microbial dehydrogenases which lead to the formation of (*R*)-alcohols are known, but they are not commercially available [47-49].

Yeast-ADH has a very narrow substrate specificity and, in general, only accepts aldehydes and methyl ketones [50, 51]. As a consequence, cyclic ketones and those, with carbon chains larger than a methyl group, are excluded as substrates. Thus, YADH is only of limited use for the preparation of chiral secondary alcohols. Similarly, other ADH´s from *Curvularia falcata*, *Mucor javanicus* and *Pseudomonas* sp. are of limited use as long as they are not commercially available. The most commonly used dehydrogenases may be displayed as shown in Figure 2.16, with reference to their preferred size of their substrates [52].

Scheme 2.105. Prelog´s rule for the asymmetric reduction of ketones.

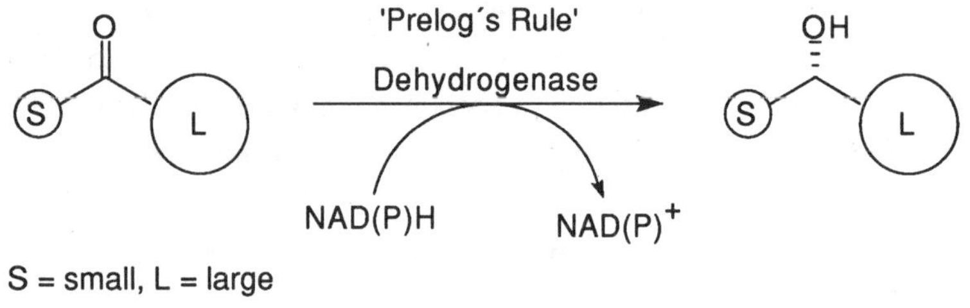

S = small, L = large

Dehydrogenase	Specificity	Cofactor	Commercially Available
yeast-ADH	Prelog	NADH	+
horse liver-ADH	Prelog	NADH	+
Thermoanaerobium brockii-ADH	Prelog[a]	NADPH	+
Hydroxysteroid-DH	Prelog	NADH	+
Curvularia falcata-ADH	Prelog	NADPH	-
Mucor javanicus-ADH	Anti-Prelog	NADPH	-
Lactobacillus kefir-ADH	Anti-Prelog	NADPH	-
Pseudomonas sp.-ADH	Anti-Prelog	NADH	-

[a] The specificity is reversed when small ketones are used as substrates.

Figure 2.16. Preferred substrate size for dehydrogenases.

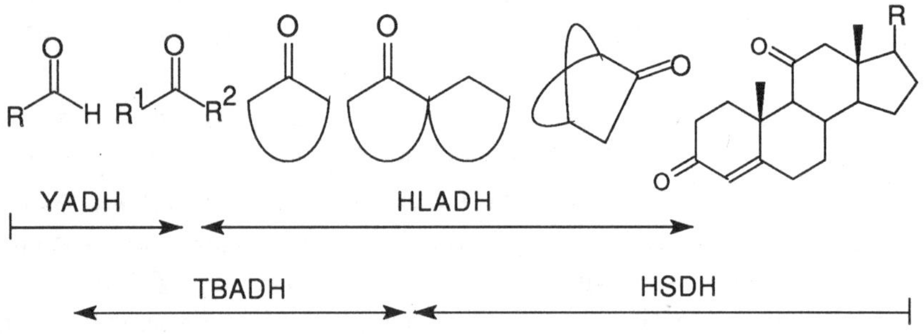

2.2 Reduction Reactions

Horse liver ADH is a very catholic enzyme with a broad substrate specificity and a narrow stereospecificity. Thus it is probably the most widely used dehydrogenase in biotransformations [53]. It is a dimer consisting of almost identical subunits each of which contains two zinc atoms. The three-dimensional structure has been elucidated by X-ray diffraction [54]. Although the primary sequence is quite different, the tertiary structure of HLADH is similar to that of YADH [55]. The most useful applications are found in the reduction of medium-ring monocyclic ketones (from four to nine-membered) and bicyclic ketones [56-58]. Sterically demanding molecules which are larger than decalines are not readily accepted and acyclic ketones are usually reduced with low enantioselectivities [59, 60].

A considerable number of monocyclic and bicyclic racemic ketones have been resolved using HLADH with fair to excellent specificities [61-63]. Even sterically demanding cage-shaped polycyclic ketones were readily accepted [64, 65]. For instance, *rac*-2-twistanone was reduced to give the *exo*-alcohol and the enantiomeric ketone in 90% and 68% e.e., respectively [66]. Also O- and S-heterocyclic ketones were shown to be good substrates [67-69]. Thus (±)-bicyclo[4.3.0]nonan-3-ones bearing either an O or S atom in position 8 were resolved with good selectivities [70]. Attempted reduction of N-heterocyclic ketones lead to deactivation of the enzyme via complexation of the essential Zn^{2+} in the active site [71].

Scheme 2.106. Resolution of bi- and polycyclic ketones using HLADH.

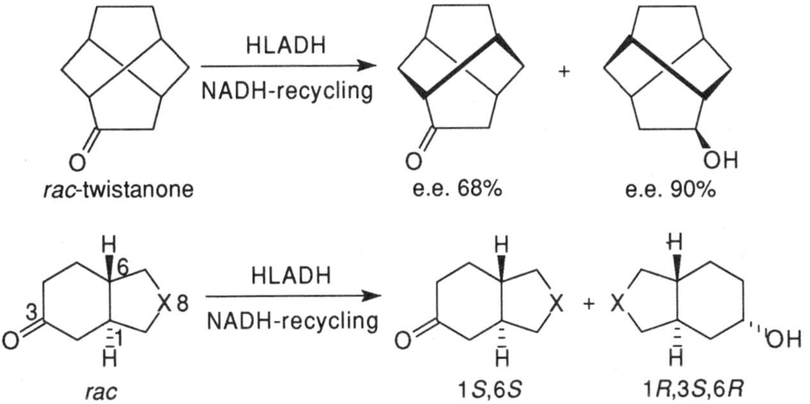

X	e.e. Ketone [%]	e.e. Alcohol [%]
O	60	>97
S	53	>97

The HLADH catalysed reduction of 2-alkylthiapyran-4-ones did not stop at 50% conversion, as one would expect for a kinetic resolution, but proceeded in a diastereoselective fashion to give a mixture of *cis*- and *trans*-(*S*)-alcohols (Scheme 2.107) [72]. The latter were chromatographically separated and subsequently oxidized to provide the optically pure ketones.

Scheme 2.107. Diastereoselective reduction of alkylthiapyranones using HLADH.

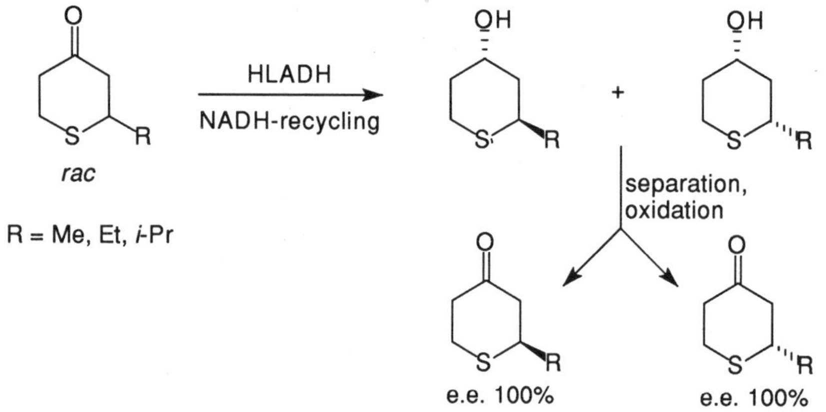

Scheme 2.108. Asymmetrisation of prochiral diketones using HLADH.

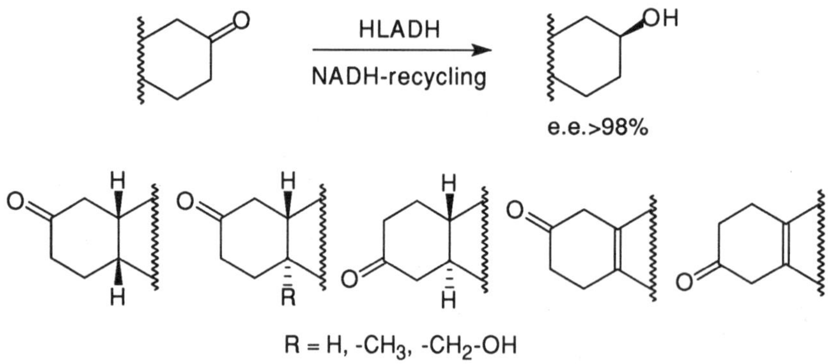

R = H, -CH$_3$, -CH$_2$-OH

Every kinetic resolution of bi- and polycyclic ketones suffers from one particular drawback because the bridgehead carbon atoms make it impossible to recycle the undesired 'wrong' enantiomer via racemisation. Hence the asymmetrisation of prochiral diketones making use of the enantioface- or

2.2 Reduction Reactions

enantiotopos-specificity of HLADH is of advantage. For instance, both the *cis*- and *trans*-forms of decalinediones shown in Scheme 2.108 were reduced to give (*S*)-alcohols with excellent optical purity. Similar results were obtained with unsaturated derivatives [73, 74].

The wide substrate tolerance of HLADH is demonstrated by the resolution of organometallic derivatives possessing axial chirality [75]. For instance, the racemic tricarbonyl cyclopentadienyl manganese aldehyde shown in Scheme 2.109 was enantioselectively reduced to give the (*R*)-alcohol and the (*S*)-aldehyde with excellent optical purities [76].

Scheme 2.109. Reduction of an organometallic aldehyde using HLADH.

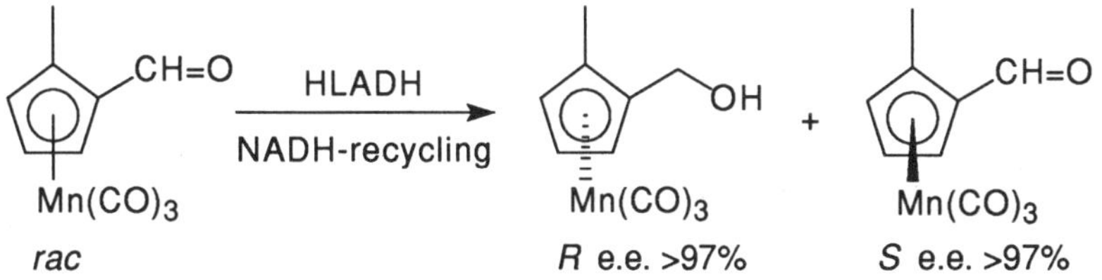

In order to predict the stereochemical outcome of HLADH catalysed reductions a number of models have been developed, each of which has its own merits. The first rationale emerged from the 'diamond lattice model' of V. Prelog, which was actually developed for *Curvularia falcata* [77]. It was refined later for HLADH using molecular graphics [78]. A more recently developed cubic-space descriptor is particularly useful for ketones bearing chirality centre(s) remote from the location of the carbonyl group [79]. Alternatively, a quadrant rule may be applied [80]. A useful substrate model based on a flattened cyclohexanone ring is shown in Figure 2.17 [81]. It shows the Zn^{2+} in the catalytic site which coordinates to the carbonyl oxygen and the nucleophilic attack of the hydride occurring from the bottom. The preferred orientation of the substrate relative to the chemical operator - the hydride ion - can be estimated by placing the substituents into the 'allowed' and 'forbidden' zones.

Since YADH and HLADH are not useful for the asymmetric reduction of open-chain ketones, it is noteworthy that a recently isolated alcohol dehydrogenase from the bacterium *Thermoanaerobium brockii* (TBADH) has been shown to fill this gap (Scheme 2.110) [82]. It is remarkably thermostable (up to 85 °C) and also can tolerate the presence of organic solvents such as *iso*-propanol, which is used for NADP-recycling in a coupled-substrate approach.

Figure 2.17. Substrate model for HLADH.

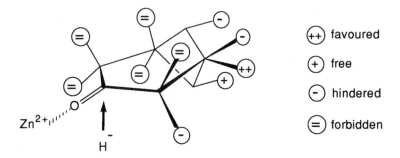

Scheme 2.110. Enantioselective reduction of ketones using TBADH.

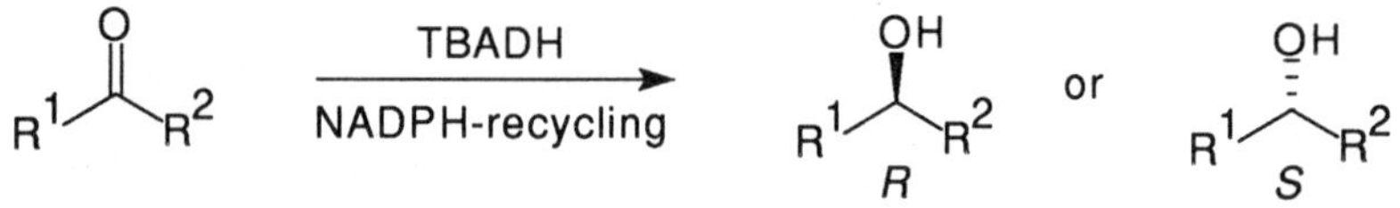

R¹	R²	Specificity	Configuration	e.e. [%]
CH_3	C_2H_5	Anti-Prelog	R	48
CH_3	$CH(CH_3)_2$	Anti-Prelog	R	86
CH_3	$c\text{-}C_3H_5$	Anti-Prelog	R	44
CH_3	$n\text{-}C_3H_7$	Prelog	S	79
CF_3	Ph	Prelog	R[a]	94
CH_3	$C{\equiv}CH$	Prelog	S	86
CH_3	$CH_2\text{-}CH(CH_3)_2$	Prelog	S	95
CH_3	$(CH_2)_3\text{-}Cl$	Prelog	S	98
CH_3	$(CH_2)_5\text{-}Cl$	Prelog	S	>99
$Cl\text{-}CH_2\text{-}$	$CH_2\text{-}CO_2Et$	Prelog	R[a]	90
C_2H_5	$(CH_2)_2\text{-}CO_2Me$	Prelog	S	98
CH_3	$n\text{-}C_5H_{11}$	Prelog	S	99
CH_3	$n\text{-}C_7H_{15}$	Prelog	S	98
C_2H_5	$n\text{-}C_3H_7$	Prelog	S	97
C_2H_5	$(CH_2)_3\text{-}Cl$	Prelog	S	>99
$n\text{-}C_3H_7$	$n\text{-}C_3H_7$	--	no reaction	--
CH_3	$E\text{-}CH{=}CH\text{-}CH(CH_3)_2$	--	no reaction	--

[a] Sequence rule order reversed.

2.2 Reduction Reactions

Open-chain methyl- and ethyl-ketones are readily reduced by TBADH into the corresponding secondary alcohols, generally with excellent specificities [83]. ω-Haloalkyl- [84, 85] and methyl- or trifluoromethyl-ketones possessing heterocyclic substituents were converted into the corresponding secondary alcohols with excellent optical purities [86, 87]. However, α,β-unsaturated ketones and those bearing substituents on both sides which are larger than ethyl, are not accepted. In general TBADH obeys Prelog's rule with 'normal-sized' ketones leading to (S)-alcohols, but the stereoselectivity was found to be reversed with small substrates.

Hydroxysteroid dehydrogenases (HSDH) are ideally suited enzymes for the reduction of bulky mono- [88] and bicyclic ketones (Scheme 2.111) [89]. This is not surprising, if one thinks of the steric requirements of their natural substrates - steroids [90, 91].

Scheme 2.111. Resolution of sterically demanding ketones using HSDH.

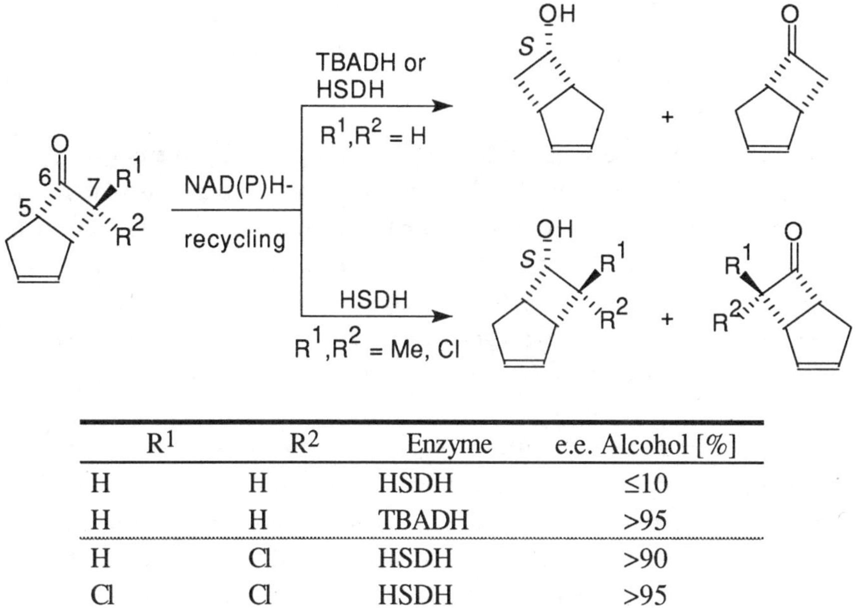

R^1	R^2	Enzyme	e.e. Alcohol [%]
H	H	HSDH	≤10
H	H	TBADH	>95
H	Cl	HSDH	>90
Cl	Cl	HSDH	>95
Me	Cl	HSDH	>98
Me	Me[a]	HSDH	>95

[a] No reaction was observed with HLADH or TBADH.

For instance, bicyclo[3.2.0]heptan-6-one systems were reduced with HSDH with very low selectivity when substituents in the adjacent 7-position were small (R^1, R^2 = H). On the other hand, TBADH showed an excellent enantioselectivity. When the steric requirements of the substrate were increased by additional methyl- or chloro-substituents, the situation changed. Then, HSDH became a very specific catalyst and TBADH (or HLADH) proved to be unable to accept the bulky substrate [92, 93]. The switch in the stereochemical preference is not surprising and may be explained by Prelog´s rule: i.e. with the unsubstituted ketone, the position 5 is 'larger' than position 7. However, when the hydrogen atoms on carbon atom 7 are replaced by sterically demanding chlorine- or methyl-groups, the situation is reversed.

The majority of synthetically useful ketones can be transformed into the correponding chiral secondary alcohols by choosing the appropriate dehydrogenase from the above mentioned set of enzymes. Other enzymes, which have been shown to be useful for specific types of substrates, are mentioned below.

α-Ketoacids may be transformed into either (R)- or (S)-2-hydroxyacids using NADH-dependent lactate dehydrogenases (LDH) of different origins. Both D- and L-specific enzymes, isolated from rabbit muscle [94] and microorganisms such as *Staphylococcus epidermidis* [95], and *Bacillus stearothermophilus* [96], have been shown to be useful for such transformations.

The natural rôle of glycerol dehydrogenase (GDH) is the interconversion of glycerol and dihydroxyacetone. The enzyme is commercially available from different sources and has been used for the stereoselective reduction of α-hydroxyketones [97]. The enzyme has been found to tolerate some structural variation of its natural substrate - dihydroxyacetone - including cyclic derivatives. A glycerol dehydrogenase (GDH) from *Geotrichum candidum* was shown to reduce not only α- but also β-ketoesters with high selectivity [98].

Enzymes from thermophilic organisms (which grow in the hostile environment of hot springs with temperatures ranging from 70-100 °C) have recently received much attention [99-102]. Thermostable enzymes are not only stable to heat but, in general, show enhanced stability in the presence of common protein denaturants and organic solvents. Since they are not restricted to work in the narrow temperature range which is set for mesophilic, 'normal' enzymes (20-40 °C), an influence of the temperature on the selectivity can be studied over a wider range. For instance, it was demonstrated that the diastereoselectivity of the HLADH-catalysed reduction of 3-cyano-4,4-dimethyl-cyclohexanone is diminished at 45 °C (the upper operational limit for HLADH) when compared with that observed at 5 °C [103]. On the other hand, a temperature-dependent reversal of the enantiospecificity of an alcohol dehydrogenase derived from *Thermoanaerobacter ethanolicus* was observed when the temperature was raised to 65 °C [104].

2.2.3 Reduction of Aldehydes and Ketones using Whole Cells

Instead of isolated dehydrogenases, which require costly cofactor-recycling, whole microbial cells can be employed. They contain multiple dehydrogenases which are able to accept non-natural substrates, all the necessary cofactors and the metabolic pathways for their regeneration. Thus, cofactor-recycling can be omitted since it is automatically done by the cell. As a consequence, cheap carbon-sources such as saccharose or glucose can be used as auxiliary substrates for asymmetric reduction reactions. Furthermore, all the enzymes and cofactors are well protected within their natural cellular environment.

These distinct advantages have to be taken into consideration alongside some significant drawbacks. The productivity of microbial conversions is usually low since the majority of non-natural substrates are toxic to living organisms and are therefore only tolerated at low concentrations (~0.1-0.3%). The high amount of biomass present in the reaction medium causes low overall yields and makes product recovery troublesome, particularly when the product is not excreted from the cells. Since only a minor fraction (typically 0.5 to 2%) of the carbon source is used for coenzyme-recycling, the bulk of it is metabolized forming byproducts which often impede product purification. Chiral transport phenomena into and out of the cell may influence the specificity of the reaction, particularly when racemic substrates are used. Finally, different strains of a microorganism may possess different specificities, thus it is important to use exactly the same culture to obtain comparable results [105].

Low stereoselectivities may be due to a variety of reasons:
- A substrate may be reduced by a single oxidoreductase with different transition states for the two enantiomers or the enantiotopic faces. In other words, insufficient chiral recognition takes place allowing an alternative fit of the substrate.
- If two enzymes, each with high but opposite stereochemical preference compete for the same substrate, the relative rates of the individual reactions, which determine the optical purity of the product, depend on the substrate concentration. Under physiological conditions (i.e. the substrate concentration is below saturation), the relative rates of the Michaelis-Menten system are given by the ratio V_{max}/K_m for the individual reactions. When saturation

is reached using increased substrate concentrations, the relative rates depend on the ratio of k_{cat} of the two reactions. Consequently, the relative rates of the reaction of enantiomers and hence the optical purity of product (e.e.p) depend on the substrate concentration. With yeasts, it is a well known phenomenon that lower substrate concentrations give higher e.e.p's [106].

The following general techniques can be applied to enhance the selectivity of microbial reduction reactions:
- Substrate modification by variation of protecting groups which can be removed after the transformation [107-109],
- variation of the metabolic parameters by immobilization [110-112],
- variation of the fermentation conditions [113],
- sceening of microorganisms to obtain strains with the optimum properties (a hard task for non-microbiologists) [114, 115], and
- selective inhibition of one of the competing enzymes (see below).

Reduction of Aldehydes and Ketones by Baker's Yeast

Baker's yeast (*Saccharomyces cerevisiae*) is by far the most widely used microorganism for the asymmetric reduction of ketones [116-119]. It is ideal for non-microbiologists, since it is easily available and it does not require sterile fermentation equipment and can be handled using standard laboratory equipment. Thus it is not surprising that yeast-catalysed transformations leading to chiral products have been reported from the beginning of this century [120] and that the first comprehensive review which covers almost all the different strategies of yeast-reductions dates back to 1949! [121].

Simple aliphatic and aromatic ketones are reduced by fermenting yeast according to Prelog's rule to give the corresponding (*S*)-alcohols in good optical purities (Scheme 2.112) [122]. Long-chain ketones such as *n*-propyl-*n*-butylketone and several bulky phenyl ketones are not accepted. One long alkyl chain is tolerated, if the other group is methyl [123, 124]. As might be expected, high chiral discrimination was achieved with groups of greatly different size. A wide range of functional groups can be tolerated adjacent to the ketone including chloro- [125], bromo- [126], perfluoroalkyl- [127], nitro- [128, 129], hydroxyl- [130, 131], dithianyl- [132] and silyl-groups [133].

Acyclic β-ketoesters shown in Scheme 2.113 are readily reduced by yeast to yield β-hydroxyesters [134-139], which serve as chiral starting material for the synthesis of β-lactams [140], insect pheromones [141] and carotenoids [142].

2.2 Reduction Reactions

Scheme 2.112. Reduction of aliphatic and aromatic ketones using baker's yeast.

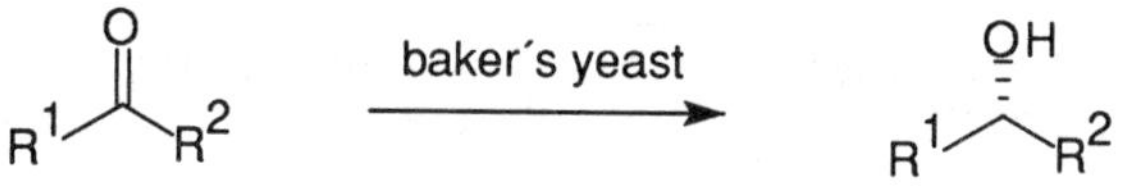

$$R^1-CO-R^2 \xrightarrow{\text{baker's yeast}} R^1-CH(OH)-R^2$$

R^1	R^2	Configuration	e.e. [%]
Me	Et	S	67
Me	n-Bu	S	82
Me	Ph	S	89
Me	CF_3	S	>80
CF_3	CH_2-Br	S	>80
Me	$C(CH_3)_2$-NO_2	S	>96
Me	CH_2-OH	S	91
Me	$(CH_2)_2$-CH=$C(CH_3)_2$	S	94
Me	c-C_6H_{11}	S	>95

It is obvious that the enantioselectivity and the stereochemical preference of the β-hydroxyester depends on the relative size of the alkoxy moiety and the ω-substituent of the ketone, with the nucleophilic attack of the hydride occurring according to Prelog's rule, as shown in Scheme 2.113. As a consequence, the absolute configuration of the alcoholic center may be directed by substrate-modification using either the corresponding short- or long-chain alkyl ester [143].

The reason for this divergent behaviour is not due to an alternative fit of the substrates in a single enzyme, but rather to the presence of two dehydrogenases possessing an opposite stereochemical preference which compete for the substrate [144]. The D-specific enzyme - belonging to the fatty acid synthetase complex - shows a higher activity towards substrates having a short-chain alcohol moiety, and the L-enzyme is more active on long-chain counterparts. As a consequence, the stereochemical direction of the reduction may be controlled by careful design of the substrate, or by selective inhibition of one of the competing dehydrogenases.

Scheme 2.113. Reduction of acyclic β-ketoesters using baker's yeast.

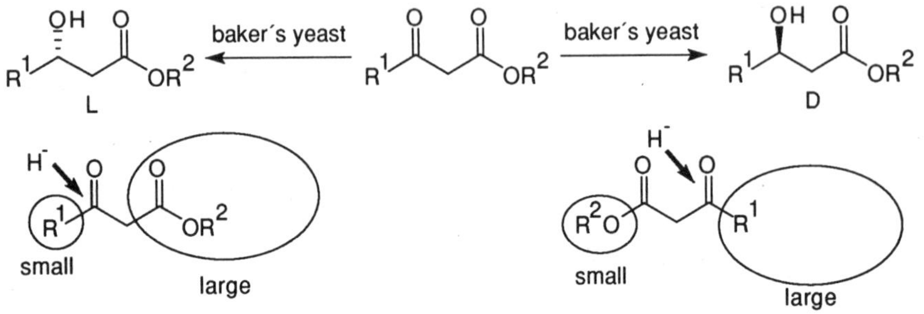

R^1	R^2	Configuration	e.e. [%]
Cl-CH$_2$-	CH$_3$	D	64
Cl-CH$_2$-	C$_2$H$_5$	D	54
Cl-CH$_2$-	n-C$_3$H$_7$	D	27
Cl-CH$_2$-	n-C$_5$H$_{11}$	L	77
Cl-CH$_2$	n-C$_8$H$_{17}$	L	97[a]
(CH$_3$)$_2$C=CH-(CH$_2$)$_2$-	CH$_3$	D	92
CCl$_3$	C$_2$H$_5$	D	85
CH$_3$	CH$_3$	L	87
CH$_3$	C$_2$H$_5$	L	>96
t-Bu-O-CH$_2$-	C$_2$H$_5$	L	97
N$_3$-CH$_2$-	C$_2$H$_5$	L	80
Br-CH$_2$-	n-C$_8$H$_{17}$	L	100
C$_2$H$_5$-	n-C$_8$H$_{17}$	L	95

[a] Low yield.

Inhibition of the L-enzyme which leads to the increased formation of D-β-hydroxyesters, was accomplished by addition of unsaturated compounds such as allyl alcohol [145] or methyl vinyl ketone (Table 2.2, Scheme 2.113) [146]. The same effect was observed when the yeast cells were immobilized by entrapment into a polyurethane gel [147, 148]. As expected, L-enzyme inhibitors led to a considerable increase in the optical purity of D-β-hydroxyesters; substrates, which normally form L-enantiomers, give products of lower optical purity.

Various haloacetates have been found to be selective inhibitors for the D-enzyme (Table 2.3, Scheme 2.113) [149]: this leads to an increased formation of the L-enantiomer. In one case shown in Table 2.3, the effect is drastic enough to afford an inversion of the configuration of the formed β-hydroxyester. As can be seen without an inhibitor, the D-enantiomer was obtained in moderate e.e., but the L-counterpart was formed with the addition of ethyl chloroacetate.

Table 2.2. Selectivity enhancement via L-enzyme inhibition (see Scheme 2.113).

R^1	R^2	Conditions	Configuration	e.e. [%]
Cl-CH$_2$-	C$_2$H$_5$	standard	D	43
Cl-CH$_2$-	C$_2$H$_5$	+ allyl alcohol	D	85
C$_2$H$_5$	CH$_3$	standard	D	37
C$_2$H$_5$	CH$_3$	+ CH$_3$-CO-CH=CH$_2$	D	89
CH$_3$	C$_2$H$_5$	standard	L	>98
CH$_3$	C$_2$H$_5$	PU-immobilized	L	60
C$_2$H$_5$	CH$_3$	standard	D	5
C$_2$H$_5$	CH$_3$	PU-immobilized	D	86

Table 2.3. Selectivity enhancement via D-enzyme inhibition (see Scheme 2.113).

R^1	R^2	Conditions	Configuration	e.e. [%]
C$_2$H$_5$	CH$_3$	standard	L	15
C$_2$H$_5$	CH$_3$	+ Cl-CH$_2$-CO$_2$-n-Bu	L	69
C$_2$H$_5$	CH$_3$	+ Cl-CH$_2$-CO$_2$-C$_2$H$_5$	L	91
Cl-CH$_2$-	C$_2$H$_5$	standard	D	43
Cl-CH$_2$-	C$_2$H$_5$	+ Cl-CH$_2$-CO$_2$-C$_2$H$_5$	L	80

The asymmetric reduction of β-ketoacids (handled as their stable potassium salts) was reported to give chiral β-hydroxyacids with even better optical purity compared to when the corresponding esters were used as substrates [150, 151]. However, low yields were encountered through decarboxylation of the liberated β-ketoacid in the acidic culture medium [152].

Yeast-reduction of α-monosubstituted β-ketoesters leads to the formation of diastereomeric *syn*- and *anti*-β-hydroxyesters [153-156]. The ratio between them is not 1:1, as may be expected from the fact that enolisation causes the interconversion of the enantiomers of the substrate [157]. Thus, *syn-anti* ratios of up to 95:5 have been observed [158]. With small α-substituents, the formation of *syn*-diastereomers predominates, but the selectivity is reversed when the substituents are increased in size.

Scheme 2.114. Reduction of acyclic α-substituted β-ketoesters using baker's yeast.

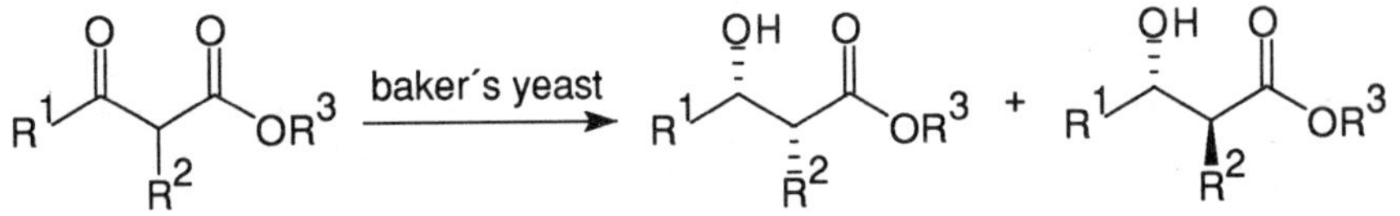

R^1	R^2	R^3	e.e [%] syn	anti	Ratio syn/anti
CH$_3$	CH$_3$	C$_2$H$_5$	100	100	83:17
CH$_3$	CH$_3$	Ph-CH$_2$-	100	80	67:33
Ph-CH$_2$-O-CH$_2$-	CH$_3$	(CH$_3$)$_2$CH-	96^a	50^a	24:76
CH$_3$	CH$_2$=CH-CH$_2$-	C$_2$H$_5$	100	100	25:75
CH$_3$	Ph-CH$_2$-	C$_2$H$_5$	199	100	33:67
CH$_3$	Ph-S-	CH$_3$	>96	>96	17:83

^{a} Due to a change in the sequence rule order the (2*R*,3*R*)-*syn*- and the (2*S*,3*R*)-*anti*-enantiomer is obtained.

These observations have led to the construction of a simple model which allows one to predict the diastereoselectivity (*syn/anti*-ratio) of yeast-catalysed reductions of α-substituted β-ketoesters [159]. Thus, when α-substituents are smaller than the carboxylate moiety, they fit well into the small pocket (S), but substrates bearing space-filling groups on the α-carbon have to occupy both of the pockets in an inverted orientation (i.e. the carboxylate group occupies the pocket S).

Figure 2.18. Model for predicting the diastereoselectivity in yeast-reductions.

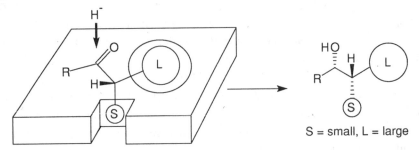

S = small, L = large

2.2 Reduction Reactions

As may be predicted by this model, the yeast-reduction of cyclic ketones bearing an α-alkoxycarbonyl group always leads to the corresponding *syn*-β-hydroxyesters [160-162]. The corresponding *anti*-diastereomers cannot be formed because rotation around the α,β-carbon bond is impossible with such cyclic structures. Furthermore, the reductions are generally more enantioselective than the corresponding acyclic substrate due to the enhanced rigidity of the system. Thus it can be worthwhile to create a ring in the substrate temporarily and to re-open it after the biotransformation in order to make use of this benefit.

Scheme 2.115. Yeast-reduction of cyclic β-ketoesters.

R^1	R^2	e.e. [%]
-(CH$_2$)$_2$-	C$_2$H$_5$	>98
-(CH$_2$)$_3$-	C$_2$H$_5$	86-99[a]
-S-CH$_2$-	CH$_3$	85
-CH$_2$-S-	CH$_3$	>95

[a] Depending on the yeast strain used.

In the same way, α-ketoesters and α-amides can be transformed into the correponding α-hydroxy derivatives. Thus, following Prelog's rule, (S)-lactate [163] and (R)-mandelate esters [164] were obtained from pyruvate and α-ketophenylacetic esters by fermenting baker's yeast in excellent optical purity (e.e. 91-100%).

Cyclic β-diketones are selectively reduced to give β-hydroxyketones without the formation of dihydroxy products (Scheme 2.116) [165-168]. It is important, however, that the highly acidic protons on the α-carbon are replaced by substituents in order to avoid the chemical condensation of the substrate with acetaldehyde, which is always present in fermenting yeast [169]. Again, with small size rings, the corresponding *syn*-products are predominantly formed, usually with excellent optical purity. However, the diastereoselectivity becomes less predictable and the yields drop when the rings are enlarged.

Scheme 2.116. Yeast-reduction of cyclic β-diketones.

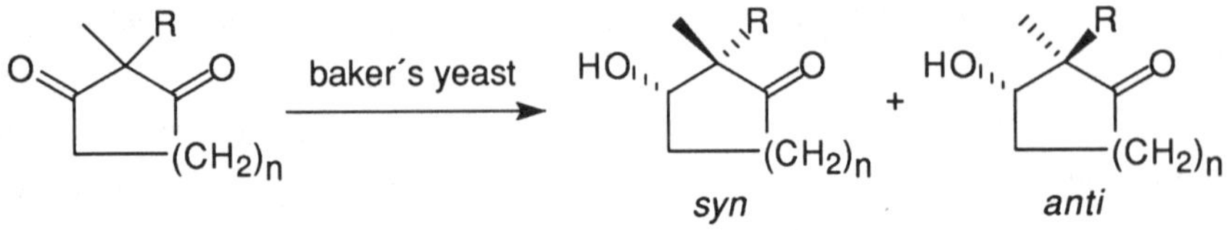

R	n	syn/anti	e.e. syn [%]
$CH_3\text{-}C(=CH_2)\text{-}CH_2\text{-}$	1	100:0	98
$CH_2=CH\text{-}CH_2\text{-}$	1	90:10	>98
$HC\equiv C\text{-}CH_2\text{-}$	1	100:0	>90
$MeO_2C\text{-}CH_2\text{-}$	1	100:0	>98
$NC\text{-}(CH_2)_2\text{-}$	1	96:4	>98
$CH_3\text{-}C(=CH_2)\text{-}CH_2\text{-}$	2	24:76	>98
$CH_2=CH\text{-}CH_2\text{-}$	2	45:55	>98
$HC\equiv C\text{-}CH_2\text{-}$	2	27:73	>98
$NC\text{-}(CH_2)_2\text{-}$	2	30:70	>98
$CH_2=CH\text{-}CH_2\text{-}$	3	100:0	>98
$CH_2=CH\text{-}CH_2\text{-}$	4	82:18	>98
$CH_2=CH\text{-}CH_2\text{-}$	5	no reaction	>98

In contrast, the reduction of α-diketones does not stop at the keto-alcohol stage (Scheme 2.117). In general, the less hindered carbonyl group is quickly reduced to give the (S)-α-hydroxyketone, but further reduction of the remaining (more sterically hindered) carbonyl group finally yields the corresponding diols predominantly in the *anti*-configuration [170, 171].

Scheme 2.117. Yeast-reduction of α-diketones.

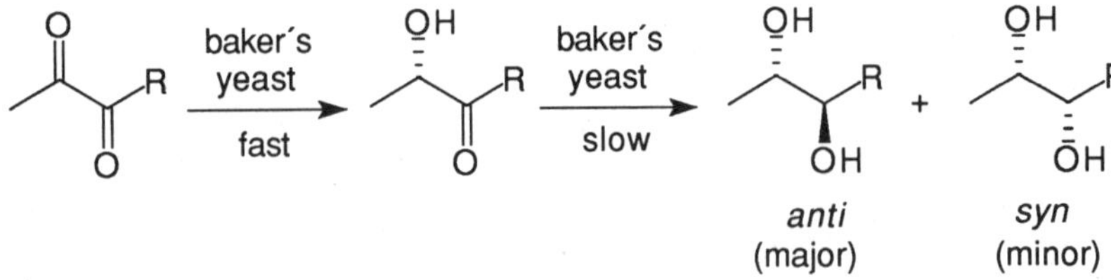

R	anti/syn	e.e. anti-diol [%]
Ph-	>95:<5	94
1,3-dithian-2-yl-	95:5	97
Ph-S-CH_2-	86:14	>97

2.2 Reduction Reactions

The use of fermenting yeast for the stereoselective reduction of ketones and aldehydes is of course not limited to the few examples shown above but has also been used for the transformation of substrates possessing nitro- [172], fluoro- [173-176], sulphur- [177-179] and heterocyclic functional groups [180, 181] and even metalloorganic compounds [182, 183]. In many cases, the use of microorganisms other than yeast may be of an advantage, particularly if the stereoselectivity of yeast-catalysed reaction is low or if the product of opposite configuration is desired [184-189]. However, in this case the help of a microbiologist is strongly recommended for organic chemists.

2.2.4 Reduction of C=C-Bonds using Whole Cells

The enzyme-catalysed enantioface-selective reduction of prochiral C=C double bonds, which is difficult to perform with conventional methods, has been shown to proceed with high specificities and to be applicable to a wide range of substrates (see Scheme 2.100). The enzymes responsible for these reactions are NADH-dependent enoate reductases, which can be found in many microorganisms such as *Clostridia* [190] and methanogenic *Proteus* sp. [191] and even in baker's yeast. Although enoate reductases have been isolated and characterized [192, 193], the majority of transformations on a preparative scale have been performed by using whole microbial cells [194]. This was mainly due to problems of cofactor-recycling and due to the sensitivity of enoate reductases to traces of oxygen. The stereochemical course of enoate reduction has been elucidated and is made up of a *trans*-addition of hydrogen across the C=C bond [195-197]. To be effectively reduced by an enoate reductase, the double bond has to be 'activated' by an electron withdrawing substituent X (Scheme 2.100).

The capacity of baker's yeast to asymmetrically reduce activated C=C bonds was discovered in 1934 [198]. The data collected since this time form a set of rules for the asymmetric hydrogenation of carbon-carbon double bonds by fermenting baker's yeast, which is more or less applicable also to other microorganisms, such as *Beauveria sulfurescens* [199].
- Only C=C-bonds which are 'activated' by electron-withdrawing substituents are reduced [200]. Isolated double bonds and acetylenic triple bonds are not accepted [201]. The following functional groups may serve as 'activating' groups:
- α,β-Unsaturated carboxylic acids and -esters are readily transformed into their saturated analogues [202]. In some cases, an ester may be enzymatically hydrolysed to give the corresponding acid prior to reduction.

- α,β-Unsaturated carboxaldehydes are generally transformed in a two-step sequence [203, 204]. Firstly, fast reduction of the carbonyl group by dehydrogenases yields allylic alcohols, which in turn are converted by enoate reductases via a second (slow) step into the respective chiral alkanols. Alternatively, allylic alcohols may be used as substrates [205].
- A reaction sequence in the reverse order is observed with α,β-unsaturated ketones. In this case, the C=C bond is reduced first to give the saturated ketone, which - in a second slow step - is further reduced by dehydrogenase enzymes to yield a chiral secondary alcohol [206]. Only in rare cases, when the resonance of the α,β-unsaturated ketone is destabilized by electron-withdrawing groups (e.g. halogens) being attached to the alkene unit the ketone is reduced first [207].
- α,β-Unsaturated nitro compounds can be transformed into chiral nitroalkanes. To avoid racemisation of the product, only hydrogen is allowed in the α-position [208].
- With activated, conjugated 1,3-dienes only the α,β-bond is selectively reduced, leaving the γ,δ-bond behind.
- In a similar manner, cumulated 1,2-dienes (allenes) mainly give the corresponding 2-alkenes. A rearrangement of the allene to give an acetylene may be observed occasionally [209].
- Sometimes the absolute configuration (*R/S*) of the product can be controlled by starting with *E*- or *Z*-alkenes. However, this is not always the case.
- Haloalkyl substituents attached to the alkene unit are not sufficiently powerful activating groups alone, but they may increase the reaction rate in the presence of other 'activators' [210].

Yeast-mediated reduction of methyl 2-chloro-2-alkenoates gave chiral 2-chloroalkanoic acids with high optical purity [211] (Scheme 2.118). In this reduction the absolute configuration of the product could be controlled by using either the (*E*)- or (*Z*)-isomer of the starting alkene, which gave rise to the (*R*)- and (*S*)-alkanoic acid, resp. Whereas the chiral recognition of the (*Z*)-alkenes was perfect, the (*E*)-isomers led to products with lower e.e. In addition it was shown that the microbial reduction took place on the carboxylic acids, which were formed enzymatically by hydrolysis of the starting esters. The (*R*)-dichloromethyl derivative was subsequently transformed into the rare amino acid L-armentomycin [212].

Scheme 2.118. Yeast-reduction of α,β-unsaturated acids and esters.

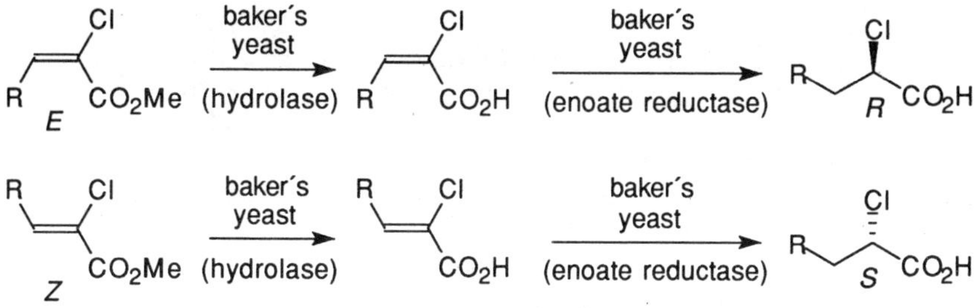

R	Configuration	e.e. [%]
C_2H_5-	$(E) \rightarrow (R)$	47
$(CH_3)_2CH$-	$(E) \rightarrow (R)$	68
$CHCl_2$-	$(E) \rightarrow (R)$	92
C_2H_5-	$(Z) \rightarrow (S)$	>98
$(CH_3)_2CH$-	$(Z) \rightarrow (S)$	>98
$CHCl_2$-	$(Z) \rightarrow (S)$	98
CCl_3-	$(Z) \rightarrow (S)$	>98

(2S,6R)- And (2S,6S)-2,6-dimethyl-1,8-octanediol (both synthetic precursors for vitamins E and K_1 and insect pheromones) were prepared by baker's yeast reduction of the corresponding (6R)- and (6S)-α,β-unsaturated aldehydes with excellent stereoselectivity (d.e. >90%, Scheme 2.119). Since the chirality of the starting material was remote from the site of the reduction (position 6), it did not influence the formation of the newly generated (2S)-centre [213]. The prochiral diene which was used for a total synthesis of (E)-(7R,11R)-phytol, contains two activated C=C bonds. The α,β-unsaturated aldehyde was selectively reduced by baker's yeast with excellent selectivity [214]. As mentioned above, the final product of these transformations was always the corresponding saturated primary alcohol which was formed via reduction of the aldehyde.

Alternatively, α- and β-substituted allylic alcohols could be reduced to give chiral alkanols. Thus, geraniol gave (R)-citronellol in >97% e.e. by conserving the isolated double bond [215]. Similarly, with conjugated 1,3-dienes, only the α,β-bond was reduced [216].

Scheme 2.119. Yeast-reduction of α,β-unsaturated aldehydes and allylic alcohols.

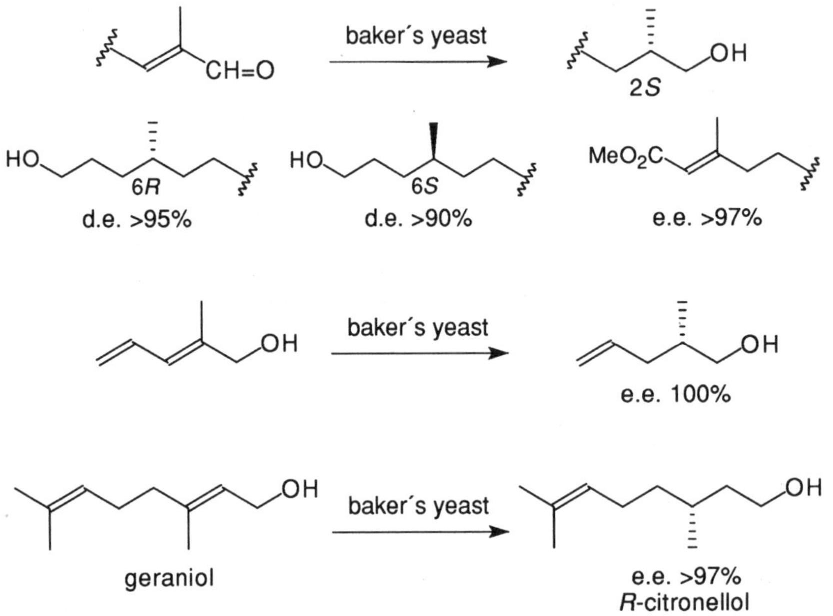

Optically active chlorohydrins were prepared by yeast-reduction of α,β-unsaturated α-chloroketones [217] (Scheme 2.120). During the course of the fermentation, (S)-chloroketones are initially formed which are subsequently reduced in a highly selective second step according to Prelog's rule to give a mixture of *syn*- and *anti*-halohydrins. As a consequence, the *syn/anti*-ratio reflects the optical purity of the intermediate ketone which was initially formed.

Scheme 2.120. Yeast-reduction of α,β-unsaturated ketones.

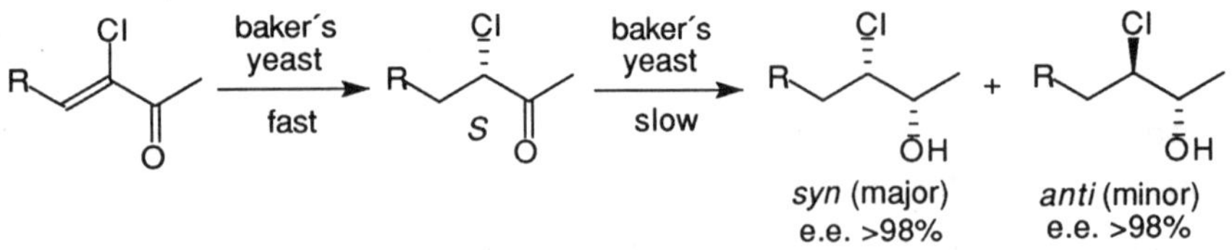

R	e.e. Ketone [%]	syn/anti
C_2H_5	44	72:28
$n\text{-}C_5H_{11}$	82	89:11
$n\text{-}C_8H_{17}$	84	95:5

2.2 Reduction Reactions

A chiral cyclohexane-1,4-dione, which was needed as the central precursor for the synthesis of various naturally occurring carotenoids, was obtained as the major product via a yeast-mediated reduction of 2,2,6-trimethylcyclohex-5-ene-1,4-dione (Scheme 2.121). Two other products arising from a stereoselective reduction of the 1- and 6-keto-moiety were formed in minor amounts (6% and 7%, respectively). The viability of such transformations was demonstrated by scale-up of the reaction. An 80% yield of dione was obtained from 13 kg of starting ene-dione [218].

Scheme 2.121. Yeast-reduction of an α,β-unsaturated diketone.

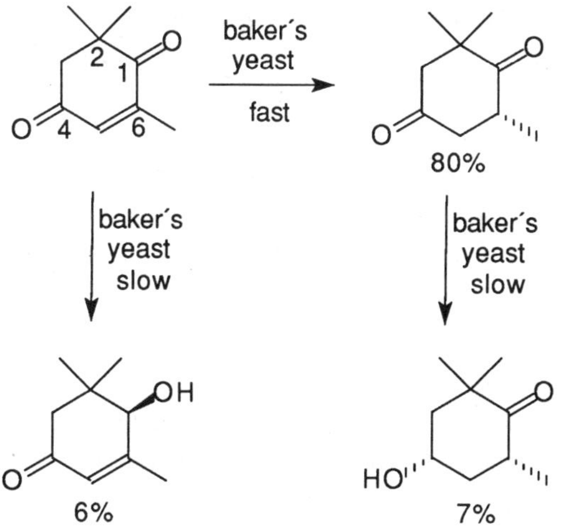

Scheme 2.122. Yeast-reduction of nitro-olefins.

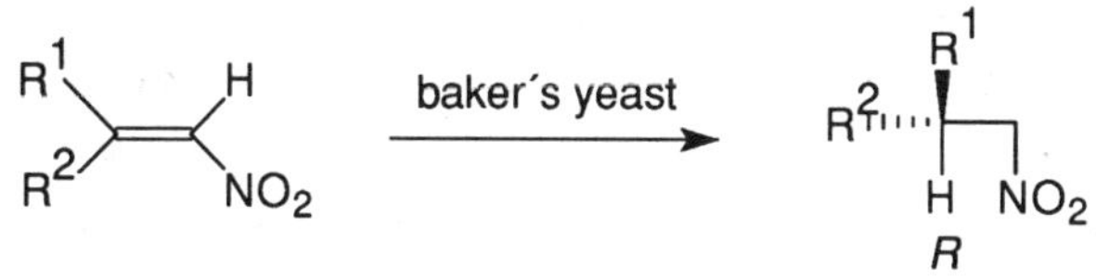

R^1	R^2	e.e. [%]
Ph	CH_3	98
Ph	n-C_3H_7	89
Ph	n-C_6H_{13}	no reaction
n-C_6H_{13}	CH_3 [a]	83
CH_3	n-C_6H_{13} [b]	66[c]

[a] (E)-olefin; [b] (Z)-olefin; [c] opposite configuration as shown, but still (R).

Nitro-olefins are readily reduced by yeast to form chiral nitro-alkanes (Scheme 2.122) [219]. In the case of 2-methyl-1-nitro-1-octene it was shown that the chirality of the product could not be controlled by using the pure (E)- and (Z)-alkenes, since only the (R)-alkanes were formed in varying optical purity. It was assumed that this phenomenon was due to isomerisation of the double bond.

Special techniques of cofactor-recycling, which offer some advantages over the 'classic' recycling methods, have been developed during the studies on enoate reductases (Scheme 2.123) [220, 221].

The ultimate source of redox equivalents in microbial reduction reactions is usually a carbohydrate. Since the majority of it is metabolized by the cells and only a minor fraction (typically 0.5-2%) is used for the delivery of redox equivalents onto the substrate, the productivity of such processes is usually low and side-reactions are common. In order to avoid the undesired metabolism of the auxiliary substrate, non-degradable organic dye molecules such as viologens have been used as 'mediators' for the electron-transport from the donor to the oxidized cofactor. As a consequence, the productivities were improved by one to three orders of magnitude.

The reduced form of the mediator (a radical cation) can either directly or indirectly regenerate NADH. The oxidized mediator (a di-cation) can be enzymatically regenerated using a cheap reducing agent such as hydrogen, formate or carbon monoxide. The redox enzymes for these reactions - hydrogenase, formate dehydrogenase and carbon monoxide dehydrogenase, respectively - are usually present together with the enoate reductase in the same microorganism. Alternatively, electrons from a cathode of an electrochemical cell can be used in a non-enzymic reaction. The course of the reaction can be easily monitored by the consumption of hydrogen or carbon monoxide, or by following the current in the electrochemical cell. Furthermore, the equilibrium of the reaction can be shifted from reduction to oxidation by choosing a mediator with the appropriate redox potential. The disadvantage of this method is the considerable toxicity of the commonly used mediators, e.g. methyl- or benzyl-viologen [222].

Although the use of a selected microorganism containing an enoate-reductase and the sophisticated cofactor recycling described above offers a considerable benefit, it represents a complicated task for non-specialists. Again, a simple alternative for organic chemists exists in the form of baker's yeast.

2.2 Reduction Reactions

Scheme 2.123. Cofactor-recycling *via* a mediator.

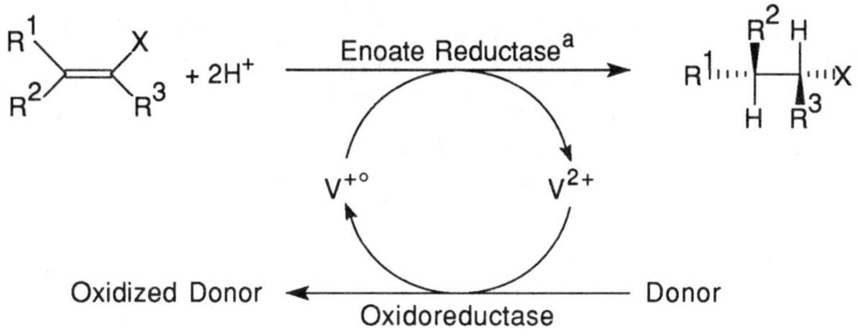

$$R^1R^2C=CR^3X + 2H^+ \xrightarrow{\text{Enoate Reductase}^a} R^1(H)C-C(H)R^3 \text{ with } R^2, X$$

Cycle: $V^{+\circ}$ ⇌ V^{2+} (Oxidized Donor ← Oxidoreductase ← Donor)

$$H_2 + 2V^{2+} \xrightarrow{\text{hydrogenase}} 2H^+ + 2V^{+\circ}$$

$$HCO_2H + 2V^{2+} \xrightarrow{\text{formate DH}} CO_2 + H^+ + 2V^{+\circ}$$

$$CO + H_2O + 2V^{2+} \xrightarrow{\text{carbon monoxide DH}} CO_2 + 2H^+ + 2V^{+\circ}$$

$$e^- + V^{2+} \xrightarrow{\text{non-enzymic}} V^{+\circ}$$

[a] accepts the mediator as cofactor
X = activating group (-CH=O, -CO$_2$R, -CH$_2$-OH, -NO$_2$)
V = viologen (*e.g.* 1,1'-dimethyl-4,4'-dipyridinium cation)

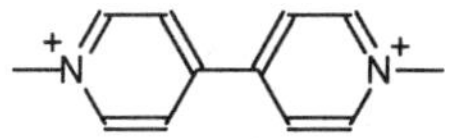

References

1. May SW, Padgette SR (1983) Biotechnology 677
2. Jones JB, Beck JF (1976) Asymmetric Syntheses and Resolutions using Enzymes, In: Applications of Biochemical Systems in Organic Chemistry, Jones JB, Sih CJ, Perlman D (eds), p 236, Wiley, New York
3. Hummel W, Kula M-R (1989) Eur. J. Biochem. 184: 1
4. The current (1988) prices for one mole are: NAD$^+$ 710 $, NADH 3000 $, NADP$^+$ 25.000 $ and NADPH 215.000 $.
5. Willner I, Mandler D (1989) Enzyme Microb. Technol. 11: 467
6. Chenault HK, Whitesides GM (1987) Appl. Biochem. Biotechnol. 14: 147
7. Chenault HK, Simon ES, Whitesides GM (1988) Biotechnol. Gen. Eng. Rev. 6: 221
8. Jones JB, Sneddon DW, Higgins W, Lewis AJ (1972) J. Chem. Soc., Chem. Commun. 856
9. Ramio R, Lilius E-M (1971) Enzymologia 40: 360
10. Jensen MA, Elving PJ (1984) Biochim. Biophys. Acta 764: 310
11. Wienkamp R, Steckhan E (1982) Angew. Chem., Int. Ed. Engl. 21: 782
12. Simon H, Bader J, Günther H, Neumann S, Thanos J (1985) Angew. Chem., Int. Ed. Engl. 24: 539
13. Mandler D, Willner I (1986) J. Chem. Soc., Perkin Trans. II, 805
14. Jones JB, Taylor KE (1976) Can. J. Chem. 54: 2069
15. Legoy M-D, Laretta-Garde V, LeMoullec J-M, Ergan F, Thomas D (1980) Biochimie 62: 341

16. Julliard M, Le Petit J, Ritz P (1986) Biotechnol. Bioeng. 28: 1774
17. van Eys J (1961) J. Biol. Chem. 236: 1531
18. Gupta NK, Robinson WG (1966) Biochim. Biophys. Acta 118: 431
19. Wang SS, King CK (1979) Adv. Biochem. Eng. Biotechnol. 12: 119
20. Karabatsos GL, Fleming JS, Hsi N, Abeles RH (1966) J. Am. Chem. Soc. 88: 849
21. Sund H, Theorell H (1963) The Enzymes, Boyer PD, Lardy H, Myrback K (eds) vol 7, p 25, Academic Press, New York
22. Levy HR, Loewus FA, Vennesland B (1957) J. Am. Chem. Soc. 79: 2949
23. Tischer W, Tiemeyer W, Simon H (1980) Biochimie 62: 331
24. Wichmann R, Wandrey C, Bückmann AF, Kula M-R (1981) Biotechnol. Bioeng. 23: 2789
25. Shaked Z, Whitesides GM (1980) J. Am. Chem. Soc. 102: 7104
26. Hummel W, Schütte H, Schmidt E, Wandrey C, Kula M-R (1987) Appl. Microbiol. Biotechnol. 26: 409
27. Vandecasteele J-P, Lemal J (1980) Bull. Soc. Chim. Fr. 101
28. Wong C-H, Drueckhammer DG, Sweers HM (1985) J. Am. Chem. Soc. 107: 4028
29. Pollak A, Blumenfeld H, Wax M, Baughn RL, Whitesides GM (1980) J. Am. Chem. Soc. 102: 6324
30. Wong C-H, Whitesides GM (1981) J. Am. Chem. Soc. 103: 4890
31. Hirschbein BL, Whitesides GM (1982) J. Am. Chem. Soc. 104: 4458
32. Wong C-H, Gordon J, Cooney CL, Whitesides GM (1981) J. Org. Chem. 46: 4676
33. Guiseley KB, Ruoff PM (1961) J. Org. Chem. 26: 1248
34. Dodds DR, Jones JB (1982) J. Chem. Soc., Chem. Commun. 1080
35. Wang SS, King C-K (1979) Adv. Biochem. Eng. 12: 119
36. Mansson MO, Larsson PO, Mosbach K (1982) Methods Enzymol. 89: 457
37. Wong C-H, Whitesides GM (1983) J. Am. Chem. Soc. 105: 5012
38. Danielsson B, Winquist F, Malpote JY, Mosbach K (1982) Biotechnol. Lett. 4: 673
39. Payen B, Segui M, Monsan P, Schneider K, Friedrich CG, Schlegel HG (1983) Biotechnol. Lett. 5: 463
40. Lee LG, Whitesides GM (1986) J. Org. Chem. 51: 25
41. Carrea G, Bovara R, Longhi R, Riva S (1985) Enzyme Microb. Technol. 7: 597
42. Matos JR, Wong C-H (1986) J. Org. Chem. 51: 2388
43. Bednarski MD, Chenault HK, Simon ES, Whitesides GM (1987) J. Am. Chem. Soc. 109: 1283
44. Lemière GL, Lepoivre JA, Alderweireldt FC (1985) Tetrahedron Lett. 4527
45. Lemière GL (1986) Alcohol Dehydrogenase catalysed Oxidoreduction Reactions in Organic Chemistry; In: Enzymes as Catalysts in Organic Synthesis, NATO ASI Series C, vol. 178, Schneider MP (ed) p 17, Reidel, Dordrecht
46. Prelog V (1984) Pure Appl. Chem. 9: 119
47. Shen G-J, Wang Y-F, Bradshaw C, Wong C-H (1990) J. Chem. Soc., Chem. Commun. 677
48. Hummel W (1990) Appl. Microbiol. Biotechnol. 34: 15
49. Hou CT, Patel R, Barnabe N, Marczak I (1981) Eur. J. Biochem. 119: 359
50. Nakamura K, Miyai T, Kawai J, Nakajima N, Ohno A (1990) Tetrahedron Lett. 1159
51. MacLeod R, Prosser H, Fikentscher L, Lanyi J, Mosher HS (1964) Biochemistry 3: 838
52. Roberts SM (1988) Enzymes as Catalysts in Organic Synthesis; In: The Enzyme Catalysis process, Cooper A, Houben JL, Chien LC (eds) NATO ASI Series A, vol 178, p 443, Reidel, Dordrecht
53. Lepoivre JA (1984) Janssen Chim. Acta 2: 20
54. Cedergen-Zeppezauer ES, Andersson I, Ottonello S (1985) Biochemistry 24: 4000
55. Ganzhorn AJ, Green DW, Hershey AD, Gould RM, Plapp BV (1987) J. Biol. Chem. 262: 3754
56. Jones JB, Schwartz HM (1981) Can. J. Chem. 59: 1574
57. Van Osselaer TA, Lemière GL, Merckx EM, Lepoivre JA, Alderweireldt FC (1978) Bull. Soc. Chim. Belg. 87: 799
58. Jones JB, Takemura T (1984) Can. J. Chem. 62: 77
59. Davies J, Jones JB (1979) J. Am. Chem. Soc. 101: 5405

60. Lam LKP, Gair IA, Jones JB (1988) J. Org. Chem. 53: 1611
61. Krawczyk AR, Jones JB (1989) J. Org. Chem. 54: 1795
62. Irwin AJ, Jones JB (1976) J. Am. Chem. Soc. 98: 8476
63. Sadozai SK, Merckx EM, Van De Val AJ, Lemière GL, Esmans EL, Lepoivre JA, Alderweireldt FC (1982) Bull. Soc. Chim. Belg. 91: 163
64. Nakazaki M, Chikamatsu H, Fujii T, Sasaki Y, Ao S (1983) J. Org. Chem. 48: 4337
65. Nakazaki M, Chikamatsu H, Naemura K, Suzuki T, Iwasaki M, Sasaki Y, Fujii T (1981) J. Org. Chem. 46: 2726
66. Nakazaki M, Chikamatsu H, Sasaki Y (1983) J. Org. Chem. 48: 2506
67. Matos JR, Smith MB, Wong C-H (1985) Bioorg. Chem. 13: 121
68. Takemura T, Jones JB (1983) 48: 791
69. Haslegrave JA, Jones JB (1982) J. Am. Chem. Soc. 104: 4666
70. Lam LKP, Gair IA, Jones JB (1988) J. Org. Chem. 53: 1611
71. Fries RW, Bohlken DP, Plapp BV (1979) J. Med. Chem. 22: 356
72. Davies J, Jones JB (1979) J. Am. Chem. Soc. 101: 5405
73. Dodds DR, Jones JB (1982) J. Chem. Soc., Chem. Commun. 1080
74. Dodds DR, Jones JB (1988) J. Am. Chem. Soc. 110: 577
75. Yamazaki Y, Hosono K (1988) Tetrahedron Lett. 5769
76. Yamazaki Y, Hosono K (1989) Tetrahedron Lett. 5313
77. Prelog V (1964) Colloqu. Ges. Physiol. Chem. 14: 288
78. Horjales E, Brändén C-I (1985) J. Biol. Chem. 260: 15445
79. Jones JB, Jakovac IJ (1982) Can. J. Chem. 60: 19
80. Nakazaki M, Chikamatsu H, Naemura K, Sasaki Y, Fujii T (1980) J. Chem. Soc., Chem. Commun. 626
81. Lemière GL, Van Osselaer TA, Lepoivre JA, Alderweireldt FC (1982) J. Chem. Soc., Perkin Trans. II, 1123
82. Keinan E, Hafeli EK, Seth KK, Lamed R (1986) J. Am. Chem. Soc. 108: 162
83. Drueckhammer DG, Sadozai SK, Wong C-H, Roberts SM (1987) Enzyme Microb. Technol. 9: 564
84. Keinan E, Seth KK, Lamed R (1986) J. Am. Chem. Soc. 108: 3474
85. Wong C-H, Drueckhammer DG, Sweers HM (1985) J. Am. Chem. Soc. 107: 4028
86. De Amici M, De Micheli C, Carrea G, Spezia S (1989) J. Org. Chem. 54: 2646
87. Drueckhammer DG, Barbas III CF, Nozaki K, Wong C-H (1988) J. Org. Chem. 53: 1607
88. De Amici M, De Micheli C, Molteni G, Pitrè D, Carrea G, Riva S, Spezia S, Zetta L (1991) J. Org. Chem. 56: 67
89. Butt S, Davies HG, Dawson MJ, Lawrence GC, Leaver J, Roberts SM, Turner MK, Wakefield BJ, Wall WF, Winders JA (1987) J. Chem. Soc., Perkin Trans. I, 903
90. Riva S, Ottolina G, Carrea G, Danieli B (1989) J. Chem. Soc., Perkin Trans. I, 2073
91. Carrea G, Colombi F, Mazzola G, Cremonesi P, Antonini E (1979) Biotechnol. Bioeng. 21: 39
92. Butt S, Davies HG, Dawson MJ, Lawrence GC, Leaver J, Roberts SM, Turner MK, Wakefield BJ, Wall WF, Winders JA (1985) Tetrahedron Lett. 5077
93. Leaver J, Gartenmann TCC, Roberts SM, Turner MK (1987) Biocatalysis in Organic Media, Laane C, Tramper J, Lilly MD (eds) p 411, Elsevier, Amsterdam
94. Kim M-J, Whitesides GM (1988) J. Am. Chem. Soc. 110: 2959
95. Kim M-J, Kim JY (1991) J. Chem. Soc., Chem. Commun. 326
96. Luyten MA, Bur D, Wynn H, Parris W, Gold M, Frieson JD, Jones JB (1989) J. Am. Chem. Soc. 111: 6800
97. Lee LG, Whitesides GM (1986) J. Org. Chem. 51: 25
98. Nakamura K, Yoneda T, Miyai T, Ushio K, Oka S, Ohno A (1988) Tetrahedron Lett. 2453
99. Fontana A (1984) Thermophilic Enzymes and their Potential Use in Biotechnology, p 221, Dechema, Weinheim
100. Bryant FO, Wiegel J, Ljungdahl LG (1988) Appl. Environ. Microbiol. 54: 460
101. Pham VT, Phillips RS, Ljungdahl LG (1989) J. Am. Chem. Soc. 111: 1935
102. Daniel RM, Bragger J, Morgan HW (1990) Enzymes from Extreme Environments; In: Biocatalysis, Abramowicz DA (ed) p 243, Van Nostrand Reinhold, New York

103. Willaert JJ, Lemière GL, Joris LA, Lepoivre JA, Alderweideldt FC (1988) Bioorg. Chem. 16: 223
104. Pham VT, Phillips RS (1990) J. Am. Chem. Soc. 112: 3629
105. Chen C-S, Zhou B-N, Girdaukas G, Shieh W-R, VanMiddlesworth F, Gopalan AS, Sih CJ (1984) Bioorg. Chem. 12: 98
106. Shieh W-R, Gopalan AS, Sih CJ (1985) J. Am. Chem. Soc. 107: 2993
107. Hoffmann RW, Ladner W, Helbig W (1984) Liebigs Ann. Chem. 1170
108. Nakamura K, Ushio K, Oka S, Ohno A (1984) Tetrahedron Lett. 3979
109. Fuganti C, Grasselli P, Casati P, Carmeno M (1985) Tetrahedron Lett. 101
110. Nakamura K, Kawai Y, Oka S, Ohno A (1989) Tetrahedron Lett. 2245
111. Christen M, Crout DHG (1987) Bioreactors and Biotransformations, Moody GW, Baker PB (eds) p 213, Elsevier, London
112. Sakai T, Nakamura T, Fukuda K, Amano E, Utaka M, Takeda A (1986) Bull. Chem. Soc. Jpn. 59: 3185
113. Ushio K, Inoue K, Nakamura K, Oka S, Ohno A (1986) Tetrahedron Lett. 2657
114. Buisson D, Azerad R (1986) Tetrahedron Lett. 2631
115. Seebach D, Züger MF, Giovannini F, Sonnleitner B, Fiechter A (1984) Angew. Chem., Int. Ed. Engl. 23: 151
116. Servi S (1990) Synthesis 1
117. Sih CJ, Chen C-S (1984) Angew. Chem., Int. Ed. Engl. 23: 570
118. Ward OP, Young CS (1990) Enzyme Microb. Technol. 12: 482
119. Csuk R , Glänzer B (1991) Chem. Rev. 91: 49
120. Neuberg C, Lewite A (1918) Biochem. Z. 91: 257
121. Neuberg C (1949) Adv. Carbohydr. Chem. 4: 75
122. Macleod R, Prosser H, Fikentscher L, Lanyi J, Mosher HS (1964) Biochemistry 3: 838
123. Le Drian C, Greene AE (1982) J. Am. Chem. Soc. 104: 5473
124. Belan A, Bolte J, Fauve A, Gourcy JG, Veschambre H (1987) J. Org. Chem. 52: 256
125. Takano S, Yanase M, Sekiguchi Y, Ogasawara K (1987) Tetrahedron Lett. 1783
126. Bucchiarelli M, Forni A, Moretti I, Torre G (1983) Synthesis 897
127. Kitazume T, Lin JT, (1987) J. Fluorine Chem. 34: 461
128. Fujisawa T, Hayashi H, Kishioka Y (1987) Chem. Lett. 129
129. Seebach D, Roggo S, Maetzke T, Braunschweiger H, Cerkus J, Krieger M (1987) Helv. Chim. Acta 70: 1605
130. Guette J-P, Spassky N (1972) Bull. Soc. Chim. Fr. 4217
131. Levene PA, Walti A (1943) Org. Synth., Coll. Vol. II: 545
132. Bernardi R, Ghiringhelli D (1987) J. Org. Chem. 52: 5021
133. Syldatk C, Andree H, Stoffregen A, Wagner F, Stumpf B, Ernst L, Zilch H, Tacke R (1987) Appl. Microbiol. Biotechnol. 27: 152
134. Hirama M, Nakamine T, Ito S (1984) Chem. Lett. 1381
135. Seebach D, Eberle M (1986) Synthesis 37
136. Seebach D, Züger MF, Giovannini F, Sonnleitner B, Fiechter A (1984) Angew. Chem., Int. Ed. Engl. 23: 151
137. Fuganti C, Grasselli P, Seneci PF, Casati P (1986) Tetrahedron Lett. 5275
138. Seebach D, Renaud P, Schweizer WB, Züger MF (1984) Helv. Chim. Acta 67: 1843
139. Deshong P, Lin M-T, Perez JJ (1986) Tetrahedron Lett. 2091
140. Tschaen DM, Fuentes LM, Lynch JE, Laswell WL, Volante RP, Shinkai I (1988) Tetrahedron Lett. 2779
141. Mori K (1989) Tetrahedron 45: 3233
142. Kramer A, Pfader H (1982) Helv. Chim. Acta 65: 293
143. Sih CJ, Zhou B, Gopalan AS, Shieh W-R, VanMiddlesworth F (1983) Strategies for Controlling the Stereochemical Course of Yeast Reductions; In: Selectivity - a Goal for Synthetic Efficiency, Bartmann W, Trost BM (eds) Proc. 14th Workshop Conference Hoechst, p 250, Verlag Chemie, Weinheim
144. Shieh W-R, Gopalan AS, Sih CJ (1985) J. Am. Chem. Soc. 107: 2993
145. Nakamura K, Inoue K, Ushio K, Oka S, Ohno A (1987) Chem. Lett. 679
146. Nakamura K, Kawai Y, Oka S, Ohno A (1989) Bull. Chem. Soc. Jpn. 62: 875
147. Nakamura K, Higaki M, Ushio K, Oka S, Ohno A (1985) Tetrahedron Lett. 4213

148. Chibata I, Tosa T, Sato T (1974) Appl. Microbiol. 27: 878
149. Nakamura K, Kawai Y, Ohno A (1990) Tetrahedron Lett. 267
150. Hirama M, Uei M (1982) J. Am. Chem. Soc. 104: 4251
151. Hirama M, Nakamine T, Ito S (1984) Chem. Lett. 1381
152. Feichter C, Faber K, Griengl H (1990) Biocatalysis 3: 145
153. Fujisawa T, Itoh T, Sato T (1984) Tetrahedron Lett. 5083
154. Buisson D, Henrot S, Larcheveque M, Azerad R (1987) Tetrahedron Lett. 5033
155. Akita H, Furuichi A, Koshoji H, Horikoshi K, Oishi T (1983) Chem. Pharm. Bull. 31: 4376
156. Frater G, Müller U, Günther W (1984) Tetrahedron 40: 1269
157. Deol B, Ridley D, Simpson G (1976) Aust. J. Chem. 29: 2459
158. Nakamura K, Miyai T, Nozaki K, Ushio K, Ohno A (1986) Tetrahedron Lett. 3155
159. VanMiddlesworth F, Sih CJ (1987) Biocatalysis 1: 117
160. Hoffmann RW, Helbig W, Ladner W (1982) Tetrahedron Lett. 3479
161. Chenevert R, Thiboutot S (1986) Can. J. Chem. 64: 1599
162. Ohta H, Ozaki K, Tsuchihashi G (1986) Agric. Biol. Chem. 50: 2499
163. Nakamura K, Inoue K, Ushio K, Oka S, Ohno A (1988) J. Org. Chem. 53: 2598
164. Deol BS, Ridley DD, Simpson GW (1976) Aust. J. Chem. 29: 2459
165. Brooks DW, Mazdiyasni H, Chakrabarti S (1984) Tetrahedron Lett. 1241
166. Brooks DW, Woods KW (1987) J. Org. Chem. 52: 2036
167. Brooks DW, Mazdiyasni H, Grothaus PG (1987) J. Org. Chem. 52: 3223
168. Brooks DW, Mazdiyasni H, Sallay P (1985) J. Org. Chem. 50: 3411
169. Eichberger G, Faber K (1984) unpublished results.
170. Fujisawa T, Kojima E, Sato T (1987) Chem. Lett. 2227
171. Takeshita M, Sato T (1989) Chem. Pharm. Bull. 37: 1085
172. Nakamura K, Inoue Y, Shibahara J, Oka S, Ohno A (1988) Tetrahedron Lett. 4769
173. Bucchiarelli M, Forni A, Moretti I, Prati F, Torre G, Resnati G, Bravo P (1989) Tetrahedron 45: 7505
174. Bernardi R, Bravo P, Cardillo R, Ghiringhelli D, Resnati G (1988) J. Chem. Soc., Perkin Trans. I, 2831
175. Kitazume T, Kobayashi T (1987) Synthesis 87
176. Kitazume T, Nakayama Y (1986) J. Org. Chem. 51: 2795
177. Takano S (1987) Pure Appl. Chem. 59: 353
178. Itoh T, Yoshinaka A, Sato T, Fujisawa T (1985) Chem. Lett. 1679
179. Kozikowski AP, Mugrage BB, Li CS, Felder L (1986) Tetrahedron Lett. 4817
180. Ticozzi C, Zanarotti A (1988) Tetrahedron Lett. 6167
181. Dondoni A, Fantin G, Fogagnolo M, Mastellari A, Medici A, Nefrini E, Pedrini P (1988) Gazz. Chim. Ital. 118: 211
182. Top S, Jaouen G, Gillois J, Baldoli C, Maiorana S (1988) J. Chem. Soc., Chem. Commun. 1284
183. Yamazaki Y, Hosono K (1988) Agric. Biol. Chem. 52: 3239
184. Kieslich K (1976) Microbial Transformations of Non-Steroid Cyclic Compounds. Thieme, Stuttgart
185. Adlercreutz P (1991) Enzyme Microb. Technol. 13: 9
186. Ohta H, Kato Y, Tsuchihashi G (1987) J. Org. Chem. 52: 2735
187. Bernardi R, Cardillo R, Ghiringhelli D, de Pavo V (1987) J. Chem. Soc., Perkin Trans. I, 1607
188. Trost BM, Lynch J, Renaut P, Steinman DH (1986) J. Am. Chem. Soc. 108: 284
189. Neef G, Petzoldt K, Wieglepp H, Weichert R (1985) Tetrahedron Lett. 5033
190. Simon H, White H, Lebertz H, Thanos I (1987) Angew. Chem., Int. Ed. Engl. 26: 785
191. Bader J, Günther H, Rambeck B, Simon H (1978) Hoppe-Seyler´s Z. Physiol. Chem. 359: 19
192. Tischer W, Bader J, Simon H (1979) Eur. J. Biochem. 97: 103
193. Tischer W, Tiemeyer W, Simon H (1980) Biochimie 62: 331
194. Simon H, Bader J, Günther H, Neumann S, Thanos J (1985) Angew. Chem., Int. Ed. Engl. 24: 539

195. Bartl K, Cavalar C, Krebs T, Ripp E, Retey J, Hull WE, Günther H, Simon H (1977) Eur. J. Biochem. 72: 247
196. Giesel H, Machacek G, Bayeri J, Simon H (1981) FEBS Lett. 123: 107
197. Fuganti C, Ghiringhelli D, Grasselli P (1975) J. Chem. Soc., Chem. Commun. 846
198. Fischer FG, Wiedemann O (1934) Liebigs Ann. Chem. 513: 260
199. Desrut M, Kergomard A, Renard MF, Veschambre H (1981) Tetrahedron 37: 3825
200. Fuganti C, Grasselli P (1989) Baker's Yeast Mediated Synthesis of Natural Products. Whitaker JR, Sonnet PE (eds) Biocatalysis in Agricultural Biotechnology, ACS Symp. Ser. 389, p 359, Am. Chem. Soc., Washington
201. Suemune H, Hayashi N, Funakoshi K, Akita H, Oishi T, Sakai K (1985) Chem. Pharm. Bull. 33: 2168
202. Ferraboschi P, Grisenti P, Casati R, Fiecchi A, Santaniello E (1987) J. Chem. Soc., Perkin Trans. I, 1743
203. Fuganti C, Grasselli P (1982) J. Chem. Soc., Chem. Commun. 205
204. Fuganti C, Grasselli P, Servi S (1983) J. Chem. Soc., Perkin Trans. I, 241
205. Sato T, Hanayama K, Fujisawa T (1988) Tetrahedron Lett. 2197
206. Kergomard A, Renard MF, Veschambre H (1982) J. Org. Chem. 47: 792
207. Sih CJ, Heather JB, Sood R, Price P, Peruzzotti G, Lee HFH, Lee SS (1975) J. Am. Chem. Soc. 97: 865
208. Ohta H, Ozaki K, Tsuchihashi G (1987) Chem. Lett. 191
209. Gil G, Ferre E, Barre M, Le Petit J (1988) Tetrahedron Lett. 3797
210. Kitazume T, Ishikawa N (1984) Chem. Lett. 587
211. Utaka M, Konishi S, Mizuoka A, Ohkubo T, Sakai T, Tsuboi S, Takeda A (1989) J. Org. Chem. 54: 4989
212. Utaka M, Konishi S, Okubo T, Tsuboi S, Takeda A (1987) Tetrahedron Lett. 1447
213. Gramatica P, Manitto P, Poli L (1985) J. Org. Chem. 50: 4625
214. Gramatica P, Manitto P, Monti D, Speranza G (1987) Tetrahedron 43: 4481
215. Gramatica P, Manitto P, Ranzi BM, Delbianco A, Francavilla M (1983) Experientia 38: 775
216. Gramatica P, Manitto P, Monti D, Speranza G (1988) Tetrahedron 44: 1299
217. Utaka M, Konishi S, Takeda A (1986) Tetrahedron Lett. 4737
218. Leuenberger HG, Boguth W, Widmer E, Zell R (1976) Helv. Chim. Acta 59: 1832
219. Ohta H, Kobayashi N, Ozaki K (1989) J. Org. Chem. 54: 1802
220. Simon H (1988) GIT Fachz. Lab. 458
221. Simon H, Günther H, Bader J, Neumann S (1985) Chiral Products from Non-pyridine Nucleotide Dependent Reductases and Methods fro NAD(P)H Regeneration. Porter R, Clark S (eds) In: Enzymes in Organic Synthesis, Ciba Foundation Symp. 111, p 97, Pitman, London
222. Salts of methyl viologen (1,1'-Dimethyl-4,4'-dipyridinium dication, Paraquat) have been used as a total herbicide.

2.3 Oxidation Reactions

2.3.1 Oxidation of Alcohols and Aldehydes

Oxidations of primary and secondary alcohols to furnish aldehydes and ketones, respectively, are common chemical reactions that rarely present insurmountable problems to the synthetic organic chemist. In contrast to the corresponding reduction reactions, oxidation reactions using isolated dehydrogenase enzymes have been scarcely reported [1]. The reasons for this situation are as follows.
- Oxidations of alcohols are thermodynamically unfavourable; this makes the recycling of the oxidized cofactor a complicated issue.
- Product inhibition is a common phenomenon to such reactions, due to the fact that the aldehydic or ketonic products are often more tightly bound onto the hydrophobic active site of the enzyme than the more polar alcohol [2].
- Enzymatic oxidations usually work best at elevated pH (8-9) where cofactors and (particularly aldehydic) products are unstable.
- Oxidation of a secondary alcohol involves the destruction of an asymmetric centre.

Regioselective Oxidation of Polyols

The enzyme-catalysed oxidation of alcohols is only of practical interest to the synthetic organic chemist if complex molecules such as polyols are involved (Scheme 2.124). Such compounds present problems of selectivity with conventional chemical oxidants. The selective oxidation of a single hydroxyl group in a polyol requires a number of protection-deprotection steps if performed by chemical means. In contrast, numerous sugars and related polyhydroxy compounds have been selectively oxidized in a single step into the corresponding keto-ols or keto-acids using a variety of microorganisms, for example *Acetobacter* or *Pseudomonas* sp.

Resolution of Alcohols by Oxidation

An interesting example for the deracemisation of a secondary alcohol involving a microbial oxidation-reduction sequence is shown in Scheme 2.125 [3]. Washed cells of *Rhodococcus erythropilis* oxidize L-pantoyl lactone to the α-ketolactone, which in turn is reduced by another dehydrogenase present in the organism to yield the corresponding (R)-D-pantoyl lactone, a key intermediate for the synthesis of pantothenic acid [4]. During this inversion process a formal theoretical yield of 100% is achieved together with an in situ regeneration of the cofactor.

Scheme 2.124. Regioselective oxidation of polyols by *Acetobacter suboxydans*.

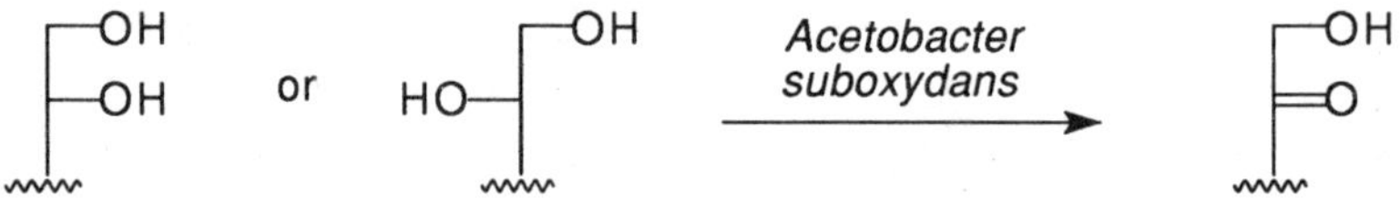

Substrate Polyol	Product Keto-alcohol	Reference
adonitol	L-adonulose	[5]
D-sorbitol	L-sorbose	[6]
L-fucitol	4-keto-L-fucose	[7]
D-gluconic acid	5-keto-D-gluconic acid	[8]
1-deoxy-D-sorbitol	6-deoxy-L-sorbose	[9]

Scheme 2.125. Microbial deracemisation of pantoyl lactone.

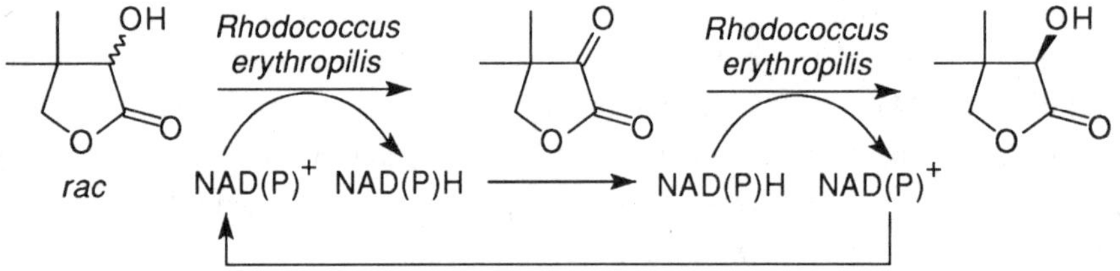

The major difficulty in the resolution of alcohols via selective oxidation of one enantiomer using isolated horse liver alcohol dehydrogenase (HLADH) is the regeneration of NAD^+. Besides the highly efficient enzymatic systems described in Section 2.2.1, a flavine mononucleotide (FMN) recycling system [10] (in which molecular oxygen is the ultimate oxidant) was used to resolve numerous mono- [11, 12] bi- [13] and polycyclic secondary alcohols [14] by HLADH. To avoid enzyme deactivation, the hydrogen peroxide produced during this process was removed using catalase [15].

Glycols having a primary hydroxyl group were enantioselectively oxidized to L-α-hydroxy acids using a co-immobilized alcohol and aldehyde dehydrogenase system. In the first step, resolution of the diol furnished a mixture of L-hydroxyaldehyde and D-diol. The former was oxidized in situ by an aldehyde dehydrogenase to yield the L-hydroxyacid in high optical purity [16, 17].

Scheme 2.126. Resolution of diols by a HLADH/aldehyde DH system.

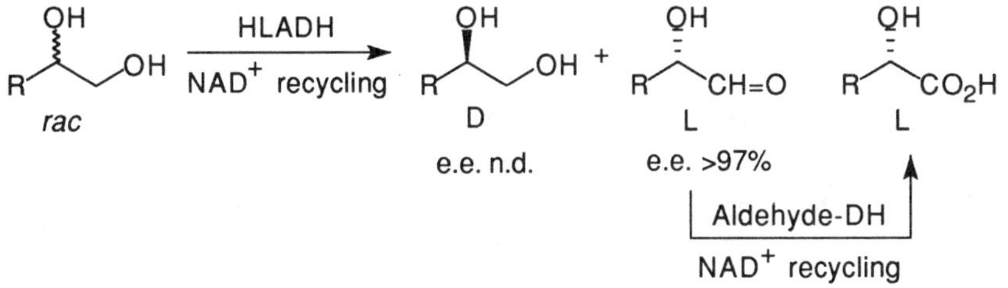

$$\underset{rac}{R\overset{OH}{\curvearrowright}OH} \xrightarrow[\text{NAD}^+ \text{ recycling}]{\text{HLADH}} \underset{\underset{\text{e.e. n.d.}}{D}}{R\overset{OH}{\curvearrowright}OH} + \underset{\underset{\text{e.e. >97\%}}{L}}{R\overset{OH}{\curvearrowright}CH=O} \xrightarrow[\text{NAD}^+ \text{ recycling}]{\text{Aldehyde-DH}} \underset{L}{R\overset{OH}{\curvearrowright}CO_2H}$$

R = -CH$_2$-OH, -CH$_2$-F, -CH$_2$-Cl, -CH$_2$-Br, -CH$_3$, -CH$_2$-NH$_2$, -CH=CH$_2$, -CH$_2$-CH$_3$.

Asymmetrisation of Prochiral or *meso*-Diols by Oxidation

In contrast to the resolution of secondary alcohols, where the more generally applicable lipase-technology is recommended instead of redox reactions, asymmetrisation of primary diols of prochiral or *meso*-structure has been shown to be a valuable method for the synthesis of chiral lactones (Scheme 2.127) [18].

As a rule of thumb, oxidation of the (*S*)- or pro-(*S*) hydroxyl group occurs selectively with HLADH (Scheme 2.127). In case of 1,4- and 1,5-diols, the intermediate hydroxyaldehydes spontaneously cyclize to form the more stable 5- and 6-membered hemiacetals (lactols). Then, in a second step, the latter are oxidized by HLADH to form γ- or δ-lactones following the same (*S*)- or pro-(*S*) specificity [19]. Both steps - asymmetrisation of the prochiral or *meso*-diol and kinetic resolution of the lactol - are often highly selective.

The two ways in which enantiospecificity can arise during this process are exemplified in Scheme 2.128 [20]. First, the enzyme oxidizes the (*S*)-hydroxymethyl group to give rise to a hemiacetal intermediate with (*S*)-chirality. Secondly, if the enzyme does not discriminate between the enantiotopic groups of the substrate, it can selectively oxidize the (*S*)-enantiomer of the intermediate hemiacetal affording a kinetic resolution. Furthermore, the final chirality may also reflect a synergistic combination of the two specificities.

Scheme 2.127. Asymmetrisation of prochiral or *meso*-diols by HLADH.

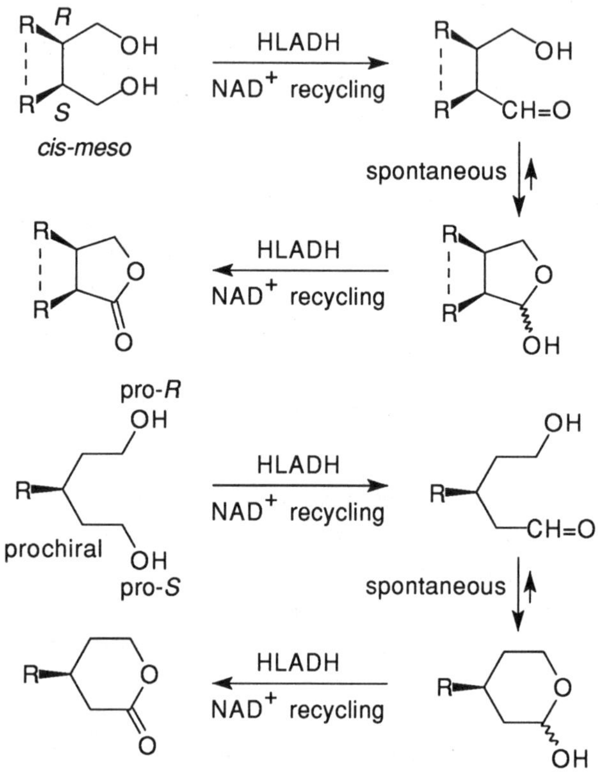

Scheme 2.128. Selectivity in asymmetrisation of *meso*-diols by HLADH.

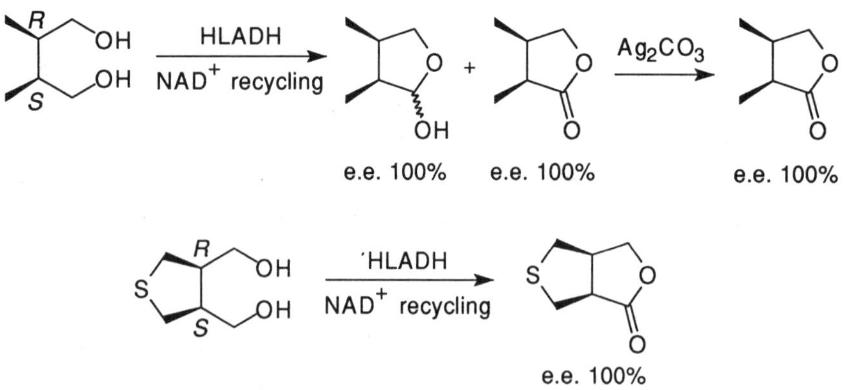

2.3 Oxidation Reactions

For the *cis*-2,3-dimethylbutane-1,4-diol and the thiacyclopentane derivative the first scenario is operating (Scheme 2.128). As a consequence, both of the lactones were obtained in optically pure form.

Scheme 2.129. Selectivity in the resolution of lactols by HLADH.

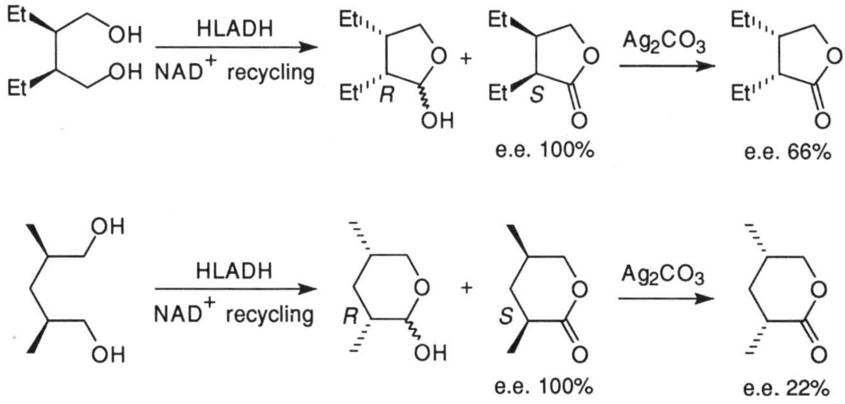

In contrast, the oxidation of the *cis*-2,4-diethylbutane-1,4-diol and the *cis*-2,4-dimethylpentane-1,5-diol proceeds non-selectively giving the corresponding racemic lactols (Scheme 2.129). The latter are subject to a kinetic resolution during which the (S)-enantiomers are oxidized into the lactones. Consequently, when the lactol-lactone mixture was chemically oxidized using silver carbonate, the lactones were obtained in low enantiomeric purity.

Scheme 2.130. Asymmetrisation of bicyclic *meso*-diols using HLADH.

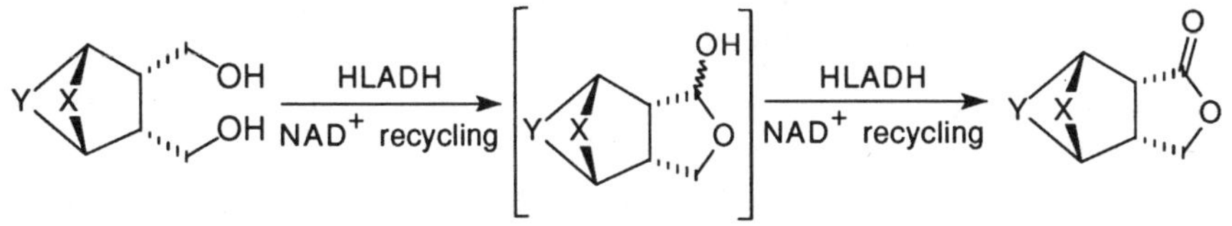

X	Y	e.e. [%]
-CH$_2$-	-CH=CH-	>97
-CH$_2$-	-CH$_2$-CH$_2$-	>97
-CH=CH-	-CH$_2$-	>97
-CH$_2$-CH$_2$-	-CH$_2$-	>97
-CH$_2$-CH$_2$-	-CH=CH-	>97
-CH$_2$-CH$_2$-	-CH$_2$-CH$_2$-	>97

Sterically demanding bicyclic *meso*-diols were transformed into optically pure lactones via the asymmetrisation process mentioned above (Scheme 2.130) [21]. Following the general trend, HLADH was completely enantiotopically specific for the hydroxymethyl group attached to an (*S*)-centre. The initially formed hydroxyaldehydes underwent in situ HLADH-promoted oxidation via their hemiacetals to yield lactone products. Only with the bicyclo[2.2.2]octane system were the intermediate hemiacetals detected in small amounts.

2.3.2 Oxygenation Reactions

Enzymes which catalyse the direct incorporation of molecular oxygen into an organic molecule are called '*oxygen*ases' [22-25]. Enzymatic oxygenation reactions are particularly intriguing since direct oxyfunctionalization of unactivated organic substrates remains a largely unresolved challenge to synthetic chemistry. This is particularly true for those cases where regio- or enantiospecifity is desired. This may be achieved in oxygenation reactions by means of enzymes.

Oxygen-transfer onto organic acceptor molecules may proceed through three different mechanisms (see Scheme 2.131).

- Mono-oxygenases incorporate one oxygen atom from molecular oxygen into the substrate, the other is reduced at the expense a donor (usually NADH or NADPH) to form water [26, 27].
- Di-oxygenases simultaneously incorporate both oxygen atoms of O_2 into the substrate, thus they are sometimes misleadingly called oxygen-transferases (although they are redox-enzymes).
- Oxidases, on the other hand, mainly catalyse the electron-transfer onto molecular oxygen. This may proceed through a two- or four-electron transfer, involving either hydrogen peroxide or water as oxygen donor, respectively. Oxidases include flavoprotein oxidases (amino acid oxidases, glucose oxidase), metallo-flavin oxidases (aldehyde oxidase) and haeme-protein oxidases (catalase, H_2O_2-specific peroxidases [28]). Some of the enzymes have been found to be very useful. For instance, D-glucose oxidase is used on a large scale, in combination with catalase, as a food antioxidant [29]; however, from a synthetic viewpoint, oxidases have not been utilized extensively [30].

2.3 Oxidation Reactions

Scheme 2.131. Enzymatic oxygenation reactions.

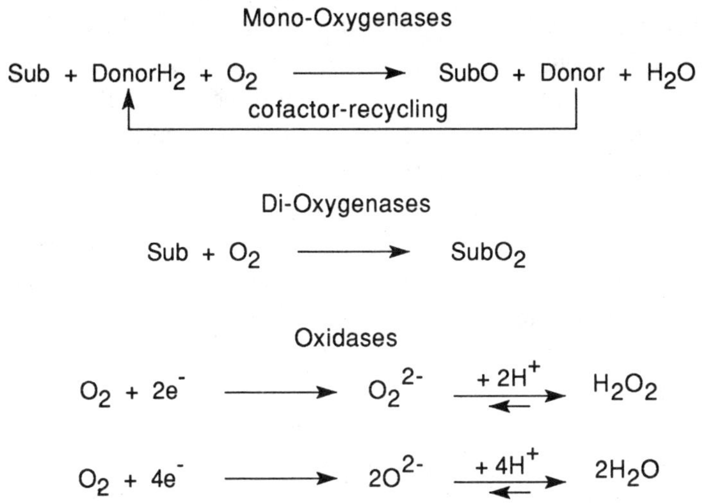

Mono-Oxygenases

$$Sub + DonorH_2 + O_2 \longrightarrow SubO + Donor + H_2O$$
$$\text{cofactor-recycling}$$

Di-Oxygenases

$$Sub + O_2 \longrightarrow SubO_2$$

Oxidases

$$O_2 + 2e^- \longrightarrow O_2^{2-} \underset{}{\overset{+ 2H^+}{\rightleftarrows}} H_2O_2$$

$$O_2 + 4e^- \longrightarrow 2O^{2-} \underset{}{\overset{+ 4H^+}{\rightleftarrows}} 2H_2O$$

Scheme 2.132. Mono-oxygenase catalysed reactions.

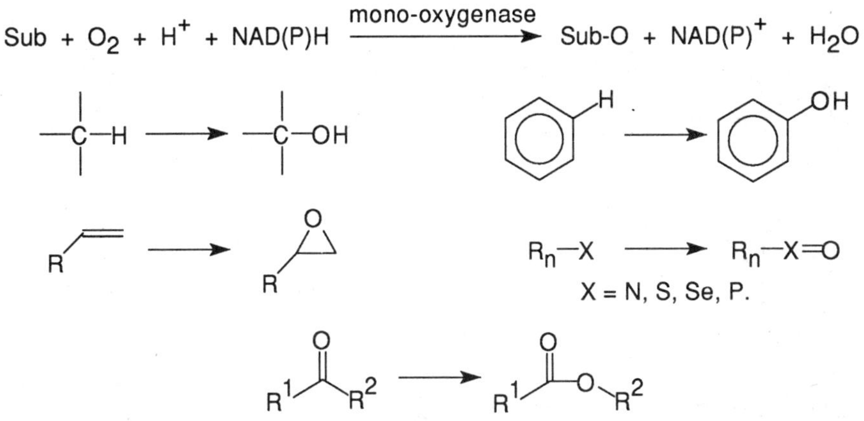

$$Sub + O_2 + H^+ + NAD(P)H \xrightarrow{\text{mono-oxygenase}} Sub\text{-}O + NAD(P)^+ + H_2O$$

$R_n\text{-}X \longrightarrow R_n\text{-}X=O$

X = N, S, Se, P.

Substrate	Product	Type of Reaction	Type of Cofactor
alkane	alcohol	hydroxylation	Fe-dependent
aromat	phenol	hydroxylation	Fe-dependent
alkene	epoxide	epoxidation	Fe-dependent
heteroatom[a]	heteroatom-oxide	heteroatom oxidation	flavin-dependent
ketone	ester/lactone	Baeyer-Villiger	flavin-dependent

[a] N, S, Se or P.

Mono-Oxygenases

Although the reaction mechanisms of different mono-oxygenases differ greatly depending on the sub-type of enzyme, their mode of oxygen-activation is the same. Molecular oxygen is activated by reduction at the expense of a reductant, usually NADH or NADPH, which serves as the ultimate source of redox equivalents (the 'donor' in Scheme 2.131). One of the oxygen atoms is transferred onto the substrate, the other is reduced to form a water molecule.

The net reaction and a number of synthetically useful oxygenation reactions which may be performed by mono-oxygenases are shown in Scheme 2.132.

The generation of the activated oxygen-transferring species is mediated either by cofactors containing a transition metal (Fe or Cu) or by a heteroaromatic system (a pteridin [31] or flavin). The catalytic cycle of the iron-depending mono-oxygenases, the majority of which belong to the cytochrome P-450 type (Cyt P-450 [32]) [33-37], has been deduced largely from studies on the camphor hydroxylase of *Pseudomonas putida* [38, 39]. A summary is depicted in Scheme 2.133.

Scheme 2.133. Catalytic cycle of cytochrome P-450 dependent mono-oxygenases.

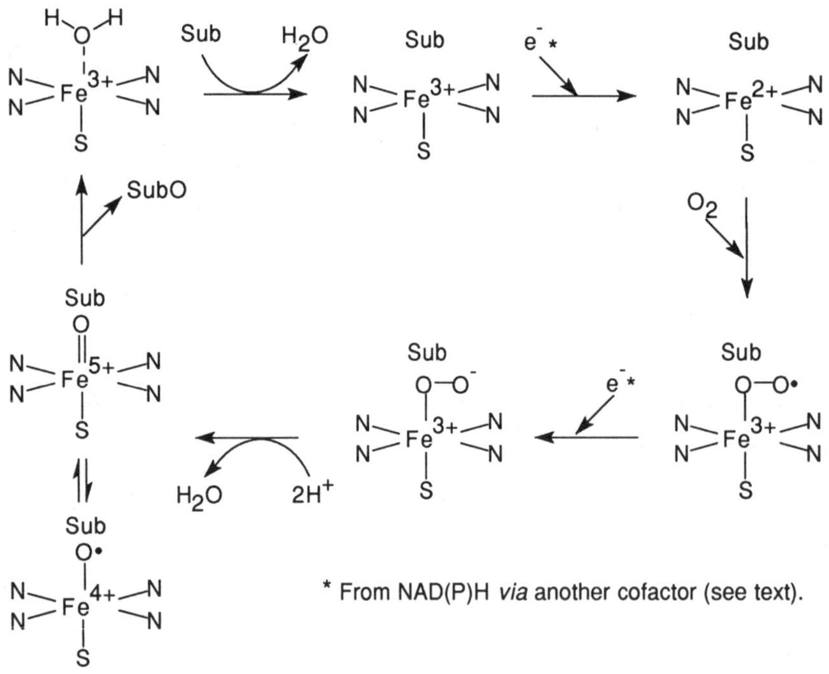

* From NAD(P)H *via* another cofactor (see text).

2.3 Oxidation Reactions

The iron species is coordinated equatorially by a haem moiety and axially by the sulphur atom of a cysteine residue. After binding of the substrate ([Sub], presumably by replacing a water molecule) in a hydrophobic pocket adjacent to the porphine molecule [40], the iron is reduced to the ferrous state. The single electron is delivered from NAD(P)H via another cofactor, which may be either a flavin-nucleotide, an iron-sulphur protein (ferredoxin) or a cytochrome b$_5$ (depending on the enzyme source). Next, molecular oxygen is bound to give a Cyt P-450 dioxygen complex. Delivery of another electron weakens the O-O bond and allows the cleavage of the oxygen molecule. One atom is expelled as water, the other forms the ultimate oxidizing Fe^{4+}- or Fe^{5+}-species, which - as a strong electrophile - attacks the substrate. Expulsion of the product [SubO] reforms the iron(III)-species and closes the catalytic cycle.

Cyt P-450 is a generic term for a wide variety of haemoproteins found in virtually all organisms. They are particularly abundant in detoxification organs such as the liver, but they are also found in microorganisms. Although the individual enzymes differ significantly, for example in their hydrogen-source (NADH or NADPH) [41], a short sequence of 26 amino acids in the proximity of the haem unit is identical in all Cyt P-450 mono-oxygenases.

Scheme 2.134. Catalytic cycle of flavin-dependent mono-oxygenases.

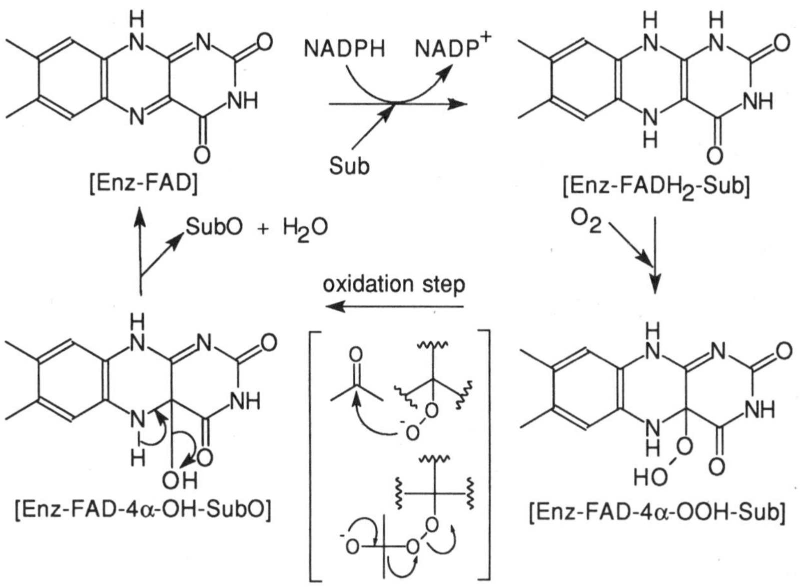

On the contrary, flavin-dependent mono-oxygenases (see Scheme 2.134 and Table of Scheme 2.132) use a different mechanism which involves a flavine cofactor [42-44]. First, NADPH reduces the Enz-FAD complex. The FADH$_2$ so formed is oxidized in the presence of molecular oxygen yielding a hydroperoxide species (FAD-4a-OOH). Deprotonation of the latter affords a peroxide anion, which undergoes a nucleophilic attack on the carbonyl group of the substrate, usually an aldehyde or a ketone. The tetrahedral intermediate collapses via rearrangement of the carbon-framework forming the product ester or lactone, respectively. Finally, water is eliminated from the FAD-4a-OH species to reform FAD.

Whereas the Cyt P-450 mechanism resembles the chemical oxidation by hypervalent transition metal oxidants, the FAD-dependent mechanism parallels the oxidation of organic compounds by peroxides or peracids [45].

Many mono-oxygenases are membrane-bound and are therefore difficult to isolate. This fact and the need for recycling of NAD(P)H makes it clear that the majority of mono-oxygenase catalysed reactions have been performed, out of necessity, by using whole microbial cells. The disadvantage of this method lies in the fact that the desired product is often obtained in low yields due to further metabolism of the product by the cell. However, selective blocking of enzymes which are responsible for the degradation of the product or using enzyme-deficient mutants has been shown to make microbial oxygenation reactions feasible on a commercial scale.

2.3.2.1 Hydroxylation of Alkanes

The hydroxylation of non-activated centres in hydrocarbons is one of the most useful biotransformations [46-51] due to the fact that this process has only very few counterparts in traditional organic synthesis [52-54]. In general, the relative reactivity of carbon atoms in bio-hydroxylation reactions declines in the order of secondary > tertiary > primary, which is in contrast to radical reactions (tertiary > secondary > primary) [55].

Intense research on the stereoselective hydroxylation of alkanes started in the late 40´s in the steroid field [56-61]. In the meantime, some of the hydroxylation processes, *e.g.* 9α- and 16α-hydroxylation of the steroid framework [62, 63], have been developed to the scale of industrial production. Nowadays, vitually any centre in a steroid can be selectively hydroxylated by choosing the appropriate microorganism [64]. For example, hydroxylation of progesterone in the 11α-position by *Rhizopus arrhizus* [65] or *Aspergillus niger* [66] made roughly half of the 37 steps of the conventional chemical

2.3 Oxidation Reactions

synthesis redundant and made 11α-hydroxyprogesterone available for therapy at a reasonable cost. A highly selective hydroxylation of lithiocholic acid in position 7β was achieved by using *Fusarium equiseti* [67]. The product (ursodeoxycholic acid) is capable of dissolving cholesterol and thus can be used in the therapy of gallstones.

Scheme 2.135. Microbial hydroxylation of steroids.

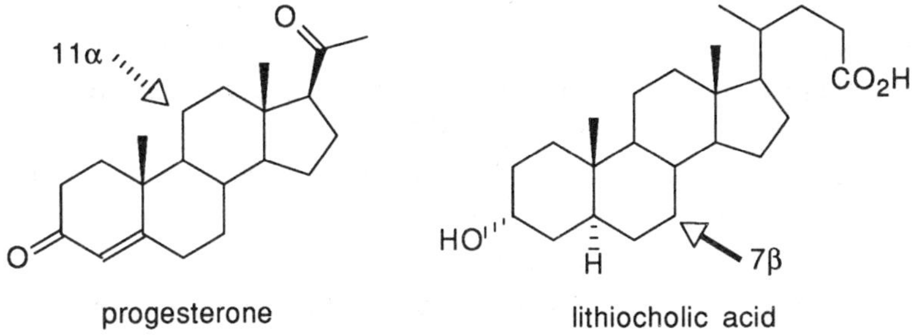

progesterone

lithiocholic acid

The production of new materials for the aroma and fragrance industry was the powerful driving force in the research on the hydroxylation of terpenes [68-70]. For instance, 1,4-cineole, a major constituent of eucalyptus oil, was regioselectively oxidized by *Streptomyces griseus* to give 8-hydroxycineole as the major product along with minor amounts of *exo*- and *endo*-2-hydroxy derivatives, with low optical purity [71]. On the other hand, when *Bacillus cereus* was used, (2R)-*exo*- and (2R)-*endo*-hydroxycineoles were exclusively formed in a ratio of 7:1, both in excellent enantiomeric excess [72].

Scheme 2.136. Microbial hydroxylation of 1,4-cineole.

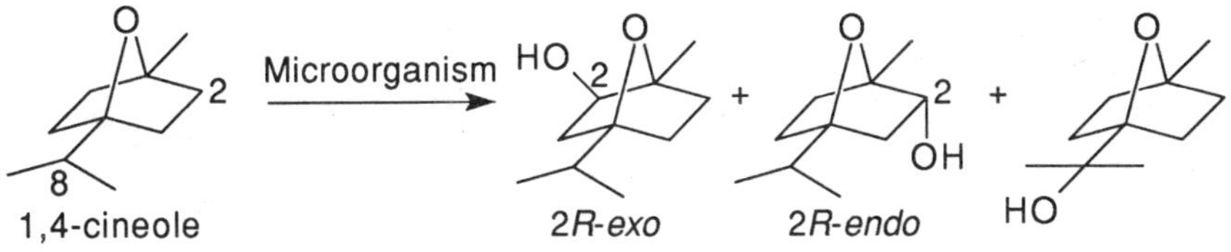

Microorganism	e.e. [%] *exo*	e.e. [%] *endo*	*exo/endo* ratio
Streptomyces griseus	46	74	1:1.7[a]
Bacillus cereus	94	94	7:1

[a] 8-Hydroxycineole was the major product.

Among the many hundreds of microorganisms tested for their capability to perform hydroxylation of non-natural aliphatic compounds, the fungus *Beauveria sulfurescens* has been studied most thoroughly [73-77]. In general the presence of a polar group in the substrate such as an acetamide, benzamide or *p*-toluene-sulphonamide moiety seems to be advantageous in order to facilitate the orientation of the substrate in the active site [78]. Hydroxylation occurs in a distance of 3.3 to 6.2 Å from the polar anchor group [79]. In cases where a competition between cycloalkane rings of different size was possible, hydroxylation preferably occurred in the order cycloheptyl > cyclohexyl > cyclopentyl.

In the majority of cases, hydroxylation by *Beauveria sulfurescens* occurs in a regioselective manner, but high enantioselectivity is not always observed. As shown in Scheme 2.137, both enantiomers of the *N*-benzyl protected bicyclic lactam are hydroxylated with high regioselectivity in position 11, but the reaction showed very low enantioselectivity. On the other hand, when the lactam moiety, which functions as the polar anchor group, was modified by replacing it by a benzoyl-amide, high enantiodifferentiation occurred. The (1*R*)-enantiomer was hydroxylated at carbon 12 and the (1*S*)-counterpart gave the 11-hydroxylated product [80]. A minor amount of 6-*exo*-alcohol was formed with low enantiomeric excess.

Scheme 2.137. Regio- and enantioselective hydroxylation by *B. sulfurescens*.

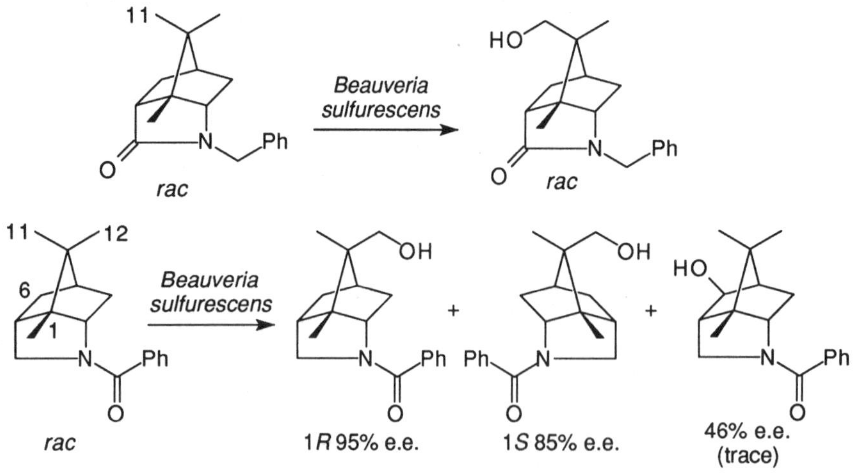

2.3 Oxidation Reactions

In order to provide a tool to predict the stereochemical outcome of hydroxylations using *Beauveria sulfurescens*, a simple model was proposed [81], which has been refined recently [82].

In summary, at present it is difficult to predict the likely site of oxidation for any novel substrate using mono-oxygenases. However, there are three strategies which can be employed to improve regio- and/or stereoselectivity in biocatalytic hydroxylation procedures: (i) Variation of the culture by stressing the metabolism of the cells, (ii) broad screening of different strains and (iii) substrate modification.

2.3.2.2 Hydroxylation of Aromatic Compounds

Regiospecific hydroxylation of aromatic compounds is notoriously difficult. There are reagents for *o*- and *p*-hydroxylation available [83, 84], but some of them are explosive and by-products are usually obtained [85]. The selective hydroxylation of aromatics in the *o*- and *p*-position to existing substituents, can be achieved by using mono-oxygenases. In contrast, *m*-hydroxylation is rather scarce [86]. Mechanistically, it has been proposed that in eukaryotic cells (fungi, yeasts and higher organisms) the reaction proceeds predominantly via epoxidation of the aromatic species which leads to an unstable arene-oxide [87]. Rearrangement of the latter involving the migration of a hydride anion (NIH-shift) forms the phenolic product [88]. Arene-oxides are responsible for toxic and mutagenic effects by reacting with DNA, RNA and proteins. An alternative explanation for flavin-dependent oxidases (which are independent of NAD(P)H) has been proposed which involves a hydroperoxide intermediate (FAD-4a-OOH) [89].

Phenolic components can be selectively oxidized by polyphenol oxidase - one of the few available isolated oxygenating enzymes [90] to give *o*-hydroxylated products (catechols) in high yields [91]. Unfortunately the reaction does not stop at this point but proceeds further to form unstable *o*-quinones, which are prone to polymerization, particularly in water.

Two techniques have been developed to solve the problem of *o*-quinone instability:

- One way is to remove the *o*-quinone from the reaction mixture by chemical reduction (e.g. by ascorbate), which leads to catechols. Ascorbate, however, like many other reductants can act as inhibitor of polyphenol oxidase. To circumvent the inhibition, the concentration of the reducing agent should be

kept at a minimum. Furthermore, a borate buffer which leads to the formation of a complex with catechols, thus preventing their oxidation, is advantageous [92].
- Polymerization, which requires the presence of water, can also be avoided if the reaction is performed in a lipophilic organic solvent such as chloroform (see Section 3) [93].

Scheme 2.138. *o*-Hydroxylation of phenols by polyphenol oxidase.

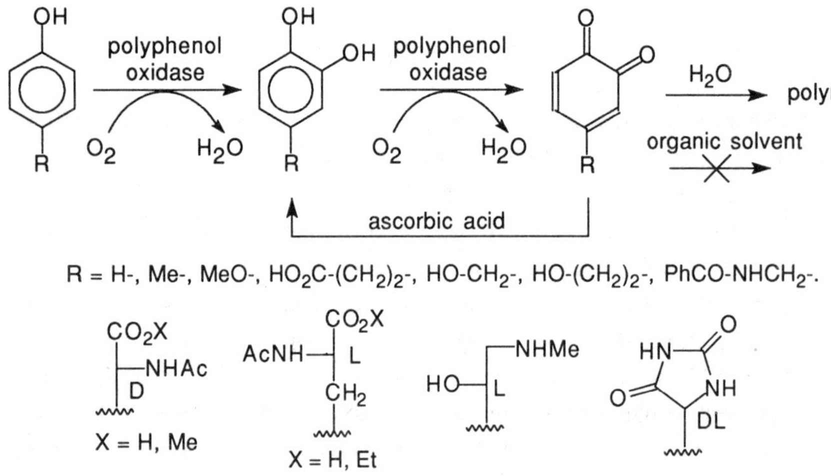

R = H-, Me-, MeO-, HO_2C-$(CH_2)_2$-, HO-CH_2-, HO-$(CH_2)_2$-, PhCO-NHCH$_2$-.

The following rules were deduced for polyphenol oxidase:

- A remarkable range of simple phenols are accepted, as long as the substituent R is in the *p*-position. *m*- And *o*-derivatives are unreactive.
- The reactivity decreases upon a transition of the nature of the group R from electron-donating to electron-withdrawing.
- Bulky phenols (*p-t*-butylphenol and 1- or 2-naphthols) are not substrates; some non-phenolic species such as *p*-toluidine were accepted.

The synthetic utility of this reaction was demonstrated by the oxidation of amino acids and -alcohols containing a *p*-hydroxyphenyl moiety. Thus L-DOPA (3,4-dihydroxyphenyl alanine) used for the treatment of Parkinson´s desease, D-3,4-dihydroxy-phenylglycine and L-epinephrine (adrenaline) were synthesized from their *p*-monohydroxy precursors without racemisation in good yield.

2.3 Oxidation Reactions

Regioselective hydroxylation of aromatic compounds can also be achieved by using whole cells [94-97]. For instance, 6-hydroxynicotinic acid is produced from nicotinic acid by a *Pseudomonas* or *Bacillus* sp. on an industrial scale [98]. Racemic prenalterol, a compound with important pharmacological activity as a β-blocker, was obtained by regioselective *p*-hydroxylation of a simple aromatic precursor using *Cunninghamella echinulata* [99].

Scheme 2.139. Microbial hydroxylation of aromatics.

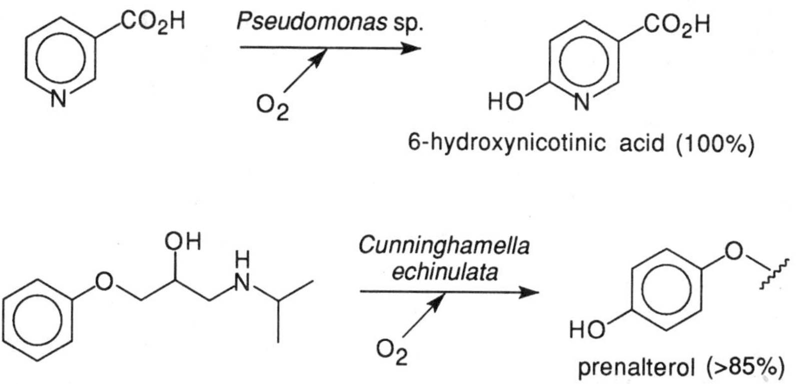

2.3.2.3 Epoxidation of Alkenes

There are well defined procedures available for the asymmetric epoxidation of olefins which bear an additional directing group, such as a hydroxyl group [100]. In contrast, the strength of enzymatic epoxidation, catalysed by mono-oxygenases, is in the preparation of small and non-functionalized epoxides, where traditional methods fail. Despite the wide distribution of mono-oxygenases within all types of organisms, their capability to epoxidize alkenes seems to be associated mainly with alkane- and alkene-utilizing bacteria [101-106].

For the epoxidation of alkenes by mono-oxygenases, NAD(P)H and molecular oxygen are required. In olefin-assimilating microorganisms, recycling of the cofactor is effectively performed since a part of the substrate olefin is degraded to carbon dioxide and water. On the other hand, for microorganisms which are unable to grow on olefins, such as methanotrophs, an additional co-substrate such as methanol or formate has to be added to the medium.

The most intensively studied microbial epoxidizing agent is the ω-hydroxylase system of *Pseudomonas oleovorans* [107, 108]. It consists of three protein components: rubredoxin, NADH-dependent rubredoxin reductase and an ω-hydroxylase (a sensitive non-haem iron protein), which catalyse not only the hydroxylation of aliphatic C-H bonds, but also the epoxidation of alkenes [109]. The following rules can be formulated for epoxidations using *Pseudomonas oleovorans*.

- Terminal, acyclic alkenes are converted into (*R*)-1,2-epoxides of high enantiomeric excess along with varying amounts of ω-en-1-ols or 1-als [110], the ratio of which depends on the chain-length of the substrate [111, 112]. Alkane-hydroxylation predominates over epoxidation for 'short' substrates (propene, 1-butene) and is a major pathway for 'long' chain olefins. Alkene-epoxidation occurs mainly with substrates of 'moderate' chain length (1-octene).
- α,ω-Dienes are transformed into the corresponding terminal (*R,R*)-di-epoxides.
- Cyclic and internal olefins, aromatic compounds and alkene units which are conjugated to an aromatic system are not epoxidized [113].
- To avoid problems arising from the toxicity of the epoxide [114] (which accumulates in the cells) a water-immiscible organic cosolvent such as hexane can be added [115, 116]. Alternatively, the alkene itself can constitute the organic phase into which the product is removed from the cells. However, the bulk apolar phase causes damage to the cell membranes, while the product epoxide reacts with cellular enzymes. Both factors reduce and eventually abolish all enzyme activity [117].

Besides *Pseudomonas oleovorans* numerous bacteria have been shown to epoxidize alkenes [118, 119]. As shown in Scheme 2.140, the optical purity of epoxides depends on the strain used, although the absolute configuration is usually *R* [120]. This concept has been recently applied to the synthesis of chiral alkyl and aryl gycidyl ethers [121, 122]. The latter are of interest for the preparation of 3-substituted 1-alkylamino-2-propanols, which are widely used as β-adrenergic receptor blocking agents [123].

More recently, the structural restrictions for substrates which were elucidated for *Pseudomonas oleovorans* (see above) could be expanded by using different microorganisms. As can be seen from Scheme 2.140, non-terminal alkenes can be epoxidized by a *Mycobacterium* or *Xanthobacter* sp. [124]. More recently, *Nocardia corallina* has been reported to convert branched alkenes into the corresponding (*R*)-epoxides in good optical purities (Scheme 2.141) [125].

Scheme 2.140. Microbial epoxidation of alkenes.

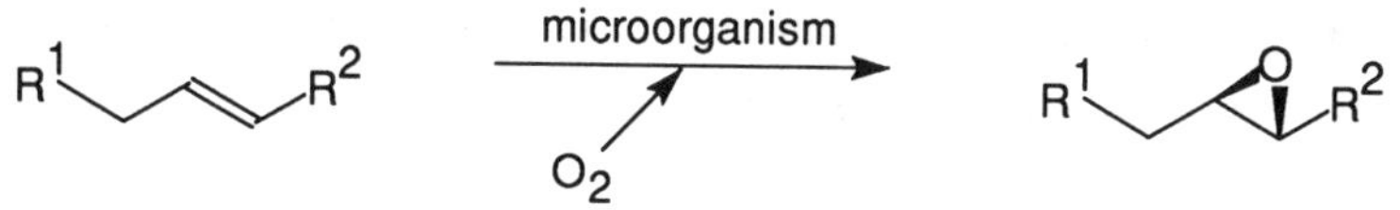

$R^1\text{-CH=CH-}R^2 \xrightarrow[O_2]{\text{microorganism}} R^1\text{-(epoxide)-}R^2$

Microorganism	R^1	R^2	Configuration	e.e. [%]
Pseudomonas oleovorans	$n\text{-}C_5H_{11}$	H	R	70-80
	$n\text{-}C_7H_{15}$	H	R	60
	H	H	R	86
	$NH_2CO\text{-}CH_2\text{-}C_6H_4\text{-}O$	H	S^a	97
	$CH_3O(CH_2)_2\text{-}C_6H_4\text{-}O$	H	S^a	98
Corynebacterium equi	CH_3	H	R	70
	$n\text{-}C_{13}H_{27}$	H	R	~100
Mycobacterium sp.	H	H	R	98
	CH_3	CH_3	R,R	74
	Ph-O	H	S^a	80
Xanthobacter Py2	Cl	H	S^a	98
	CH_3	CH_3	R,R	78
Nocardia sp. IP1	Cl	H	S^a	98
	CH_3	H	R	98

^{a} Sequence rule order reversed.

Scheme 2.141. Epoxidation of branched alkenes by *Nocardia corallina*.

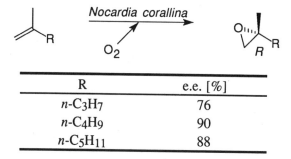

R	e.e. [%]
$n\text{-}C_3H_7$	76
$n\text{-}C_4H_9$	90
$n\text{-}C_5H_{11}$	88

At present, the disadvantages associated with mono-oxygenase catalysed epoxidation reactions - such as inherent enzyme instability, the requirement of several protein components to function together, and the dependence on cofactors [NAD(P)H] - prevent the exploitation of large-scale biocatalytic epoxidation. For certain applications on the gram-scale, however, microbial epoxidation can be a convenient method for the production of chiral epoxides. For instance, the epoxy containing phosphonic acid derivative 'fosfomycin' [126], whose enantiospecific synthesis by classical methods would have been extremely difficult, was obtained by a microbial epoxidation reaction using *Penicillium spinulosum*.

2.3.2.4 Sulphoxidation Reactions

Chiral sulphoxides have been extensively studied in the search for asymmetric units that assist stereoselective reactions. The sulphoxide functional group activates adjacent carbon-hydrogen bonds towards attack by base, and the corresponding anions can be alkylated [127] or acylated [128] with high diastereoselectivity. Similarly, thermal elimination [129] and reduction of α-keto sulphoxides [130] can proceed with transfer of chirality from sulphur to carbon. In spite of this great potential as valuable relay reagents, with rare exceptions [131], no general method is available for the synthesis of sulphoxides with high enantiomeric purities.

An alternative approach involves the use of enzymatic suphur-oxygenation reactions catalysed by mono-oxygenases [132, 133]. The main types of enzymatic sulphur oxygenation are shown in Scheme 2.142. The direct oxidation of a thioether by means of a di-oxygenase, which directly affords the corresponding sulphone, is of no synthetic use since no chirality is involved. On the other hand, the stepwise oxidation involving a chiral sulphoxide, which is catalysed by mono-oxygenases or oxidases, offers two possible ways to obtain chiral sulphoxides.

Scheme 2.142. Enzymatic sulphur oxygenation reactions.

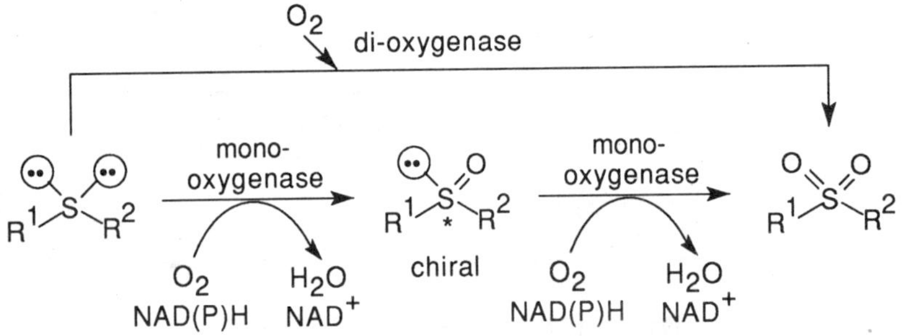

- The asymmetric mono-oxidation of a thioether leading to a chiral sulphoxide resembles an asymmetrisation of a prochiral substrate. As discussed below, it is of high synthetic value.

2.3 Oxidation Reactions

- The kinetic resolution of a racemic sulphoxide during which one enantiomer is oxidized to yield an achiral sulphone is feasible but it has been shown to proceed with low selectivities.

The first asymmetric sulphur oxygenation using cells of *Aspergillus niger* was reported in the early 1960s [134]. Since this time it was shown that the enantiomeric excess and the absolute configuration of the sulphoxide do not only depend on the species but also on the strain of microorganism used [135]. In general, the formation of (R)-sulphoxides predominates.

Thioethers can be asymmetrically oxidized both by bacteria (e.g. *Corynebacterium equi* [136]) and fungi (e.g. *Helminthosporium* sp. and *Mortierella isabellina* [137]). Even baker's yeast has this capacity [138]. As shown in Scheme 2.143, a large variety of aryl-alkyl thioethers were oxidized to yield sulphoxides with good to excellent optical purities [139-141]. The undesired second oxidation step was generally negligable, but with certain substrates the formation of the sulphone was observed.

Scheme 2.143. Microbial oxidation of aryl-alkyl thioethers.

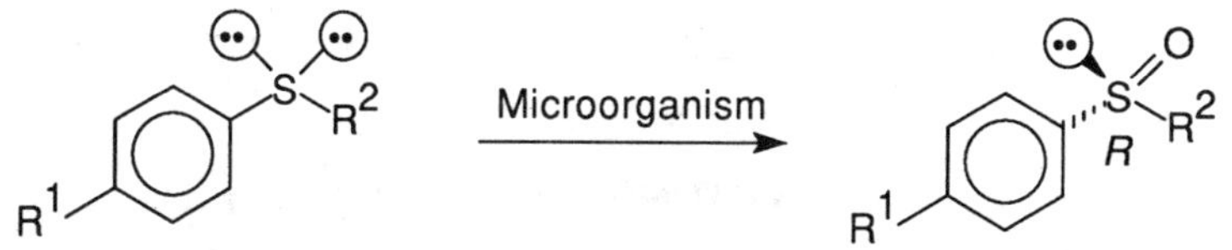

Microorganism	R^1	R^2	e.e. [%]
Mortierella isabellina	H	C_2H_5	85
	H	n-C_3H_7	~100
	H	$(CH_3)_2CH$	83
	H	$C(CH_3)_3$	60
	C_2H_5	CH_3	90
	$(CH_3)_2CH$	CH_3	82
	Cl	CH_3	90
	Br	CH_3	~100[a]
Corynebacterium equi	H	CH_3	92
	H	CH_2-$CH=CH_2$	~100
	H	n-C_4H_9	~100
	H	n-$C_{10}H_{21}$	98[a]
	CH_3	CH_3	97

[a] Sulphone was also formed in this case.

The transformation of thioacetals into mono- or bis-sulphoxides presents intriguing stereochemical possibilities. In a symmetric thioacetal of an aldehyde other than formaldehyde, the sulphur atoms are enantiotopic and each of them contains two diastereotopic non-bonded pairs of electrons (see Scheme 2.144). Unfortunately, most of the products from asymmetric oxidation of thioacetals are of low to moderate optical purity [142, 143]. Two exceptions, however, are worth mentioning. Oxidation of 2-*tert*-butyl-1,3-dithiane by *Mortierella isabellina* gave the (*R,R*)-mono-sulphoxide in 72% optical purity [144] and formaldehyde thioacetals were oxidized by *Corynebacterium equi* to yield (*R*)-sulphoxide-sulphone products [145] with excellent enantiomeric purity.

Scheme 2.144. Microbial oxidation of dithioacetals.

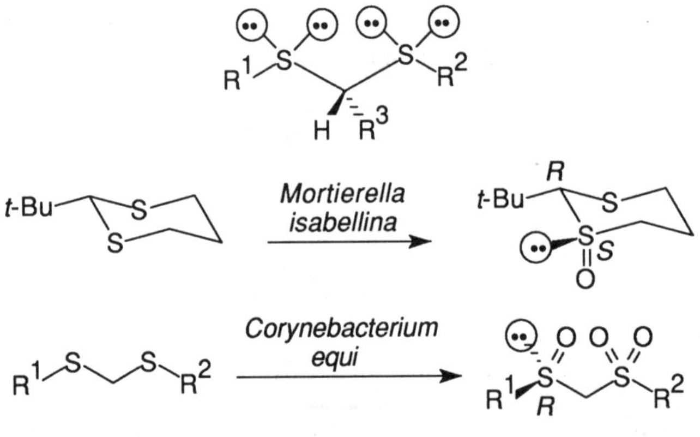

R^1	R^2	e.e. [%]
n-C$_4$H$_9$	n-C$_4$H$_9$	>95
CH$_3$	Ph	>95

Due to the problems arising from NAD(P)H-recycling, sulphoxidation reactions using isolated mono-oxygenases are not feasible on a preparative scale and therefore were only used for mechanistic studies [146]. Only recently it was found that readily available oxidases, which depend on hydrogen peroxide but not on nicotinamide cofactors, can be used for the preparation of chiral sulphoxides (Scheme 2.145) [147]. Chloroperoxidase from the marine fungus *Caldariomyces fumago* (see also Section 2.7) is a selective catalyst for the oxidation of methyl-thioethers. When the 'natural' oxidant (H$_2$O$_2$) was replaced by *tert*-butyl hydroperoxide the selectivity of the enzyme was more than doubled. Another oxidase - horseradish peroxidase - was shown to be relatively unselective.

At present microbial sulphur oxygenation reactions cannot yet be recommended as a simple method for the preparation of chiral sulphoxides mainly due to operational problems such as poor recoveries of the water-soluble products from the considerable amounts of biomass as a consequence of using whole cells as the biocatalyst. Whether the use of isolated oxidases is of an advantage, future investigations will tell.

Scheme 2.145. Oxidation of methyl thioethers by chloroperoxidase.

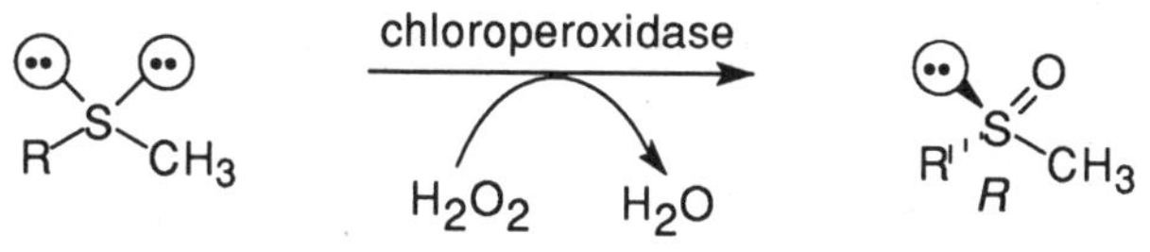

R	Oxidant	e.e. [%]
p-CH$_3$-C$_6$H$_4$	H$_2$O$_2$	35
p-CH$_3$-C$_6$H$_4$	tert-butyl-OOH	86
p-CH$_3$-O-C$_6$H$_4$	tert-butyl-OOH	92
Ph	tert-butyl-OOH	76
p-Cl-C$_6$H$_4$	tert-butyl-OOH	85
Ph-CH$_2$	tert-butyl-OOH	91

2.3.2.5 Baeyer-Villiger Reactions

Oxidation of ketones by peracids - the Baeyer-Villiger reaction [148, 149] - is a reliable and useful method to prepare esters or lactones. The mechanism is believed to be a two-step process, in which the peracid attacks the carbonyl group of the ketone to form the so called tetrahedral 'Criegee-intermediate' [150]. The breakdown of this unstable species, which proceeds via expulsion of a carboxylate ion and migration of a carbon-bond, forms an ester or a lactone. The regiochemistry of oxygen-insertion of the chemical and enzymatic Baeyer-Villiger reaction can usually be predicted by assuming that the carbon atom best able to support a positive charge will migrate preferentially [151].

Scheme 2.146. Mechanism of the Baeyer-Villiger reaction.

$$\underset{\substack{X = \text{acyl group} \\ \text{or flavin}}}{\overset{\overset{\displaystyle O-X}{\underset{\displaystyle HO}{|}}}{\underset{R^1 \quad R^2}{\bigtriangleup}}} \rightleftharpoons \underset{\text{Criegee-intermediate}}{\overset{\overset{\displaystyle O-X}{\underset{\displaystyle O}{|}}}{\underset{R^1 \quad R^2}{HO}}} \longrightarrow \underset{R^1 \quad O-R^2}{\overset{\displaystyle O}{\|}}$$

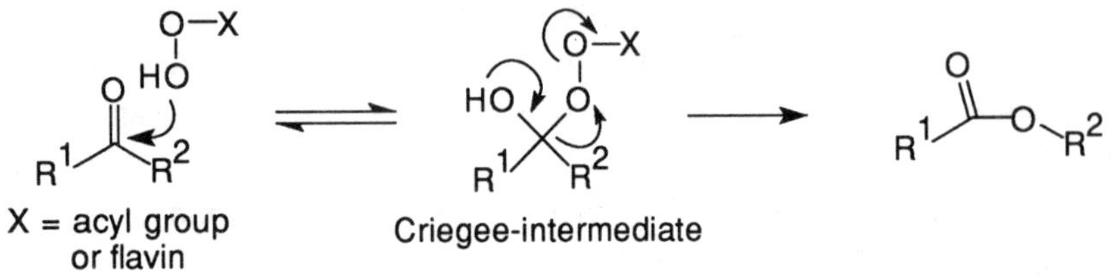

All mechanistic studies on enzymatic Baeyer-Villiger reactions support the hypothesis that conventional and enzymatic reactions are closely related [152, 153]. The oxidized flavin cofactor (FAD-4a-OOH, see Scheme 2.134) plays the rôle as a nucleophile similar to the peracid. The strength of enzyme-catalysed Baeyer-Villiger reactions resides in the recognition of chirality [154, 155], which is impossible by conventional means.

Enzymatic Baeyer-Villiger oxidation of ketones is usually catalysed by flavin-dependent mono-oxygenases and plays an important rôle in the breakdown of carbon structures containing a ketone moiety. Although a number of bacterial enzymes have been purified and characterized to date [156-159], most reactions on a preparative scale were performed by using whole microbial cells, particularly in view of the NAD(P)H-recycling problem [160]. To avoid further degradation of esters and lactones by hydrolytic enzymes and to achieve their accumulation in the culture medium, three approaches are possible:

- Blocking of the hydrolytic enzymes by selective hydrolase-inhibitors such as tetraethyl pyrophosphate (TEPP [161]) or diethyl p-nitrophenylphosphate (paraoxon),
- development of mutant strains lacking lactone-hydrolases or
- application of artificial ketones, whose lactone products are not substrates for the hydrolytic enzymes.

Prochiral (symmetric) ketones can be asymmetrically oxidized by a cyclohexanone mono-oxygenase from an *Acinetobacter* sp. to yield the corresponding lactones [162]. As depicted in Scheme 2.147, oxygen insertion occurred with high selectivity almost independent of the substitutional pattern of the substrate.

Racemic ketones can be resolved via two pathways. The 'classic' form of a kinetic resolution represents involves a transformation in which one enantiomer reacts and its counterpart remains unchanged [163]. For example, α-substituted cyclopentanones were stereospecifically oxidized by an *Acinetobacter* sp. to form the corresponding (*S*)-δ-lactones [164], which constitute valuable components of various fruit flavours (Scheme 2.148). The slower reacting (R)-ketones accumulated in the culture medium.

Scheme 2.147. Enzymatic Baeyer-Villiger oxidation of prochiral ketones.

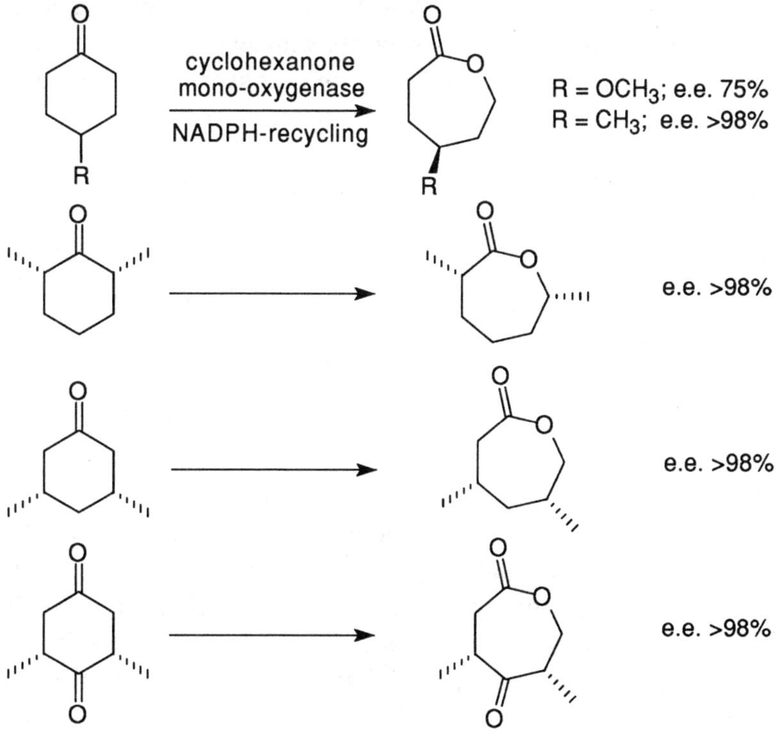

R = OCH$_3$; e.e. 75%
R = CH$_3$; e.e. >98%

e.e. >98%

e.e. >98%

e.e. >98%

Scheme 2.148. Microbial Baeyer-Villiger oxidation of monocyclic racemic ketones.

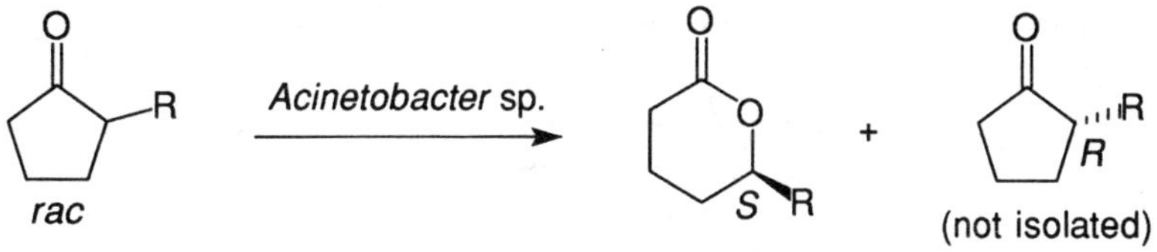

R	e.e. of lactone [%]
n-C$_5$H$_{11}$	97
n-C$_7$H$_{15}$	95
n-C$_9$H$_{19}$	85
n-C$_{11}$H$_{23}$	73[a]

[a] The (R)-ketone showed an e.e. of 36% in this case.

Bicyclic halo-ketones, which were used for the synthesis of anti-viral 6'-fluoro-carbocyclic nucleoside analogues, were resolved by using the same technique [165] (see Scheme 2.149). Both enantiomers were obtained with >95 % optical purity. The exquisite enantioselectivity of the microbial oxidation is due to the presence of the halogen atoms since the de-halogenated bicyclo[2.2.1]hepan-2-one was transformed with low selectivity. On the other hand, replacement of the halogens by methoxy- or hydroxy-groups gave rise to substrates which were not accepted as substrates.

Scheme 2.149. Microbial Baeyer-Villiger oxidation of bicyclic racemic ketones.

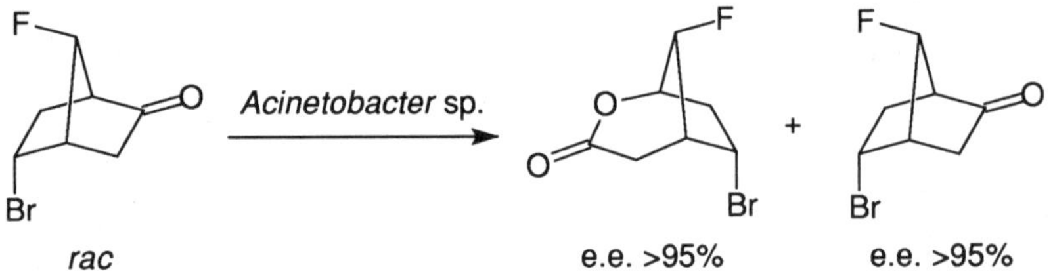

The resolution of a racemic ketone does not have to follow the 'classic' format as described above, but can proceed via an oxygen-insertion into the two opposite sides of the ketone. As shown in Scheme 2.150, both enantiomers of the bicyclo[3.2.0]heptenones were microbially oxidized, but in an enantiodivergent manner [166, 167]. Oxygen insertion on the (5R)-ketone occurred as expected, adjacent to carbon 7, forming the 3-oxabicyclic lactone. On the other hand, the (5S)-ketone underwent oxygen insertion in the 'wrong sense' towards carbon atom 5, which led to the 2-oxabicyclic species. Further results suggested the presence of two competing mono-oxygenases in the microorganisms. On a preparative scale, however, the separation of such regio-isomeric lactones is often difficult by conventional methods.

Scheme 2.150. Enantiodivergent microbial Baeyer-Villiger oxidation.

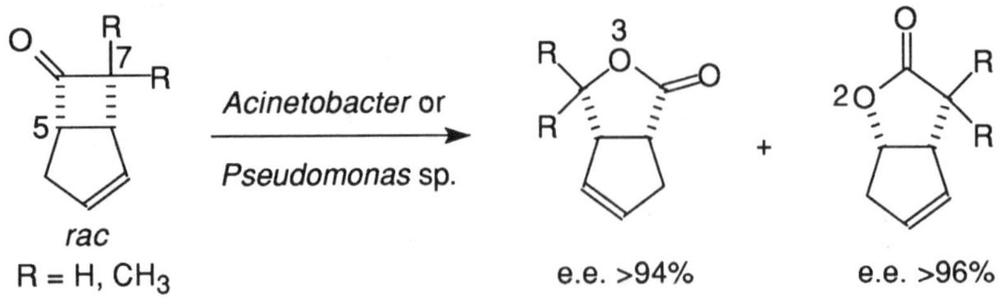

2.3 Oxidation Reactions

Such enantiodivergent transformations using microbial cells are not uncommon and they may even proceed via completely different enzymatic pathways. This was recently demonstrated by a Baeyer-Villiger oxidation of a structurally related bicyclic ketone, where one enantiomer was *oxidized* as expected to give the lactone, but the ketone of opposite configuration was *reduced* by dehydrogenase(s) to give the corresponding secondary alcohol [168].

An intriguing new concept of cofactor-recycling using isolated mono-oxygenases was developed for the Baeyer-Villiger reaction [169]. Thus, in a coupled-enzyme system, which consists of a mono-oxygenase from *Acinetobacter calcoaceticus* and a dehydrogenase from *Thermoanaeriobium brockii*, the ketone is enzymatically prepared by oxidation of the corresponding alcohol (see Section 2.2.2). As a consequence, a catalytic amount of NADPH which was consumed during the Baeyer-Villiger oxidation step, is recycled in the first step of a closed loop.

Scheme 2.151. Cofactor-recycling via a dehydrogenase-oxygenase system.

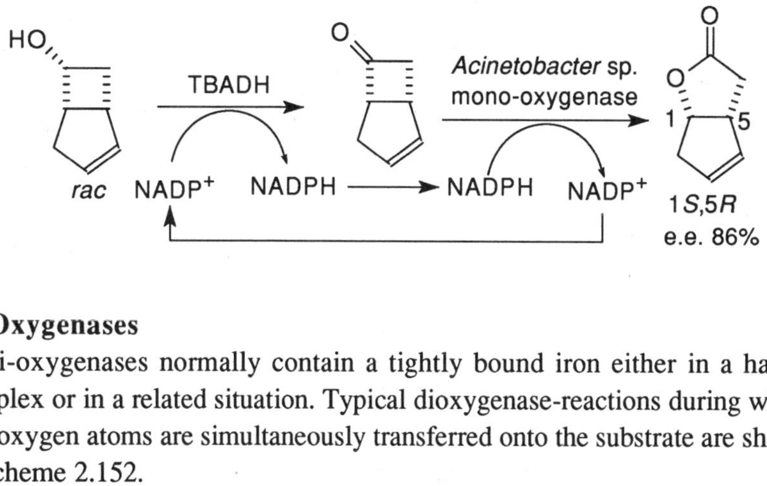

Di-Oxygenases

Di-oxygenases normally contain a tightly bound iron either in a haem-complex or in a related situation. Typical dioxygenase-reactions during which two oxygen atoms are simultaneously transferred onto the substrate are shown in Scheme 2.152.

- Alkenes may be transformed into a hydroperoxide - e.g. by a lipoxygenase - which, upon reduction (e.g. by sodium borohydride) yields an alcohol. In living systems, the formation of lipid peroxides in vivo is considered to be involved in some serious diseases and malfunctions including arteriosclerosis and cancer [170].
- Alternatively, an endo-peroxide may be formed, whose reduction leads to a diol. The latter reaction resembles the cycloaddition of singlet-oxygen onto an

unsaturated system and occurs in the biosynthesis of prostaglandins and leukotrienes. Both intermediate hydro- or endo-peroxide species may be subject to further reactions such as (enzymatic or non-enzymatic) reduction or rearrangement.
- In prokaryotic cells (bacteria) the initial step of the metabolism of aromatic compounds is the cycloaddition of oxygen catalysed by a di-oxygenase [171, 172]. In living microbial cells, the resulting endo-peroxide (dioxetan) is enzymatically reduced to yield synthetically useful *cis*-glycols.

Scheme 2.152. Di-oxygenase catalysed reactions.

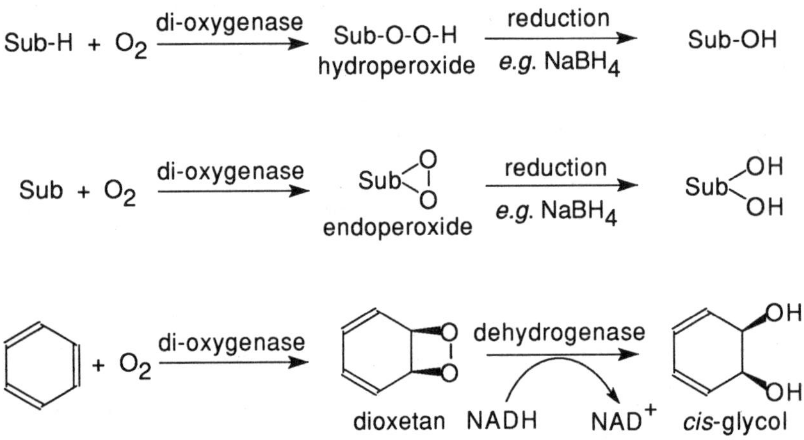

2.3.2.6 Formation of Peroxides

Lipoxygenase is a non-haeme iron di-oxygenase which catalyzes the incorporation of dioxygen into polyunsaturated fatty acids possessing a non-conjugated 1,4-diene unit by forming conjugated hydroperoxides [173-175]. The enzyme from soybean has received the most attention in terms of a detailed characterization because of its early discovery [176], ease of isolation and stability [177, 178]. The following characteristics can be given for soybean-lipoxygenase catalysed oxidations:
- The specificity of this enzyme has long been thought of being restricted to an all-Z configurated 1,4-diene unit at an appropriate location in the carbon chain of poly-unsaturated fatty acids. More recently, however, it was shown that also (*E,Z*)- and (*Z,E*)-1,4-dienes are accepted as substrates [179], albeit at slower rates.
- (*E,E*)-1,4-Dienes and conjugated 1,3-dienes are generally not oxidized.

2.3 Oxidation Reactions

- The configuration at the newly formed oxygenated chiral centre is S predominantly although not exclusively [180, 181].
- In some cases, the oxidation of non-natural substrates can be forced by increasing the pressure (up to 50 bar) of oxygen [182].

Oxidation of the natural substrate (Z,Z)-9,12-octadecadienoic acid (linoleic acid) proceeds highly selective (95% e.e.) and leads to peroxide-formation at carbon 13 (the distal region) along with traces of 9-oxygenated product [183] (the proximal region, Scheme 2.153) [184].

Scheme 2.153. Oxidation of linoleic acid by soybean lipoxygenase.

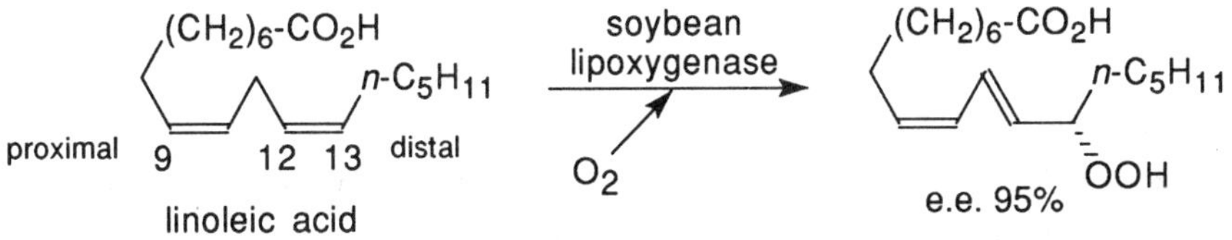

Scheme 2.154. Oxidation of 1,4-dienes by soybean lipoxygenase.

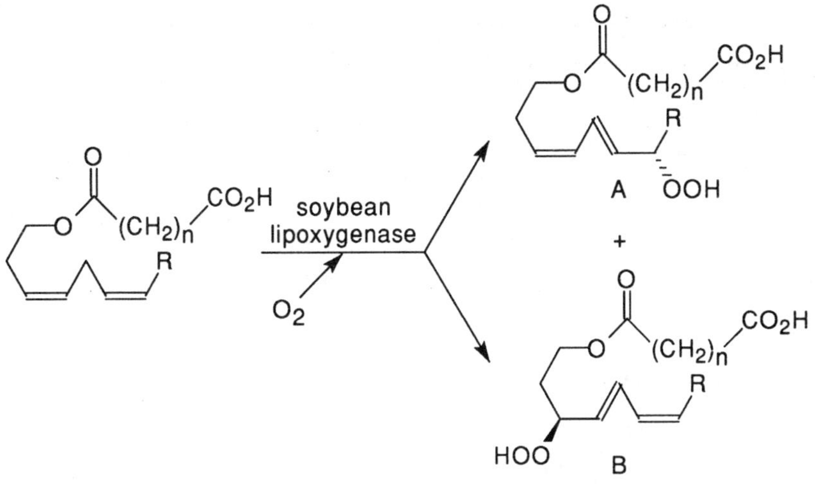

n	R	A/B	Configuration of A	e.e. of A [%]
4	n-C$_5$H$_{11}$	95:5	S	98
4	CH$_2$CH(CH$_3$)$_2$	82:18	S	96
4	CH$_2$Ph	89:11	S	98
4	CH$_2$OCH$_2$Ph	85:15	R[a]	97
4	(CH$_2$)$_3$C(O)CH$_3$	99:1	S	97

[a] Sequence rule order reversed.

Most recently, it has been shown that lipoxygenase can also be used for the oxidation of non-natural 1,4-dienes, as long as the substrate is carefully modified [185]. Thus, the (Z,Z)-1,4-diene moiety of several long-chain alcohols could be oxidized by attachment of a prostetic group, which served as a mimick of the carboxylate moiety (see Scheme 2.154). The oxidation occurred with high regioselectivity at the 'normal' (distal) site and the optical purity of the peroxides was >96%. After reduction of the hydroperoxide and removal of the prostetic group, the corresponding secondary alcohols were obtained with retention of configuration [186].

In addition, the regioselectivity of the oxidation - from 'normal' (distal) to 'abnormal' (proximal) - could be inverted by changing the lipophilicity of the modifying groups R and n (see Table 2.4 and Scheme 2.154). Increasing the lipophilicity of the distal group (n-C_5 to n-C_{10}) led to an increased reaction at the 'abnormal' site to form predominantly product B. Consequently, when the lipophilicity of the proximal prostetic group was increased by varying n from 2 to 6, the 'normal' product A was formed in favour of B.

Table 2.4. Variation of prostethic groups (see Scheme 2.154).

n	Variation	R	A/B
4	distal	n-C_5H_{11}	95:5
4	distal	n-C_8H_{17}	1:1
4	distal	n-$C_{10}H_{21}$	27:73
2	proximal	n-C_8H_{17}	20:80
4	proximal	n-C_8H_{17}	1:1
6	proximal	n-C_8H_{17}	85:15

2.3.2.7 Dihydroxylation of Aromatic Compounds

cis-Dihydroxylation by microbial di-oxygenases constitutes the major degradation pathway for aromatic compounds in lower organisms (Scheme 2.155) [187, 188]. In 'wild-type' microorganisms, the chiral *cis*-glycols are rapidly further oxidized by dihydrodiol dehydrogenase(s), involving re-aromatization of the diol intermediate with concomitant loss of chirality. The use of mutant strains with blocked dehydrogenase activity [189], however, allows the chiral glycols to accumulate in the medium, which can be isolated in good yield.

2.3 Oxidation Reactions

For a number of mutant strains of *Pseudomonas putida* the stereospecificity is high yet the substrate specificity remains low with respect to ring substituents R^1 and R^2 [190] (see Scheme 2.156). An impressive number of substituted aromatic compounds have been converted into the corresponding chiral *cis*-glycols with excellent optical purities on a commercial scale [191-193]. A general method for the determination of optical purity and absolute configuration of such compounds has recently been reported [194]. The work of Ribbons et al. has shown that polysubstituted benzene derivatives can be converted into cyclohexadienediols and that the regioselectivity of the oxygen-addition can be predicted with some accuracy [195, 196].

Scheme 2.155. Degradation of aromatics by microbial di-oxygenases.

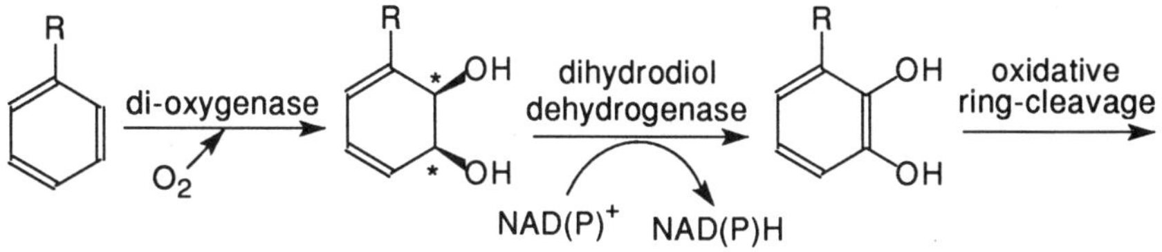

Scheme 2.156. Synthesis of *cis*-glycols.

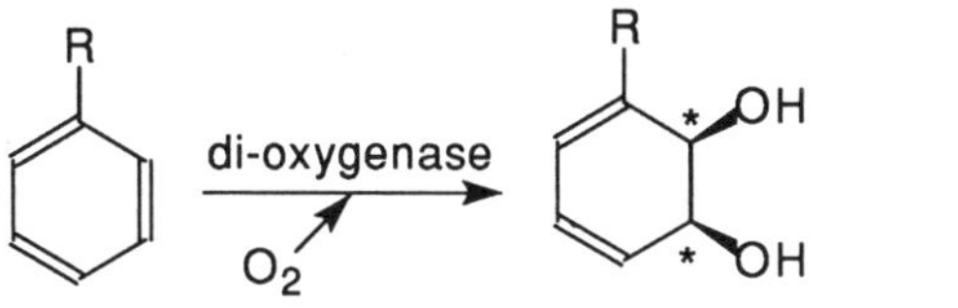

R = H, Me, Et, *n*-Pr, *i*-Pr, *n*-Bu, *t*-Bu, Et-O,
n-Pr-O, Halogen, CF_3, Ph, Ph-CH_2, Ph-CO,
CH_2=CH, CH_2=CH-CH_2, HC≡C.

The substrates need not necessarily be mono-substituted aromatic compounds such as those shown above, but may also be extended to other species including fluoro- [197], monocyclic- [198], polycyclic- [199] and heterocyclic derivatives [200].

The large synthetic potential of chiral *cis*-glycols is illustrated in Scheme 2.157. The diene may be subjected to Diels-Alder reactions, as well as Michael-type addition reactions. Both proceed in a stereocontrolled manner, guided by the chiral diol moiety. Alternatively, oxidative cleavage of the cyclohexane ring leads to open-chain products, which *inter alia* give rise to cyclopentanoids.

This synthetic potential has been exploited over the recent years to synthesize a number of bioactive compounds. Cyclohexanoids have been prepared by making use of the possibility to functionalize every carbon atom of the glycol in a stereocontrolled way. For instance, (+)-pinitol [201] and (D)-*myo*-inositol derivatives [202] were obtained using this approach. Cyclopentanoid synthons for the synthesis of prostaglandins and terpenes were prepared by a ring-opening closure sequence [203]. Rare carbohydrates such as D- and L-erythrose [204] and L-ribonolactone [205] were obtained from chlorobenzene as were pyrrolizidine alkaloids [206].

Scheme 2.157. Syntheses starting from chiral *cis*-glycols.

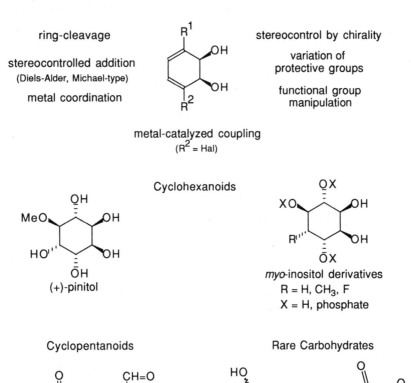

References

1. Fonken GS, Johnson RA (1972) Chemical Oxidations with Microorganisms, Belew JS (ed), Oxidation in Organic Chemistry, vol 2, p 185, Marcel Dekker, New York
2. Lemière GL, Lepoivre JA, Alderweireldt FC (1985) Tetrahedron Lett. 4527
3. Shimizu C, Hattori S, Hata H, Yamada Y (1987) Appl. Environ. Microbiol. 53: 519
4. Brown GM (1971) in: Comprehensive Biochemistry, Florkin M, Stotz EH (eds) vol 21, p 73, Elsevier, New York
5. Reichstein T (1934) Helv. Chim. Acta 17: 996
6. Sato K, Yamada Y, Aida K, Uemura T (1967) Agric. Biol. Chem. 31: 877
7. Touster O, Shaw DRD (1962) Physiol. Rev. 42: 181
8. Bernhauer K, Knobloch H (1940) Biochem. Z. 303: 308
9. Kaufmann H, Reichstein T (1967) Helv. Chim. Acta 50: 2280
10. Jones JB, Taylor KE (1976) Can. J. Chem. 54: 2969 and 2974
11. Irwin AJ, Jones JB (1977) J. Am. Chem. Soc. 99: 1625
12. Jakovac IJ, Jones JB (1985) Org. Synth. 63: 10
13. Irwin AJ, Jones JB (1976) J. Am. Chem. Soc. 98: 8476
14. Nakazaki M, Chikamatsu H, Fujii T, Sasaki Y, Ao S (1983) J. Org. Chem. 48: 4337
15. Drueckhammer DG, Riddle VW, Wong C-H (1985) J. Org. Chem. 50: 5387
16. Wong C-H, Matos JR (1985) J. Org. Chem. 50: 1992
17. Matos JR, Smith MB, Wong C-H (1985) Bioorg. Chem. 13: 121
18. Irwin AJ, Jones JB (1977) J. Am. Chem. Soc. 99: 556
19. Jones JB, Lok KP (1979) Can. J. Chem. 57: 1025
20. Ng GSY, Yuan L-C, Jakovac IJ, Jones JB (1984) Tetrahedron 40: 1235
21. Lok KP, Jakovac IJ, Jones JB (1985) J. Am. Chem. Soc. 107: 2521
22. Holland HL (1992) Organic Synthesis with Oxidative Enzymes, Verlag Chemie, Weinheim
23. Walsh C (1979) Enzymatic Reaction Mechanisms, Freeman, San Francisco, p 501
24. Dalton H (1980) Adv. Appl. Microbiol. 26: 71
25. Hayashi O (ed) (1974) The Molecular Mechanism of Oxygen Activation, Academic Press, New York
26. Gunsalus IC, Pederson TC, Sligar SG (1975) Ann. Rev. Microbiol. 377
27. Hou CT (1986) Biotechnol. Gen. Eng. Rev. 4: 145
28. Dunford HB (1982) Adv. Inorg. Biochem. 4: 41
29. Richter G (1983) Glucose Oxidase In: Industrial Enzymology, Godfrey T, Reichelt J (eds) Nature Press, New York
30. Szwajcer E, Brodelius P, Mosbach K (1982) Enzyme Microb. Technol. 4: 409
31. Dix TA, Benkovic SS (1988) Acc. Chem. Res. 21: 101
32. 450 Refers to the characteristic absorption (in nm) by the Fe^{2+}-enzyme-carbonmonoxide complex.
33. Sato R, Omura T (eds.) (1978) Cytochrome P-450, Academic Press, New York
34. Ortiz de Montellano PR (ed) (1986) Cytochrome P-450, Plenum Press, New York
35. Takemori S (1987) Trends Biochem. Sci. 12: 118
36. Dawson JH, Sono M (1987) Chem. Rev. 87: 1255
37. Korzekwa KR, Jones JP, Gillette JR (1990) J. Am. Chem. Soc. 112: 7042
38. Alexander LS, Goff HM (1982) J. Chem. Educ. 59: 179
39. Poulos TL, Finzel BC, Howard AJ (1986) Biochemistry 25: 5314
40. Poulos TL, Finzel BC, Gunsalus IC, Wagner GC, Kraut J (1985) J. Biol. Chem. 260: 16122
41. Bacterial Cyt P-450 depends on NADH, enzymes isolated from higher organisms generally use NADPH
42. Ryerson CC, Ballou DP, Walsh C (1982) Biochemistry 21: 2644
43. Visser CM (1983) Eur. J. Biochem. 135: 543
44. Ghisla S, Massey V (1989) Eur. J. Biochem. 181: 1
45. Mansuy D, Battioni P (1989) Activation and Functionalisation of Alkanes, Hill CL (ed) p 195, Wiley, New York

46. Johnson RA (1978) Oxygenations with Micro-Organisms, in: Oxidation in Organic Synthesis, part C, Trahanovsky WS (ed) p 131, Academic Press, New York
47. Fonken G, Johnson RA (1972) Chemical Oxidations with Microorganisms, Marcel Dekker, New York
48. Kieslich K (1984) in: Biotechnology, Rehm H-J, ReedG (eds) vol 6a, p 1, Verlag Chemie, Weinheim
49. Kieslich K (1980) Bull. Soc. Chim. Fr. 11: 9
50. Sariaslani FS (1989) Crit. Rev. Biotechnol. 9: 171
51. Dalton H (1980) Adv. Appl. Microbiol. 26: 71
52. Breslow R (1980) Acc. Chem. Res. 13: 170
53. Fossey J, Lefort D, Massoudi M, Nedelec J-Y, Sorba J (1985) Can. J. Chem. 63: 678
54. Barton DHR, Kalley F, Ozbalik N, Young E, Balavoine G (1989) J. Am. Chem. Soc. 111: 7144
55. Mansuy D (1990) Pure Appl. Chem. 62: 741
56. Kieslich K (1969) Synthesis 120
57. Kolot FB (1983) Process Biochem. 19
58. Marsheck WJ (1971) Progr. Ind. Microbiol. 10: 49
59. Holland HL (1984) Acc. Chem. Res. 17: 389
60. Holland HL (1982) Chem. Soc. Rev. 11: 371
61. Sedlaczek L (1988) Crit. Rev. Biotechnol. 7: 186
62. Weiler EW, Droste M, Eberle J, Halfmann HJ, Weber A (1987) Appl. Microbiol. Biotechnol. 27: 252
63. Perlman D, Titius E, Fried J (1952) J. Am. Chem. Soc. 74: 2126
64. For a complete list see: Davies HG, Green RH, Kelly DR, Roberts SM (1989) Biotransformations in Preparative Organic Chemistry, p 175, Academic Press, London
65. Peterson DH, Murray HC, Eppstein SH, Reineke LM, Weintraub A, Meister PD, Leigh HM (1952) J. Am. Chem. Soc. 74: 5933
66. Fried J, Thoma RW, Gerke JR, Herz JE, Donin MN, Perlman D (1952) J. Am. Chem. Soc. 74: 3692
67. Sawada S, Kulprecha S, Nilubol N, Yoshida T, Kinoshita S, Taguchi H (1982) Appl. Environ. Microbiol. 44: 1249
68. Ciegler A (1974) Microbial Transformations of Terpenes, in: CRC Handbook of Microbiology, vol 4, p 449-458, CRC Press, Boca Raton, Florida
69. Lamare V, Furstoss R (1990) Tetrahedron 46: 4109
70. Kieslich K (1976) Microbial Transformations of Non-Steroid Cyclic Compounds, Thieme, Stuttgart
71. Rosazza JPN, Steffens JJ, Sariaslani S, Goswami A, Beale JM, Reeg S, Chapman R (1987) Appl. Environ. Microbiol. 53: 2482
72. Liu W-G, Goswami A, Steffek RP, Chapman RL, Sariaslani FS, Steffens JJ, Rosazza JPN (1988) J. Org. Chem. 53: 5700
73. Archelas A, Furstoss R, Waegell B, le Petit J, Deveze L (1984) Tetrahedron 40: 355
74. Johnson RA, Herr ME, Murray HC, Reineke LM (1971) J. Am. Chem. Soc. 93: 4880
75. Furstoss R, Archelas A, Waegell B, le Petit J, Deveze L (1981) Tetrahedron Lett. 445
76. Johnson RA, Herr ME, Murray HC, Fonken GS (1970) J. Org. Chem. 35: 622
77. Archelas A, Fourneron JD, Furstoss R (1988) Tetrahedron Lett. 6611
78. Fonken GS, Herr ME, Murray HC, Reineke LM (1968) J. Org. Chem. 33: 3182
79. Previously this value was believed to be 5.5 Å.
80. Archelas A, Fourneron JD, Furstoss R (1988) J. Org. Chem. 53: 1797
81. Johnson RA, Herr ME, Murray HC, Fonken GS (1968) J. Org. Chem. 33: 3217
82. Furstoss R, Archelas A, Fourneron JD, Vigne B (1986) A Model for the Hydroxylation Site of the Fungus *Beauveria sulfurescens*. In: Schneider MP (ed) Enzymes as Catalysts in Organic Synthesis, p 361, NATO ASI Series C, vol 178, Reidel, Dordrecht
83. Olah GA, Ernst TD (1989) J. Org. Chem. 54: 1204
84. Komiyam AM (1989) J. Chem. Soc., Perkin Trans. I, 2031
85. Zimmer H, Lankin DC, Horgan SW (1971) Chem. Rev. 71: 229
86. Powlowski JB, Dagley S, Massey V, Ballou DP (1987) J. Biol. Chem. 262: 69
87. Wiseman A, King DJ (1982) Topics Enzymol. Ferment. Biotechnol. 6: 151

88. Boyd DR, Campbell RM, Craig HC, Watson CG, Daly JW, Jerina DM (1976) J. Chem. Soc., Perkin Trans. I, 2438
89. Schreuder HA, Hol WGJ, Drenth J (1988) J. Biol. Chem. 263: 3131
90. Also called tyrosinase, catechol oxidase, cresolase.
91. Klibanov AM, Berman Z, Alberti BN (1981) J. Am. Chem. Soc. 103: 6263
92. Doddema HJ (1988) Biotechnol. Bioeng. 32: 716
93. Kazandjian RZ, Klibanov AM (1985) J. Am. Chem. Soc. 107: 5448
94. Vigne B, Archelas A, Furstoss R (1991) Tetrahedron 47: 1447
95. Yoshioka H, Nagasawa T, Hasegawa R, Yamada H (1990) Biotechnol. Lett. 679
96. Theriault RJ, Longfield TH (1973) Appl. Microbiol. 25: 606
97. Watson GK, Houghton C, Cain RB (1974) Biochem. J. 140: 265
98. Hoeks FWJMM, Meyer H-P, Quarroz D, Helwig M, Lehky P (1990) Scale-Up of the Process for the Biotransformation of Nicotinic Acid into 6-Hydroxynicotinic Acid, in: Opportunities in Biotransformations, Copping LG, Martin RE, Pickett JA, Bucke C, Bunch AW, p 67, Elsevier, London
99. Pasutto FM, Singh NN, Jamali F, Coutts RT, Abuzar S (1987) J. Pharm. Sci. 76: 177
100. Pfenninger A (1986) Synthesis 89
101. Weijers CAGM, de Haan A, de Bont JAM (1988) Microbiol. Sci. 5: 156
102. May SW (1979) Enzyme Microb. Technol. 1: 15
103. Furuhashi K (1986) Econ. Eng. Rev. 18 (7/8): 21
104. Abraham W-R, Stumpf B, Arfmann HA (1990) J. Essent. Oil Res. 2: 251
105. Dalton H (1980) Adv. Appl. Microbiol. 26: 71
106. Habets-Ctützen AQH, Carlier SJN, de Bont JAM, Wistuba D, Schurig V, Hartmans S, Tramper J (1985) Enzyme Microb. Technol. 7: 17
107. Peterson JA, Basu D, Coon MJ (1966) J. Biol. Chem. 241: 5162
108. de Smet M-J, Witholt B, Wynberg H (1983) Enzyme Microb. Technol. 5: 352
109. Jurtshuk P, Cardini GE (1972) Crit. Rev. Microbiol. 1: 254
110. Katopodis AG, Wimalasena K, Lee J, May SW (1984) J. Am. Chem. Soc. 106: 7928
111. May SW, Abbott BJ (1973) J. Biol. Chem. 248: 1725
112. May SW (1976) Catal. Org. Synth. 4: 101
113. May SW, Schwartz RD, Abbott BJ, Zaborsky OR (1975) Biochim. Biophys. Acta 403: 245
114. Habets-Crützen AQH, de Bont JAM (1985) Appl. Microbiol. Biotechnol. 22: 428
115. Brink LES, Tramper J (1987) Enzyme Microb. Technol. 9: 612
116. Brink LES, Tramper J (1985) Biotechnol. Bioeng. 27: 1258
117. de Smet M-J, Witholt B, Wynberg H (1981) J. Org. Chem. 46: 3128
118. Hou CT, Patel R, Laskin AI, Barnabe N, Barist I (1983) Appl. Environ. Microbiol. 46: 171
119. Furuhashi K (1981) Ferment. Ind. 39: 1029
120. Ohta H, Tetsukawa H (1979) Agric. Biol. Chem. 43: 2099
121. Johnstone SL, Phillips GT, Robertson BW, Watts PD, Bertola MA, Koger HS, Marx AF (1987) Stereoselective Synthesis of (S)-β-Blockers via Microbially Produced Epoxide Intermediates; In: Laane C, Tramper J, Lilly MD (eds) Biocatalysis in Organic Media, p 387, Elsevier, Amsterdam
122. Fu H, Shen G-J, Wong C-H (1991) Recl. Trav. Chim. Pays-Bas 110: 167
123. Howe R, Shanks RG (1966) Nature 210: 1336
124. Weijers CAGM, van Ginkel CG, de Bont JAM (1988) Enzyme Microb. Technol. 10: 214
125. Takahashi O, Umezawa J, Furuhashi K, Takagi M (1989) Tetrahedron Lett. 1583
126. White RF, Birnbaum J, Meyer RT, ten Broeke J, Chemerda JM, Demain AL (1971) Appl. Microbiol. 22: 55
127. Bravo P, Resnati G, Viani F (1985) Tetrahedron Lett. 2913
128. Solladie G (1981) Synthesis 185
129. Goldberg SI, Sahli MS (1967) J. Org. Chem. 32: 2059
130. Solladie G, Demailly G, Greck C (1985) J. Org. Chem. 50: 1552
131. Kagan HB, Dunach E, Nemeck C, Pitchen P, Samuel O, Zhao S (1985) Pure Appl. Chem. 57: 1922
132. Holland HL (1988) Chem. Rev. 88: 473

133. Phillips RS, May SW (1981) Enzyme Microb. Technol. 3: 9
134. Dodson RM, Newman N, Tsuchiya HM (1962) J. Org. Chem. 27: 2707
135. Auret BJ, Boyd DR, Henbest HB, Watson CG, Balenovic K, Polak V, Johanides V, Divjak S (1974) Phytochemistry 13: 65
136. Ohta H, Okamoto Y, Tsuchihashi G (1985) Agric. Biol. Chem. 49: 2229
137. Holland HL, Carter IM (1983) Bioorg. Chem. 12: 1
138. Buist PH, Marecak DM, Partington ET, Skala P (1990) J. Org. Chem. 55: 5667
139. Ohta H, Okamoto Y, Tsuchihashi G (1985) Agric. Biol. Chem. 49: 671
140. Abushanab E, Reed D, Suzuki F, Sih CJ (1978) Tetrahedron Lett. 3415
141. Holland HL, Pöpperl H, Ninniss RW, Chenchaiah PC (1985) Can. J. Chem. 63: 1118
142. Poje M, Nota O, Balenovic K (1980) Tetrahedron 36: 1895
143. Auret BJ, Boyd DR, Breen F, Greene RME, Robinson PM (1981) J. Chem. Soc., Perkin Trans. I, 930
144. Auret BJ, Boyd DR, Cassidy ES, Turley F, Drake AF, Mason SF (1983) J. Chem. Soc., Chem. Commun. 282
145. Okamoto Y, Ohta H, Tsuchihashi G (1986) Chem. Lett. 2049
146. Light DR, Waxman DJ, Walsh C (1982) Biochemistry 21: 2490
147. Colonna S, Gaggero N, Manfredi A, Casella L, Gullotti M, Carrea G, Pasta P (1990) Biochemistry 29: 10465
148. Krow GR (1981) Tetrahedron 37: 2697
149. Mimoun H (1982) Angew. Chem., Int. Ed. Engl. 21: 734
150. Criegee R (1948) Liebigs Ann. Chem. 560: 127
151. Lee JB, Uff BC (1967) Quart. Rev. 21: 429
152. Schwab JM, Li WB, Thomas LP (1983) J. Am. Chem. Soc. 105: 4800
153. Ryerson CC, Ballou DP, Walsh C (1982) Biochemistry 21: 2644
154. Gunsalus IC, Peterson TC, Sligar SG (1975) Ann. Rev. Biochem. 44: 377
155. Walsh CT, Chen Y-CJ (1988) Angew. Chem., Int. Ed. Engl. 27: 333
156. Nealson KH, Hastings JW (1979) Microbiol. Rev. 43: 496
157. Donoghue NA, Norris DB, Trudgill PW (1976) Eur. J. Biochem. 63: 175
158. Britton LN, Markavetz AJ (1977) J. Biol. Chem. 252: 8561
159. Trower MK, Buckland RM, Griffin M (1989) Eur. J. Biochem. 181: 199
160. Abril O, Ryerson CC, Walsh C, Whitesides GM (1989) Bioorg. Chem. 17: 41
161. Alphand V, Archelas A, Furstoss R (1990) J. Org. Chem. 55: 347
162. Taschner MJ, Black DJ (1988) J. Am. Chem. Soc. 110: 6892
163. Ouazzani-Chahdi J, Buisson D, Azerad R (1987) Tetrahedron Lett. 1109
164. Alphand V, Archelas A, Furstoss R (1990) Biocatalysis 3: 73
165. Levitt MS, Newton RF, Roberts SM, Willetts AJ (1990) J. Chem. Soc., Chem. Commun. 619
166. Alphand V, Archelas A, Furstoss R (1989) Tetrahedron Lett. 3663
167. Carnell AJ, Roberts SM, Sik V, Willetts AJ (1990) J. Chem. Soc., Chem. Commun. 1438
168. Königsberger K, Alphand V, Furstoss R, Griengl H (1991) Tetrahedron Lett. 499
169. Willetts AJ, Knowles CJ, Levitt MS, Roberts SM, Sandey H, Shipston NF (1991) J. Chem. Soc., Perkin Trans. I, 1608
170. Yagi K (ed) (1982) Lipid Peroxides in Biology and Medicine, Academic Press, New York
171. Subramanian V, Sugumaran M, Vaidyanathan CS (1978) J. Ind. Inst. Sci. 60: 143
172. Jeffrey H, Yeh HJC, Jerina DM, Patel TR, Davey JF, Gibson DT (1975) Biochemistry 14: 575
173. Axelrod B (1974) ACS Adv. Chem. Ser. 136: 324
174. Vick BA, Zimmerman DC (1984) Plant. Physiol. 75: 458
175. Corey EJ, Nagata R (1987) J. Am. Chem. Soc. 109: 8107
176. Theorell H, Holman RT, Akeson A (1947) Acta Chem. Scand. 1: 571
177. Finnazzi-Agro A, Avigliano L, Veldink GA, Vliegenhart JFG, Boldingh J (1973) Biochim. Biophys. Acta 326: 462
178. Axelrod B, Cheesbrough TM, Laakso TM (1981) Methods Enzymol. 71: 441
179. Funk Jr MO, Andre JC, Otsuki T (1987) Biochemistry 26: 6880

2.3 Oxidation Reactions

180. Van Os CPA, Vente M, Vliegenhart JFG (1979) Biochim. Biophys. Acta 547: 103
181. Corey EJ, Albright JO, Burton AE, Hashimoto S (1980) J. Am. Chem. Soc. 102: 1435
182. Corey EJ, Nagata R (1987) Tetrahedron Lett. 5391
183. Iacazio G, Langrand G, Baratti J, Buono G, Triantaphylides C (1990) J. Org. Chem. 55: 1690
184. Gunstone FD (1979) In: Comprehensive Organic Chemistry, Barton DHR, Ollis WD, Haslam E (eds) p 587, Pergamon Press, New York
185. Datcheva VK, Kiss K, Solomon L, Kyler KS (1991) J. Am. Chem. Soc. 113: 270
186. Zhang P, Kyler KS (1989) J. Am. Chem. Soc. 111: 9241
187. Smith MR, Ratledge C (1989) Appl. Microbiol. Biotechnol. 32: 68
188. De Frank JJ, Ribbons DW (1977) J. Bacteriol. 129: 1356 and 1365
189. Gibson DT, Koch JR, Kallio RE (1968) Biochemistry 7: 3795
190. Ribbons DW, Evans CT, Rossiter JT, Taylor SCJ, Thomas SD, Widdowson DA, Williams DJ (1990) In: Biotechnology and Biodegradation, Kamely D, Chakrabarty A, Omenn GS (eds) Advances in Applied Biotechnology Series, vol 4, p 213, Gulf Publ. Co., Houston
191. Widdowson DA, Ribbons DW (1990) Janssen Chim. Acta 8 (3): 3
192. Ballard DHG, Courtis A, Shirley IM, Taylor SC (1988) Macromolecules 21: 294
193. Ballard DHG, Courtis A, Shirley IM, Taylor SC (1983) J. Chem. Soc., Chem. Commun. 954
194. Boyd DR, Dorrity MRJ, Hand MV, Malone JF, Sharma ND, Dalton H, Gray DJ, Sheldrake GN (1991) J. Am. Chem. Soc. 113: 666
195. Taylor SC, Ribbons DW, Slawin AMZ, Widdowson DA, Williams DJ (1987) Tetrahedron Lett. 6391
196. Rossiter JT, Williams SR, Cass AEG, Ribbons DW (1987) Tetrahedron Lett. 5173
197. Ribbons DW, Cass AEG, Rossiter JT, Taylor SJC, Woodland MP, Widdowson DA, Williams SR, Baker PB, Martin RE (1987) J. Fluorine Chem. 37: 299
198. Geary PJ, Pryce RJ, Roberts SM, Ryback G, Winders JA (1990) J. Chem. Soc., Chem. Commun. 204
199. Deluca ME, Hudlicky T (1990) Tetrahedron Lett. 13
200. Wackett LP, Kwart LD, Gibson DT (1988) Biochemistry 27: 1360
201. Ley SV, Sternfeld F (1989) Tetrahedron 45: 3463
202. Ley SV, Parra M, Redgrave AJ, Sternfeld F (1990) Tetrahedron 46: 4995
203. Hudlicky T, Luna H, Barbieri G, Kwart LD (1988) J. Am. Chem. Soc. 110: 4735
204. Hudlicky T, Luna H, Price JD, Rulin F (1989) Tetrahedron Lett. 4053
205. Hudlicky T, Price JD (1990) Synlett. 159
206. Hudlicky T, Luna H, Price JD, Rulin F (1990) J. Org. Chem. 55: 4683

2.4 Formation of Carbon-Carbon Bonds

The majority of enzymatic reactions exploited to date for biotransformations catalyse degradation reactions. Two enzymatic systems belonging to the class of lyases, which are capable of *forming* carbon-carbon bonds in a highly stereoselective manner, are known and are gaining increasing attention. Aldol reactions catalysed by aldolases are useful for the elongation of aldehydes by a three-carbon unit. A C_2-fragment is transferred via transketolase-reactions or via yeast-mediated acyloin condensations. For the addition of cyanide (a C_1-synthon) to aldehydes by oxynitrilase enzymes see Section 2.5.1.

2.4.1 Aldol-Reactions

Asymmetric C-C bond formation based on catalytic aldol addition reactions remains one of the most challenging subjects in synthetic organic chemistry. Although many successful non-biological strategies have been developed [1, 2], most of them are not without drawbacks. They are often stoichiometric in auxiliary reagent and require the use of a metal or metal-like enolate complex to achieve stereoselectivity. Due to the instability of such complexes in aqueous solutions, the aldol reactions are carried out in organic solvents at low temperature. This requirement limits the application of conventional aldol reactions to the synthesis of molecules soluble in organic solvents. For those compounds containing polyfunctional groups, such as carbohydrates, the employment of conventional aldol reactions requires extensive protection to be carried out in order to make them lipophilic. On the other hand, enzymatic aldol reactions catalysed by aldolases, which are performed in aqueous solution at neutral pH, can be achieved without extensive protection methodology and have therefore attracted increasing interest [3-5].

Aldolases were first recognized some sixty years ago as an ubiquitous class of enzymes, that catalyse the interconversion of hexoses and their three-carbon subunits [6]. It is now known that aldolases operate on a much wider range of substrates than hexoses, and a variety of enzymes has been described that add a one-, two- or three-carbon fragment onto a carbonyl group of an aldehyde or a ketone with high stereospecificity via an aldol reaction. Since glycolysis and glyconeogenesis is a *conditio sine qua non*, all organisms possess aldolase enzymes. Two distinct groups, using different mechanisms have been recognized [7].

2.4 Formation of Carbon-Carbon Bonds

Type I aldolases, found predominantly in higher plants and animals, require no metal cofactor, and catalyse the aldol reaction through a Schiff-base intermediate [8].

Scheme 2.158. Mechanism of type I aldolases.

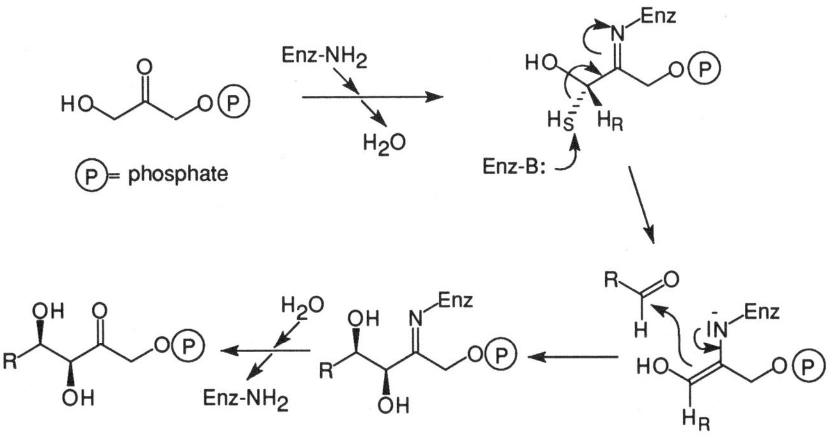

The donor (DHAP) is covalently linked to the enzyme probably involving the ε-amino group of a lysine to form a Schiff-base. Next, abstraction of H_S leads to the formation of an enamine species, which performs a nucleophilic attack on the carbonyl group of the aldehydic acceptor in an asymmetric manner. Consequently, the two new chiral centres are formed stereospecifically in a *threo*-configuration. Finally, hydrolysis of the Schiff-base liberates the diol and regenerates the enzyme.

Class II aldolases are found predominantly in bacteria and fungi, and are Zn^{2+} dependent enzymes [9]. Bearing in mind that the natural substrates of aldolases are carbohydrates, most successful enzyme-catalysed aldol reactions have been performed with carbohydrate-like (poly)hydroxy compounds as substrates. With the exception of reactions catalysed by transketolase, the carbon-chain elongation always involves a three-carbon unit. Both *threo-* or *erythro*-diols can be formed through the use of the appropriate enzyme. However, explorations of substrate specificity are limited in some cases and work in this field has been impeded by the non-availability of the majority of these enzymes. This problem should be surmountable by modern techniques in molecular biology. To date more than 15 aldolases have been isolated, but only the most useful and commercially available enzymes are described in this chapter.

Fructose-1,6-Diphosphate Aldolase

Fructose-1,6-diphosphate (FDP) aldolase from rabbit muscle, also commonly known as rabbit muscle aldolase (RAMA), catalyses the addition of dihydroxyacetone phosphate (DHAP) to D-glyceraldehyde-3-phosphate to form fructose-1,6-diphosphate [10]. The equilibrium of the reaction is predominantly on the product side and the specificity on C-3 and C-4 adjacent to the newly formed vicinal diol bond is absolute - always *threo* (see Scheme 2.159). However, if the α-carbon atom in the aldehyde component is chiral (C-5 in the product), only low chiral recognition takes place. Consequently, if an α-substituted aldehyde is employed in racemic form, a pair of diastereomeric products will be obtained.

RAMA accepts a wide range of aldehydes in place of its natural substrate, allowing the synthesis of carbohydrates [11-14] and analogues such as nitrogen-containing sugars [15, 16], deoxysugars [17], fluorosugars and rare eight- and nine-carbon sugars [18]. As depicted in Scheme 2.159, numerous aldehydes which are structurally quite unrelated to the natural acceptor substrate (D-glyceraldehyde-3-phosphate) are freely accepted [19-22].

Scheme 2.159. Aldol reactions catalysed by FDP aldolase from rabbit muscle.

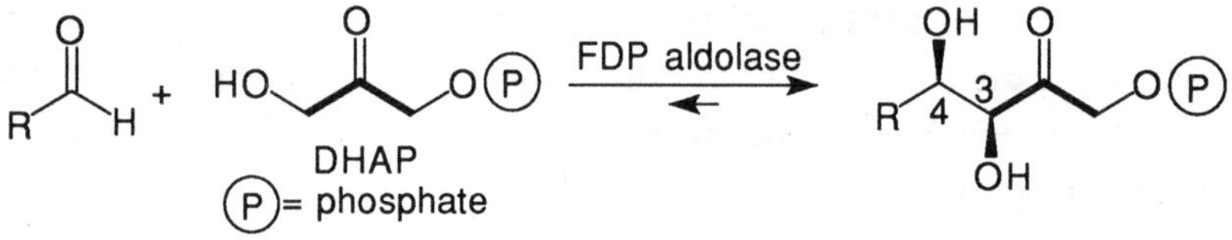

Natural Substrates R =	Non-natural Substrates R = 
D- or L-threose	H, Me, Et, *n*-Pr, *i*-Pr, CH=O, CO_2H, CH_2F,
D- or L-Erythrose	CH_2OCH_2Ph, $(CH_2)_2OH$, $CH(OCH_3)CH_2OH$,
L-arabinose, D-ribose,	$(CH_2)_2OMe$, CHOHMe, $CHOHCH_2OMe$,
D-lyxose, D-xylose	CH_2Ph, COPh, $(CH_2)_3CH=O$, $CH_2P(O)(OEt)_2$,
	2-pyridyl, 3-pyridyl, 4-cyclohexenyl, CH_2O(P)

2.4 Formation of Carbon-Carbon Bonds

For RAMA the following rules apply to the aldehyde component.

- In general, unhindered aliphatic, α-heterosubstituted, and protected alkoxy aldehydes are accepted as substrates.
- Severely hindered aliphatic aldehydes such as pivaldehyde do not react with RAMA, nor do α,β-unsaturated aldehydes or compounds that can eliminate to form α,β-unsaturated aldehydes.
- Aromatic aldehydes are either poor substrates or are unreactive.
- ω-Hydroxy compounds that are phosphorylated at the terminal hydroxyl group are accepted at enhanced rates relative to the non-phosphorylated species.

The requirement for the donor component DHAP is much more stringent. Several analogues have been tested as substitutes for DHAP; two are accepted (see Scheme 2.160) but the reaction rate is reduced by about one order of magnitude [23, 24].

Scheme 2.160. Substitutes for DHAP.

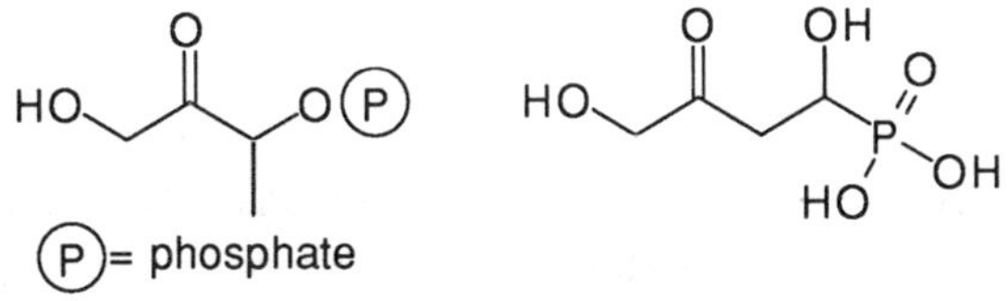

(P)= phosphate

One problem associated with FDP aldolase catalysed reactions is the need for the preparation of DHAP. This molecule is not stable in solution (t ~20 h at pH 7), and its synthesis is not trivial [25]. It may be generated in situ from fructose-1,6-diphosphate (FDP) using triosephosphate isomerase [26], from the dimer of DHAP by chemical phosphorylation with $POCl_3$ [27], or from dihydroxyacetone by enzymatic phosphorylation at the expense of ATP and glycerol kinase [28] (see Chapter 2.1.4). The in situ generation from FDP is the most convenient of the above listed methods, but the presence of excess FDP may complicate the isolation of product(s).

An alternative solution is to use a mixture of dihydroxyacetone and a small amount of inorganic arsenate to replace DHAP [29, 30]. Mechanistic studies indicate that in aqueous solution, dihydroxyacetone reacts with inorganic arsenate spontaneously to form dihydroxyacetone arsenate, which is a mimic of DHAP and is accepted by FDP aldolase as a substrate. However, the toxicity of arsenate limits the attractiveness of this method. Furthermore, the presence of the phosphate group in the aldol adducts facilitates their purification by ion-exchange chromatography or by precipitation as the barium or silver salts. Cleavage of this group is accomplished either by chemical or enzymatic methods using acid or alkaline phosphatase (see Section 2.1.4). The use of inorganic vanadate instead of arsenate is prohibited by a redox reaction involving dihydroxyacetone.

Scheme 2.161. Use of dihydroxy arsenate as a DHAP mimic.

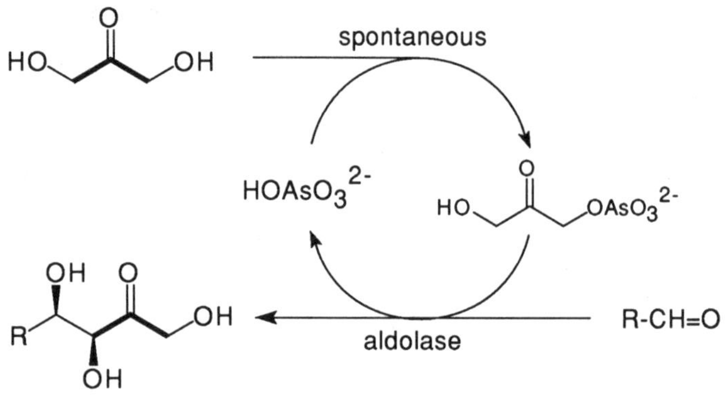

FDP aldolase from rabbit muscle has been extensively used for the synthesis of biologically active sugar analogues on a preparative scale (Scheme 2.162). For example, nojirimycin and derivatives thereof, which have been shown to be potent anti-AIDS agents with no cytotoxicity, have been obtained by a chemo-enzymatic approach using RAMA in the key step [31, 32].

An elegant synthesis of (+)-*exo*-brevicomin, the sex pheromone of the bark beetle was reported recently (Scheme 2.163) [33]. RAMA-catalysed condensation of DHAP to a keto-aldehyde gave, after enzymatic dephosphorylation, a *threo*-keto-diol, which was cyclised to form a precursor of the pheromone. Finally, the side-chain was modified in four subsequent steps to give (+)-*exo*-brevicomin.

Despite the fact that enzymatic aldol reactions are becoming useful in synthetic carbohydrate chemistry, the preparation of aldehyde substrates containing chiral centres remains a problem. Many α-substituted aldehydes racemise in aqueous solution and this would result in the production of a diastereomeric mixture, which is often not readily separable.

2.4 Formation of Carbon-Carbon Bonds

Scheme 2.162. Synthesis of nitrogen-containing sugar analogues.

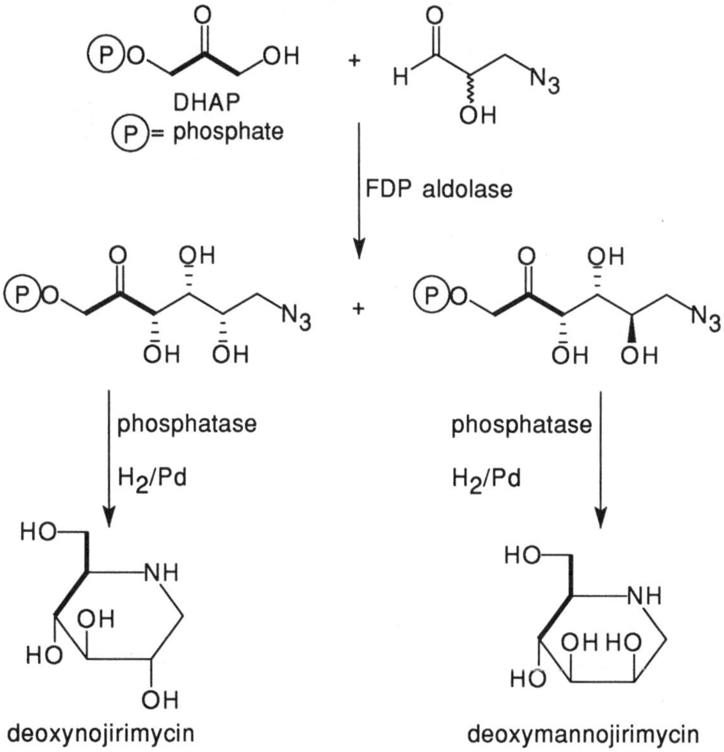

Scheme 2.163. Synthesis of (+)-*exo*-brevicomin.

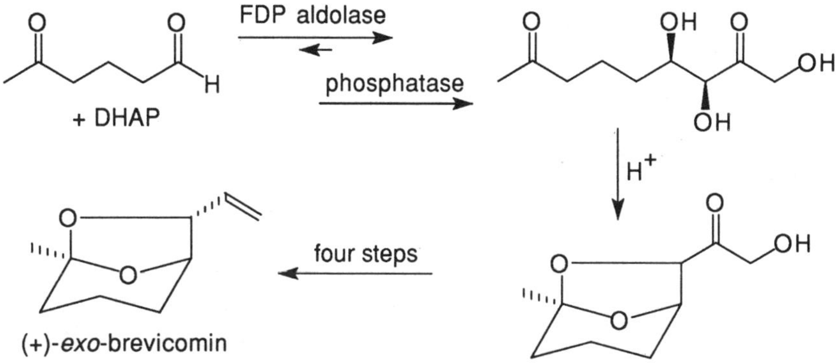

The following methods have been used to avoid the (often tedious) separation of diastereomeric products [34].

- In some cases, a stereoselective aldol reaction can be accomplished in a kinetically controlled process. If the reaction is stopped before it reaches equilibrium, a single diastereomer is obtained; this occurs if one enantiomer of the aldehyde reacts appreciably faster than the other. As mentioned above, the selectivities of aldolases for such kinetic resolutions involving the chirality on the α-carbon atom of the aldehyde is usually low.
- In cases wherein one diastereomer of the product is more stable than the other, one can utilize a thermodynamically controlled process (Scheme 2.164). For example, in the aldol reaction of 2-allyl-3-hydroxypropanal, two diastereomeric products are formed. Due to the ring-formation of the aldol product and because of the reversible nature of the aldol reaction, only the more stable product (with the 5-allyl substituent in an equatorial position) is produced when the reaction reaches equilibrium.
- Another approach to the separation of diastereomeric products is to subject the mixture to the action of glucose isomerase, whereby the D-ketose is converted into the corresponding D-aldose leaving the L-ketose component unchanged [35].

Scheme 2.164. Thermodynamic control in aldolase reactions.

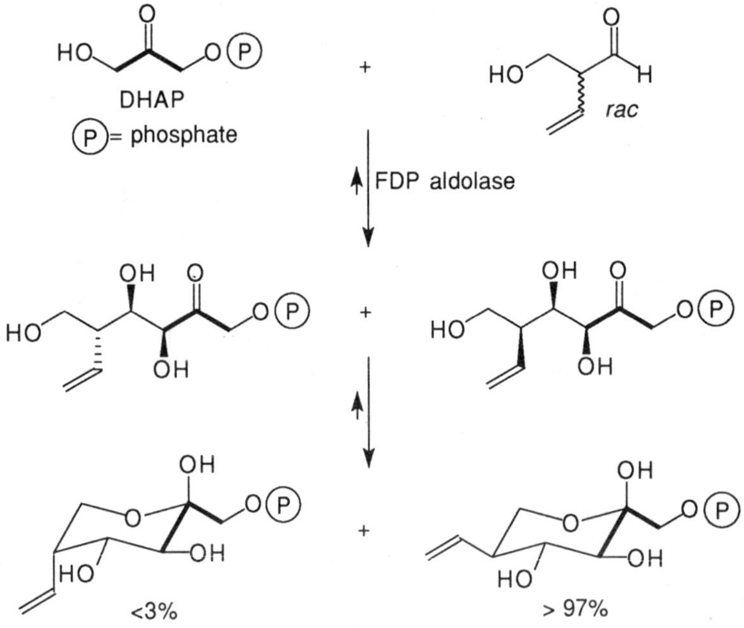

2.4 Formation of Carbon-Carbon Bonds

A potential limitation with the use of FDP aldolases for the synthesis of monosaccharides is that the products are always ketoses with fixed stereochemistry at C-3 and C-4. There are, however, methods of establishing this stereochemistry at other centres and to obtain aldoses instead of ketoses. This technique makes use of a mono-protected dialdehyde as the acceptor substrate (see Scheme 2.165). After the RAMA-catalysed aldol reaction, the resulting ketone is reduced stereospecifically with polyol dehydrogenase. The remaining aldehyde is then deprotected to yield a new aldose.

Scheme 2.165. Synthesis of aldoses using FDP aldolase.

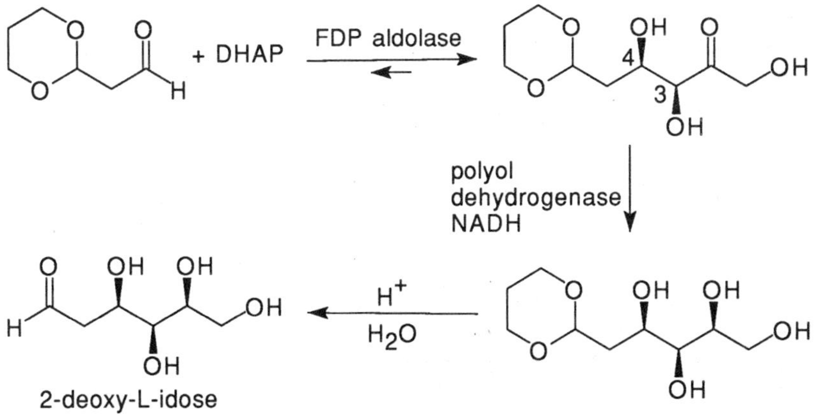

Sialyl Aldolase

N-Acetylneuraminic acid (NeuAc, also termed sialic acid) aldolase catalyses the reversible addition of pyruvate and N-acetylmannosamine to form N-acetylneuraminic acid [36, 37]. Since the equilibrium for this reaction is near unity, an excess of pyruvate must be used in synthetic reactions to drive the reaction towards completion. The synthesis of NeuAc itself is only of moderate interest, because it may be isolated from natural sources such as cow's milk. The production of analogues, however, is of significance since neuraminic acid derivatives play an important rôle in biochemical recognition processes [38]. The cloning of the enzyme has reduced its cost [39].

In parallel with the substrate requirements of FDP aldolase, the specificity of sialic acid aldolase appears to be absolute for pyruvate (the donor), but relaxed for the acceptor. As may be seen from Scheme 2.166, a range of mannosamine derivatives have been used to synthesize derivatives of NeuAc [40-44]. Substitution at C-2 of N-acetylmannosamine is readily tolerated, and the enzyme exhibits only a slight preference for defined stereochemistry at other centres.

Scheme 2.166. Aldol reactions catalysed by sialic acid aldolase.

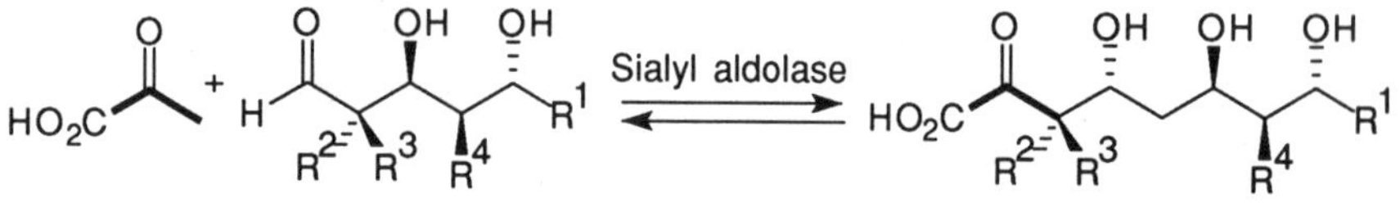

$$HO_2C-CO-CH_3 + H-CO-CHR^2-CHR^3-CHR^4-R^1 \underset{}{\overset{Sialyl\ aldolase}{\rightleftharpoons}} HO_2C-CO-CH_2-CR^2(OH)-CHR^3-CHR^4-R^1$$

R^1	R^2	R^3	R^4
CH$_2$OH	NH-Ac	H	OH[a]
CH$_2$N$_3$	NH-Ac	H	OH
CH$_2$OCH$_3$	NH-Ac	H	OH
CH$_2$O-CO-CHOH-CH$_3$	NH-Ac	H	OH
CH$_2$OAc	NH-CO-CH$_2$OH	H	OH
CH$_2$OH	H	H	OH
CH$_2$OH	OH	H	OH
CH$_2$OH	N$_3$	H	OH
H	OH	H	OH
H	H	OH	OH
CH$_2$OH	NH-Ac	H	OCH$_3$
CH$_2$OH	OH	H	H

[a] The natural substrate NeuAc.

2-Deoxyribose-5-Phosphate Aldolase

The enzyme 2-deoxyribose-5-phosphate (DER) aldolase catalyses the reversible aldol reaction of acetaldehyde and D-glyceraldehyde-3-phosphate to form 2-deoxyribose-5-phosphate. This aldolase is unique in that it is the only aldolase that condenses two aldehydes (instead of an aldehyde and a ketone) to form aldoses (Scheme 2.167). Interestingly, the enzyme (which has been overproduced [45]) shows a relaxed substrate specificity not only on the acceptor side, but also on the donor side. Thus, besides acetaldehyde it accepts also acetone, fluoroacetone and propionaldehyde as donors. Like other aldolases, it converts a variety of aldehydes in addition to D-glyceraldehyde-3-phosphate.

Other aldolases such as KDO synthetase (3-deoxy-D-*manno*-2-octulosonate-8-phosphate) [46], DAHP synthetase (3-deoxy-D-*arabino*-heptulosonate-7-phosphate) [47], serine-hydroxymethyl transferase [48] and fuculose-1-

phosphate aldolase [49] have been used for aldol reactions, but they are of limited importance since they are not commercially available.

Scheme 2.167. Aldol reactions catalysed by 2-deoxyribose-5-phosphate aldolase.

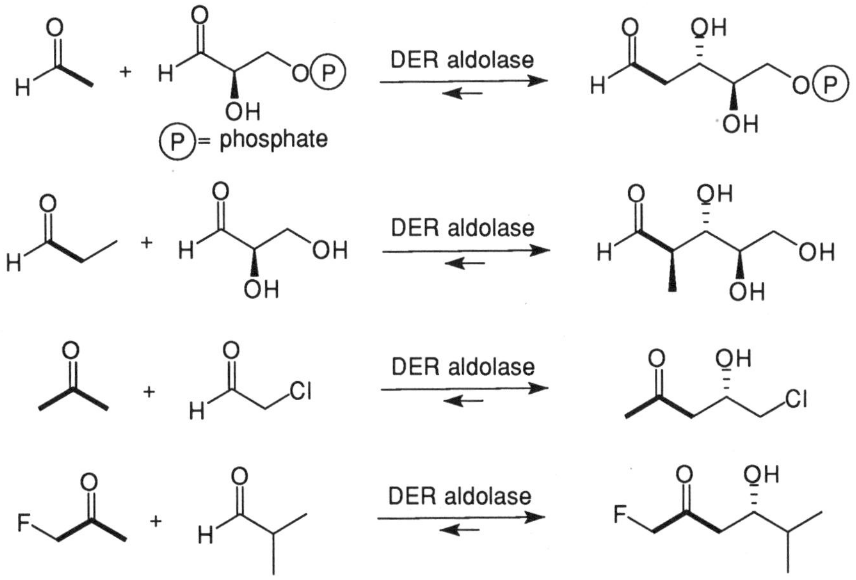

Transketolase

In the pentose pathway, transketolase (D-seduheptulose-7-phosphate: D-glyceraldehyde-3-phosphate glyco-aldehyde-transferase) catalyses the transfer of a hydroxyketo group from a ketose phosphate to an aldose phosphate [50]. It can be used to extend the chain of a variety of aldoses stereospecifically by two carbon units (i.e. a hydroxyaldehyde acyl anion equivalent). Unlike other aldolases, it requires the cofactor thiamine pyrophosphate (TPP) for the activation of the ketone.

Although the substrate specificity of transketolase has not been thoroughly explored, it appears to be a promising catalyst for use in synthesis. Hydroxypyruvate can replace the ketose, providing a C-2 hydroxyketo group after spontaneous decarboxylation. In this case, TPP is not required and the hydroxyketo group is transferred to the aldehyde acceptor in an irreversible reaction with the formation of a *threo*-diol. This method has allowed the synthesis of a number of monosaccharides on a preparative scale [51, 52].

Scheme 2.168. Aldol reactions catalysed by transketolase.

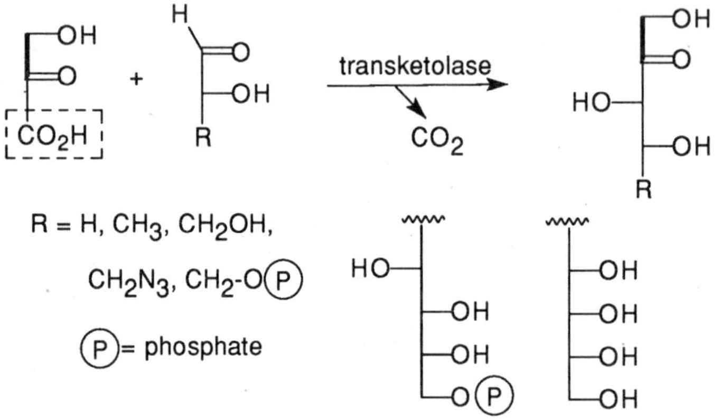

R = H, CH$_3$, CH$_2$OH, CH$_2$N$_3$, CH$_2$-O-(P)

(P) = phosphate

Interestingly, transketolase recognizes chirality in the aldehydic moiety to a greater extent than the aldolases. When (±)-3-azido-2-hydroxypropionaldehyde was chosen as acceptor, only the D-(R)-isomer reacted and the L-(S)-enantiomer remained unchanged [26].

2.4.2 Acyloin-Reactions

Closely related to the above-mentioned transketolase-reaction are acyloin condensations effected by baker's yeast [53, 54]. Although the reaction was first observed already in 1921 [55], it was only recently that the reaction pathway was elucidated in detail [56]. As shown in Scheme 2.169, the enzymic system involved in this reaction is pyruvate decarboxylase. The C$_2$-unit (an acyl anion equivalent) originates from the decarboxylation of pyruvate and is transferred to the *si*-face of the aldehydic substrate to form an (R)-α-hydroxyketone (acyloin). The Mg^{2+}-dependent reaction involves an intermediate in which the aldehyde is bound to the cofactor thiamine pyrophosphate. Since pyruvate decarboxylase accepts α-ketoacids other than pyruvate, C$_2$- through C$_4$-equivalents can be transferred in this way [57-59]. The resulting acyloin can be reduced in a subsequent step by dehydrogenase enzymes to give the *erythro*-diol. The latter reaction is a common feature of baker's yeast and the stereochemistry is guided by Prelog's rule (see Section 2.2.3). The optical purity of the diols is usually better than 90%.

There is also some tolerance by the enzyme system with respect to the structure of the aldehyde. Aromatic aldehydes, such as benzaldehyde, 2-

2.4 Formation of Carbon-Carbon Bonds

chlorobenzaldehyde, anisaldehyde [60] and *o*- and *p*-tolylaldehyde [61] are readily accepted, but *p*-chloro- and *p*-nitrobenzaldehyde are not. More significantly from a synthetic viewpoint, α,β-unsaturated aromatic and aliphatic aldehydes can be converted into the corresponding chiral diols [62, 63]. After protection they can be transformed into useful chiral carbonyl compounds with wide synthetic applications as intermediates in natural product synthesis [64-67]. For example, cinnamaldehyde and α-methylcinnamaldehyde gave the corresponding (2*S*,3*R*)-diols in 85-95% enantiomeric excess [68]. Structural requirements involving the α-position of α,β-unsaturated aldehydes tend to be a limiting factor. Although α-methyl- and α-bromo derivatives are readily converted, α-ethyl- and α-propyl substituted aldehydes are not substrates for the acyloin reaction. Furthermore, α-methylcrotonaldehyde was not accepted; however, when the terminal position was oxidized, as in ethyl 3-methyl-4-oxocrotonate, up to 35% of the diol was obtained.

Scheme 2.169. Acyloin reactions catalysed by baker's yeast.

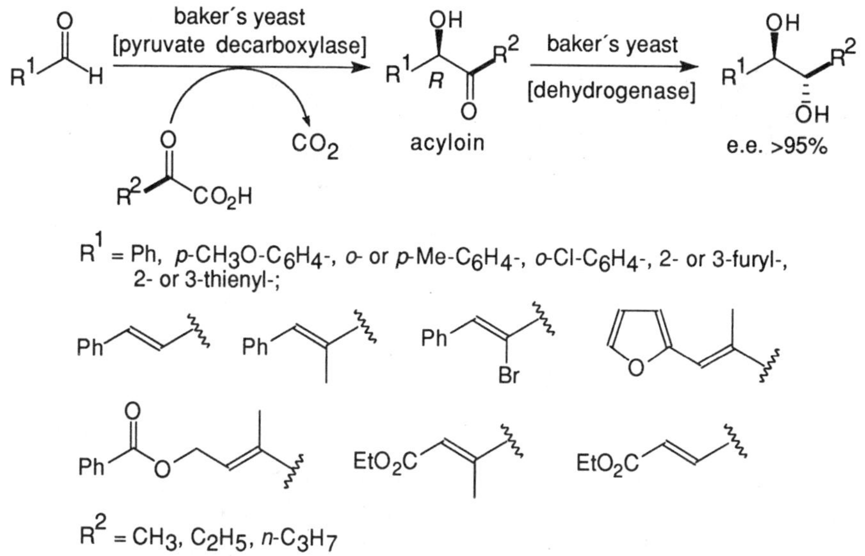

R^1 = Ph, *p*-CH$_3$O-C$_6$H$_4$-, *o*- or *p*-Me-C$_6$H$_4$-, *o*-Cl-C$_6$H$_4$-, 2- or 3-furyl-, 2- or 3-thienyl-;

R^2 = CH$_3$, C$_2$H$_5$, *n*-C$_3$H$_7$

It must be mentioned, however, that the yields of chiral diols are usually in the range of 10 to 35%, but this is offset by the ease of the reaction and the cheapness of the reagents used. Depending on the substrate structure, the reduction of the aldehyde to give the corresponding primary alcohol (catalysed by dehydrogenases, see Section 2.2.3) and saturation of the α,β-double bond (catalysed by enoate reductases, see Section 2.2.4) are the major competing reactions. This situation may sometimes be turned to advantage. As depicted in Scheme 2.170, α-methyl-β-(2-furyl)acrolein, when fermented with baker's yeast, gave the reduced acyloin in 20% yield with about 10% of the reduced furyl alcohol. These two products were then linked together in a synthesis of the C_{14} chromanyl moiety of α-tocopherol (vitamin E) [69].

Scheme 2.170. Synthesis of α-tocopherol using baker's yeast.

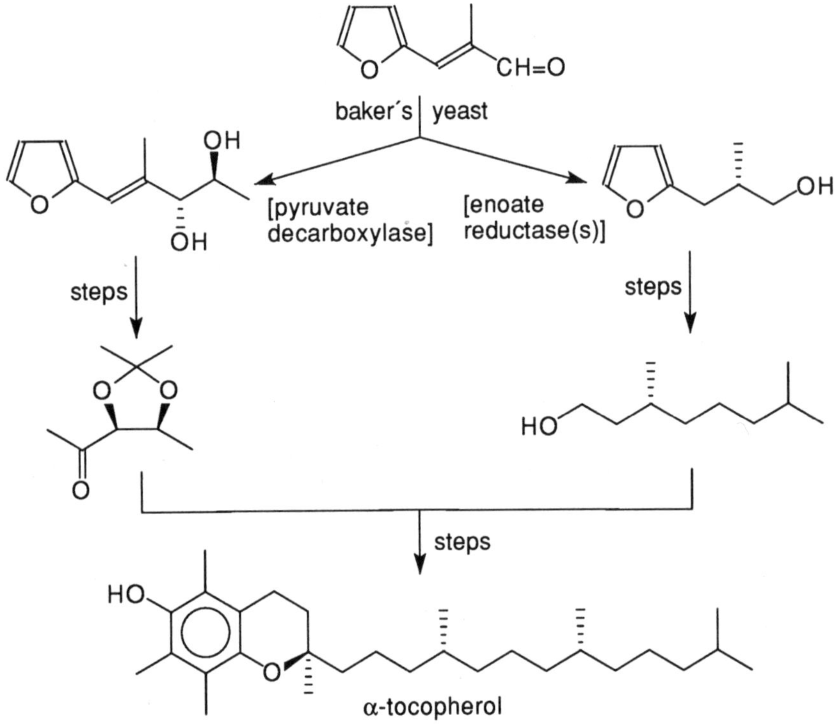

The ability to perform an acyloin reaction has also been found in other microorganisms [70, 71], but as yet they are of limited use for organic synthesis in view of the availability and ease of handling of baker's yeast.

2.4.3 Michael-Type Additions

A variant of the normal acyloin condensation involving a Michael-type addition, mediated by baker's yeast, was observed when trifluoroethanol was added to the medium (Scheme 2.171) [72]. Although the exact mechanism of this reaction has not yet been elucidated, it presumably proceeds via the

2.4 Formation of Carbon-Carbon Bonds

Scheme 2.171. Michael addition catalysed by baker's yeast.

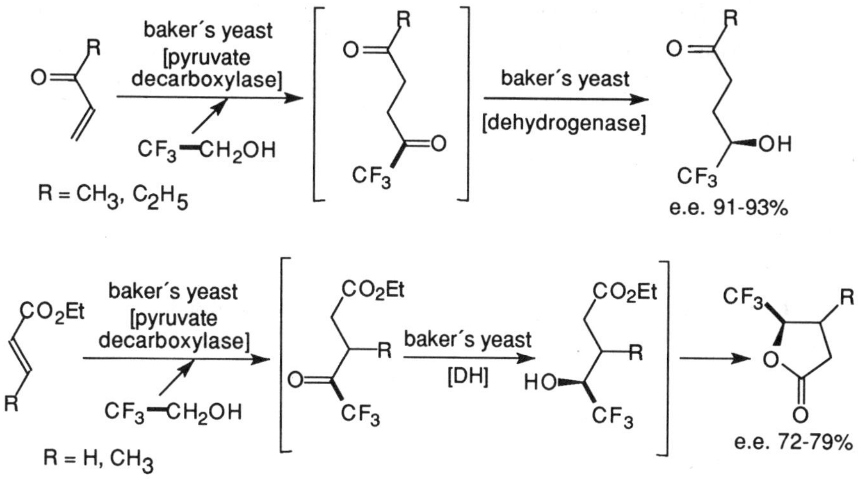

Scheme 2.172. Asymmetric Michael addition catalysed by hydrolytic enzymes.

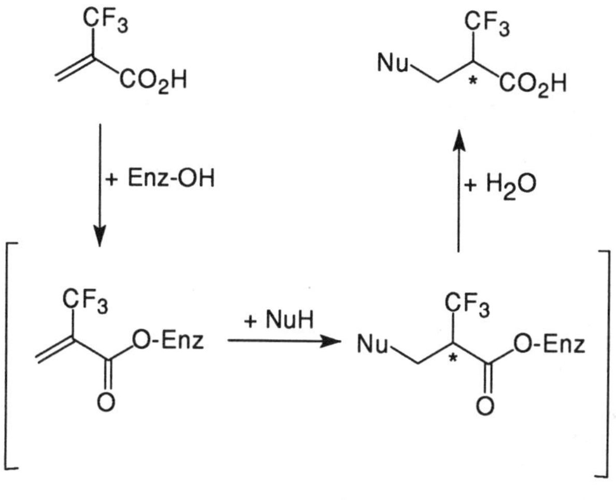

Nucleophile	Enzyme	e.e. [%]
H_2O	*Candida cylindracea* lipase	70
Et_2NH	*Candida cylindracea* lipase	71
H_2O	pig liver esterase	60
Et_2NH	pig liver esterase	69
PhSH	pig liver esterase	50

following sequence: trifluoroethanol is enzymatically oxidized to trifluoroacetaldehyde before formation of the putative thiamine pyrophosphate adduct. Then the acyl anion equivalent is added across the C=C double bond of the α,β-unsaturated carbonyl compound to form trifluoromethyl ketones as intermediates. The latter are stereoselectively reduced by dehydrogenase(s) to form chiral trifluoromethyl carbinols or the corresponding lactones, respectively. Absolute configurations were not specified, but are probably (R) as predicted by Prelog's rule.

Stereospecific Michael addition reactions may be also catalysed by hydrolytic enzymes (Scheme 2.172). When α-trifluoromethyl propenoic acid was subjected to the action of various proteases, lipases and esterases in the presence of a nucleophile (NuH), such as water, amines and thiols, chiral propanoic acids were obtained in moderate optical purity [73]. The reaction mechanism probably involves the formation of an acyl-enzyme intermediate. The latter (being more highly electrophilic than the carboxylate) undergoes a stereoselective Michael addition by the nucleophile, directed by the asymmetric environment of the enzyme.

References

1. Evans DA, Nelson JV, Taber TR (1982) Topics Stereochem. 13: 1
2. Heathcock CH (1984) Asymm. Synthesis 3: 111
3. Toone EJ, Simon ES, Bednarski MD, Whitesides, GM (1989) Tetrahedron 45: 5365
4. Wong C-H (1990) Chemtracts - Organic Chemistry 3: 91
5. Wong C-H (1990) Aldolases in Organic Synthesis, In: Biocatalysis, Abramowicz DA (ed) p 319, Van Nostrand Reinhold, New York
6. Meyerhof O, Lohmann K (1934) Biochem. Z. 271: 89
7. Horecker L, Tsolas O, Lai CY (1972) In: The Enzymes, Boyer PD (ed) vol 7, p 213, Academic Press, New York
8. Brockamp HP, Kula MR (1990) Appl. Microbiol. Biotechnol. 34: 287
9. von der Osten CH, Sinskey AJ, Barbas III CF, Pederson RL, Wang Y-F, Wong C-H (1989) J. Am. Chem. Soc. 111: 3924
10. Bednarski MD, Simon ES, Bischofberger N, Fessner W-D, Kim M-J, Lees W, Saito T, Waldmann H, Whitesides GM (1989) J. Am. Chem. Soc. 111: 627
11. Jones JKN, Sephton HH (1960) Can. J. Chem. 38: 753
12. Jones JKN, Kelly RB (1956) Can. J. Chem. 34: 95
13. Horecker BL, Smyrniotis PZ (1952) J. Am. Chem. Soc. 74: 2123
14. Huang PC, Miller ON (1958) J. Biol. Chem. 330: 805
15. Straub A, Effenberger F, Fischer P (1990) J. Org. Chem. 55: 3926
16. Kajimoto T, Chen L, Liu KK-C, Wong C-H (1991) J. Am. Chem. Soc. 113: 6678
17. Wong C-H, Mazenod FP, Whitesides GM (1983) J. Org. Chem. 3493
18. Bednarski MD, Waldmann HJ, Whitesides GM (1986) Tetrahedron Lett. 5807
19. Jones JKN, Matheson NK (1959) Can. J. Chem. 37: 1754
20. Gorin PAJ, Hough L, Jones JKN (1953) J. Chem. Soc. 2140
21. Lehninger AL, Sice J (1955) J. Am. Chem. Soc. 77: 5343
22. Charalampous FC (1954) J. Biol. Chem. 211: 249
23. Simon ES, Plante R, Whitesides GM (1989) Appl. Biochem. Biotechnol. 22: 169
24. Bischofberger N, Waldmann H, Saito T, Simon ES, Lees W, Bednarski MD, Whitesides GM (1988) J. Org. Chem. 53: 3457
25. Pederson RL, Esker J, Wong C-H (1991) Tetrahedron 47: 2643

26. Wong C-H, Whitesides GM (1983) J. Org. Chem. 48: 3199
27. Effenberger F, Straub A (1987) Tetrahedron Lett. 1641
28. Crans DC, Whitesides GM (1985) J. Am. Chem. Soc. 107: 7019
29. Lagunas R, Sols A (1968) FEBS Lett. 1: 32
30. Drueckhammer DG, Durrwachter JR, Pederson RL, Crans DC, Daniels L, Wong C-H (1989) J. Org. Chem. 54: 70
31. Pederson RL, Kim M-J, Wong C-H (1988) Tetrahedron Lett. 4645
32. Ziegler T, Straub A, Effenberger F (1988) Angew. Chem., Int. Ed. Engl. 27: 716
33. Schultz M, Waldmann H, Kunz H, Vogt W (1990) Liebigs Ann. Chem. 1010
34. Durrwachter JR, Wong C-H (1988) J. Org. Chem. 53: 4175
35. Durrwachter JR, Drueckhammer DG, Nozaki K, Sweers HM, Wong C-H (1986) J. Am. Chem. Soc. 108: 7812
36. Simon ES, Bednarski MD, Whitesides GM (1988) J. Am. Chem. Soc. 110: 7159
37. Uchida Y, Tsukada Y, Sugimori T (1985) Agric. Biol. Chem. 49: 181
38. Schauer R (1985) Trends Biochem. Sci. 10: 357
39. Ota Y, Shimosaka M, Murata K, Tsudaka Y, Kimura A (1986) Appl. Microbiol. Biotechnol. 24: 386
40. Brunetti P, Jourdian GW, Roseman S (1962) J. Biol. Chem. 237: 2447
41. Brossmer R, Rose U, Kasper D, Smith TL, Grasmuk H, Unger FM (1980) Biochem. Biophys. Res. Commun. 96: 1282
42. Augé C, David S, Gautheron C, Malleron A, Cavayre B (1988) New J. Chem. 12: 733
43. Kim M-J, Hennen WJ, Sweers HM, Wong C-H (1988) J. Am. Chem. Soc. 110: 6481
44. Augé C, Gautheron C, David S, Malleron A, Cavayé B, Bouxom B (1990) Tetrahedron 46: 201
45. Barbas III CF, Wang Y-F, Wong C-H (1990) J. Am. Chem. Soc. 112: 2013
46. Bednarski MD, Crans DC, DiCosmio R, Simon ES, Stein PD, Whitesides GM, Schneider M (1988) Tetrahedron Lett. 427
47. Reimer LM, Conley DL, Pompliano DL, Frost JW (1986) J. Am. Chem. Soc. 108: 8010
48. Lotz BT, Gasparski CM, Peterson K, Miller MJ (1990) J. Chem. Soc., Chem. Commun. 1107
49. Ozaki A, Toone EJ, von der Osten CH, Sinskey AJ, Whitesides GM (1990) J. Am. Chem. Soc. 112: 4970
50. Thunberg L, Backstrom G, Lindahl U (1982) Carbohydr. Res. 100: 393
51. Bolte J, Demuynck C, Samaki H (1987) Tetrahedron Lett. 5525
52. Mocali A, Aldinucci D, Paoletti F (1985) Carbohydr. Res. 143: 288
53. Fuganti C, Grasselli P (1989) Baker's Yeast-Mediated Synthesis on Natural Products, In Biocatalysis in Agricultural Biotechnology, Whitaker JR, Sonnet PE (eds), p 359, ACS Symp. Ser., Am. Chem. Soc., Washington DC
54. Fuganti C, Grasselli P (1977) Chem. Ind. 983
55. Neuberg C, Hirsch J (1921) Biochem. Z. 115: 282
56. Crout DHG, Dalton H, Hutchinson DW, Miyagoshi M (1991) J. Chem. Soc., Perkin Trans. I, 1329
57. Fuganti C, Grasselli P, Poli G, Servi S, Zorzella A (1988) J. Chem. Soc., Chem. Commun. 1619
58. Fuganti C, Grasselli P, Servi S, Spreafico F, Zirotti C (1984) J. Org. Chem. 49: 4087
59. Suomalainen H, Linnahalme T (1966) Arch. Biochem. Biophys. 114: 502
60. Neuberg C, Liebermann L (1921) Biochem. Z. 121: 311
61. Behrens M, Ivanoff N (1926) Biochem. Z. 169: 478
62. Fronza G, Fuganti C, Grasselli P, Poli G, Servi S (1990) Biocatalysis 3: 51
63. Fronza G, Fuganti C, Majori L, Pedrocchi-Fantoni G, Spreafico F (1982) J. Org. Chem. 47: 3289
64. Fuganti C, Servi S (1988) Bioflavour '87, p 555, Scheier P (ed) de Gruyter, Berlin
65. Fuganti C (1986) Baker's Yeast Mediated Preparation of Carbohydrate-like Chiral Synthons, In Enzymes as Catalysts in Organic Synthesis, Schneider MP (ed) p 3, NATO ASI Series, vol 178, Reidel, Dordrecht, NL

66. Fuganti C, Grasselli P (1985) Stereochemistry and Synthetic Applications of Products of Fermentation of α,β-Unsaturated Aromatic Aldehydes by Baker´s Yeast, In Enzymes in Organic Synthesis, Porter R, Clark S (eds) p 112, Pitman, London
67. Fuganti C, Grasselli P, Servi S (1983) J. Chem. Soc., Perkin Trans. I, 241
68. Fuganti C, Grasselli P, Spreafico F, Zirotti C (1984) J. Org. Chem. 49: 543
69. Fuganti C, Grasselli P (1982) J. Chem. Soc., Chem. Commun. 205
70. Abraham WR, Stumpf B (1987) Z. Naturforsch. 559
71. Bringer-Meyer S, Sahm H (1988) Biocatalysis 1: 321
72. Kitazume T, Ishikawa N (1984) Chem. Lett. 1815
73. Kitazume T, Ikeya T, Murata K (1986) J. Chem. Soc., Chem. Commun. 1331

2.5 Addition and Elimination Reactions

The asymmetric addition of small molecules such as water or ammonia to carbon-carbon double bonds, or hydrogen cyanide to C=O bonds is catalysed by lyases. During such a reaction one (or, depending on the substitution pattern of the substrate, two) chiral centres are created from a prochiral substrate. With the exception of cyanohydrin formation, the synthetic potential of these reactions is restricted due to the fact that lyases generally exhibit narrow specificities and only allow minor structural variations of their natural substrate(s).

2.5.1 Cyanohydrin Formation

Oxynitrilase enzymes catalyse the asymmetric addition of hydrogen cyanide onto a carbonyl group of an aldehyde or a ketone thus forming a chiral cyanohydrin [1, 2]. The latter are versatile starting materials for the synthesis of several types of compounds. Hydrolysis or alcoholysis of the nitrile group affords chiral α-hydroxyacids or -esters. Grignard-reactions provide acyloins [3], which in turn can be reduced to give vicinal diols [4]. Alternatively, the cyanohydrins can be subjected to reductive amination to afford chiral ethanolamines [5]. Furthermore, a number of chiral cyanohydrins are also of interest as they constitute the alcohol moieties of several commercial pyrethroid insecticides [6].

Scheme 2.173. Syntheses starting from chiral cyanohydrins.

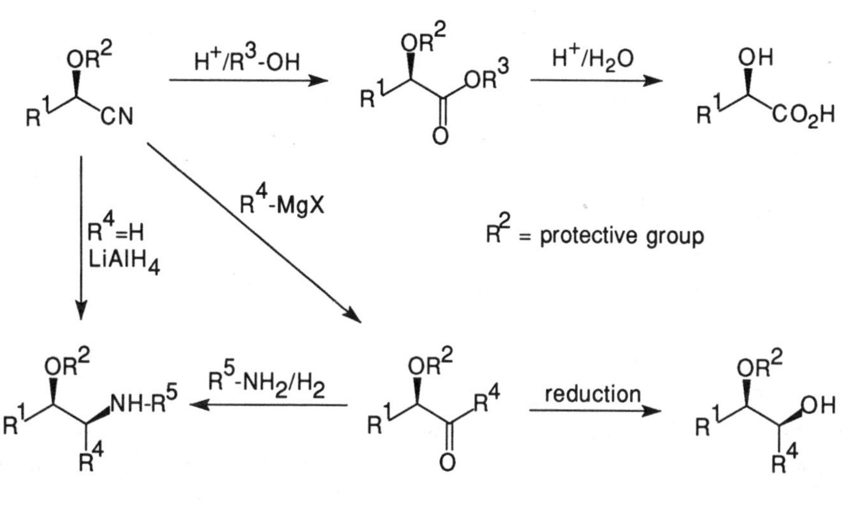

Since only a single enantiomer is produced during the reaction - asymmetrisation of a prochiral substrate - the availability of different enzymes of opposite stereochemical preference is of importance to gain access to both (*R*)- and (*S*)-cyanohydrins (Scheme 2.174). A number of oxynitrilases of opposite stereochemical preference can be isolated from cyanogenic plants [7].

(*R*)-Specific enzymes are obtained predominantly from the *Rosacea* family (almond, plum, cherry, apricot) and they have been thoroughly investigated [8-10]. They contain FAD as a prostethic group located near (but not in) the active site, but this moiety does not participate in a redox-reaction. The following set of rules for the substrate-acceptance of (*R*)-oxynitrilase was recently postulated [11].

- The best substrates are aromatic aldehydes, which may be substituted in the *meta*- or *para*-position; heteroaromatics such as furan and thiophene derivatives are well accepted [12-15].
- Straight-chain aliphatic aldehydes and α,β-unsaturated aldehydes are transformed as long as they are not longer than six carbon atoms; the α-position may be substituted with a methyl group.
- Methyl (but not ethyl) ketones are transformed into cyanohydrins [16].

(*S*)-Oxynitrilases have been used only recently [17, 18]. They are found in *Sorghum bicolor* [19] (millet), *Sambucus nigra* [20] (elder), *Ximenia americana* [21] (sandalwood), flax, clover and cassava. They do not contain FAD and they exhibit a narrow substrate tolerance, as aliphatic aldehydes are usually not accepted. The reaction rates and optical purities are sometimes lower than those which are obtained when the (*R*)-enzyme is used.

Two particular problems which are often encountered in oxynitrilase-catalysed reactions are the non-specific non-enzymatic formation of racemic cyanohydrin and racemisation of the product due to the equilibrium of the reaction. As a result, the optical purity of the product is depleted. Bearing in mind that both the chemical formation and the racemisation of cyanohydrins are pH-dependent and require water, three different techniques have been developed in order to suppress the reduction of the optical purity of the product.

- Adjusting the pH of the medium to a value below 3.5.
- Lowering the water-activity of the medium [22] by using water-miscible organic cosolvents such as ethanol or methanol. Alternatively, the reaction can be carried out in a biphasic aqueous-organic system or in a monophasic organic solvent (e.g. ethyl acetate) which contains only traces of water to preserve the enzyme´s activity.
- Most recently, acetone cyanohydrin has been proposed as a donor for hydrogen cyanide [23] instead of the in situ formation of hydrogen cyanide in

2.5 Addition and Elimination Reactions

a *trans*-cyanation reaction. Using this technique, the chemical cyanohydrin formation is negligable due to the low concentration of free hydrogen cyanide. Additionally, the use of the hazardous hydrogen cyanide is avoided.

Scheme 2.174. Asymmetric cyanohydrin formation.

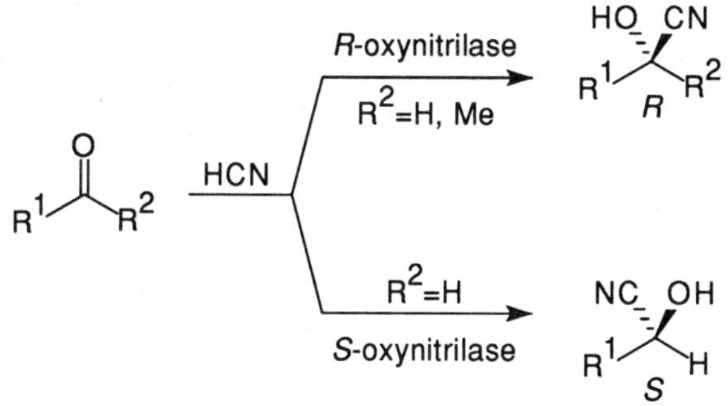

R¹	R²	Configuration	e.e. [%]
Ph-	H	R	94
Ph-CH₂-	H	R	40
p-MeO-Ph-	H	R	93
n-C₃H₇-	H	R	92-96
n-C₅H₁₁-	H	R	67
t-Bu-	H	R	73
(*E*)-CH₃-CH=CH-	H	R	69
2-furyl-	H	R	98
C₂H₅-	Me	R	76
n-C₄H₉-	Me	R	98
(CH₃)₂CH-(CH₂)₂-	Me	R	98
CH₂=CH(CH₃)-	Me	R	94
Cl-(CH₂)₃-	Me	R	84
Ph-	H	S	96-97
m-HO-Ph-	H	S	91-98
p-HO-Ph-	H	S	94-99
m-Me-Ph-	H	S	96
p-Me-Ph-	H	S	78
m-MeO-Ph-	H	S	89
m-PhO-Ph-	H	S	96
m-Br-Ph-	H	S	92
p-Cl-Ph-	H	S	54

2.5.2 Addition of Water and Ammonia

Fumarase - a 'dehydrase' - catalyses the stereospecific addition of water onto carbon-carbon double bonds conjugated with a carbonyl group (usually a carboxylic acid) [24]. The analogous addition of ammonia is catalysed by 3-methylaspartase [25]. Both reactions are mechanistically related and take place in a *trans*- or *anti*-manner [26, 27], with protonation occurring from the *re*-side (Scheme 2.175). These close similarities may be explained by the fact that both enzymes show a remarkable degree of amino acid homology [28]. Within this group of enzymes, substrate-tolerance is rather narrow, but the selectivities observed are exceptionally high.

Scheme 2.175. Lyase-catalysed addition of water and ammonia onto C=C bonds.

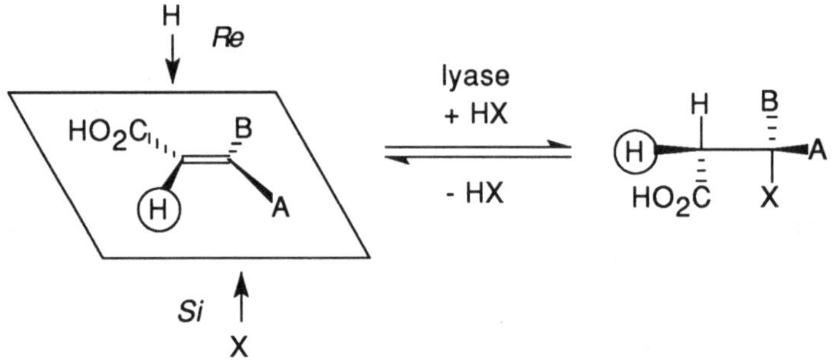

Scheme 2.176. Fumarase-catalysed addition of water.

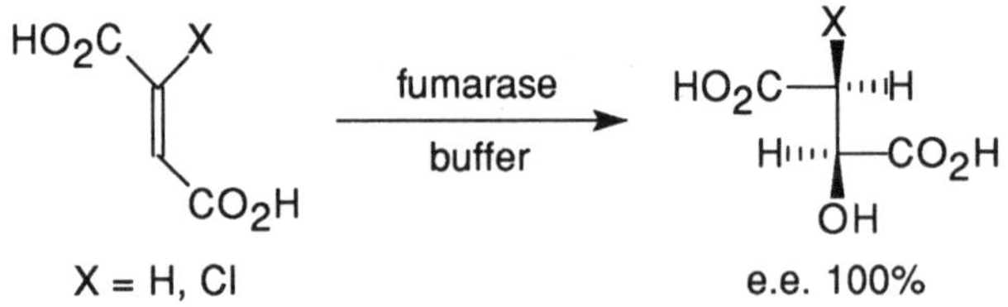

The addition of water onto fumaric acid leads to malic acid derivatives (Scheme 2.176). Only fumarate and chlorofumaric acid are well accepted, the corresponding bromo-, iodo- and methyl derivatives are transformed at exceedingly low rates. Fluoro- and 2,3-fluorofumaric acid are accepted but their transformation suffers from decomposition reactions of the first-formed

2.5 Addition and Elimination Reactions

fluoromalic acid. Replacement of one of the carboxylic groups or changing the stereochemistry of the double bond from (*E*) to (*Z*) is not tolerated by the enzyme [29].

An unexpected reaction catalysed by baker's yeast with potential synthetic utility was reported recently (Scheme 2.177) [30]. During an attempt to asymmetrically hydrogenate substituted crotonaldehyde derivatives, a lyase-catalysed addition of water was observed as a parallel reaction. Further investigation revealed that a 4-benzyloxy- or benzoyloxy substituent is required to induce this particular transformation [31].

Scheme 2.177. Addition of water by fermenting baker's yeast.

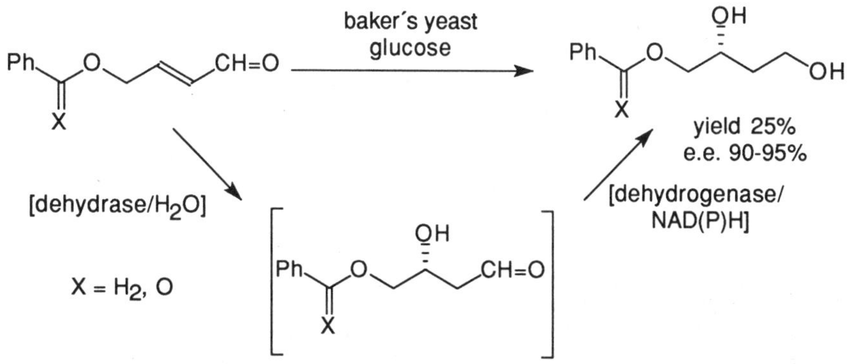

The capacity of microbial cells of different origin to perform an asymmetric hydration of C=C bonds is only poorly investigated but they show a promising synthetic potential. For instance, *Fusarium solani* cells are capable of hydrating the 'inner' (*E*)-double bond of terpene alcohols (e.g. nerolidol) or -ketones (e.g. geranyl acetone) in a highly selective manner [32]. However, side-reactions such as hydroxylation, ketone-reduction or degradation of the carbon-skeleton represent a major drawback.

Enzymic amination of fumaric acid, and derivatives thereof, leads to the formation of aspartic acid (Scheme 2.178) [33]. Some structural variations are tolerated by 3-methylaspartase. The methyl group in the natural substrate may be replaced by a chlorine atom or by other alkyl moieties, as long as they are small [34]. The fluoro- and the iodo-analogue (X = F, I) are not good substrates. Although the bromo-derivative is accepted, it irreversibly inhibits the enzyme [35].

Scheme 2.178. Aspartase-catalysed addition of ammonia.

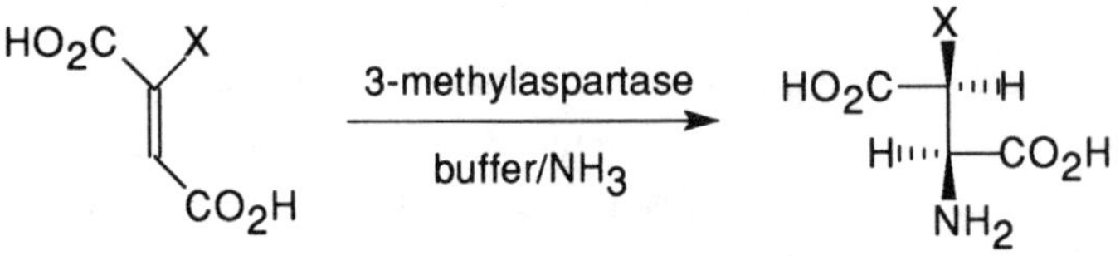

X	Yield [%]
H	90
Cl-	60
CH_3-	61
C_2H_5-	60
$(CH_3)_2CH$-	54
n-C_3H_7-	49
n-C_4H_9-	0

References

1. Smitskamp-Wilms E, Brussee J, van der Gen A, van Scharrenburg GJM, Sloothaak JB (1991) Rec. Trav. Chim. Pays-Bas 110: 209
2. Rosenthaler I (1908) Biochem. Z. 14: 238
3. Krepski LR, Jensen KM, Heilmann SM, Rasmussen JK (1986) Synthesis 301
4. Jackson WR, Jacobs HA, Matthews BR, Jayatilake GS, Watson KG (1990) Tetrahedron Lett. 1447
5. Brussee J, van Benthem RATM, Kruse CG, van der Gen A (1990) Tetrahedron: Asymmetry 1: 163
6. Matsuo T, Nishioka T, Hirano M, Suzuki Y, Tsushima K, Itaya N, Yoshioka H (1980) Pestic. Sci. 202
7. Nahrstedt A (1985) Pl. Syst. Evol. 150: 35
8. Hochuli E (1983) Helv. Chim. Acta 66: 489
9. Jorns MS, Ballenger C, Kinney G, Pokora A, Vargo D (1983) J. Biol. Chem. 258: 8561
10. Becker W, Pfeil E (1964) Naturwissensch. 51: 193
11. Brussee J, Loos WT, Kruse CG, van der Gen A (1990) Tetrahedron 46: 979
12. Effenberger F, Ziegler T, Förster S (1987) Angew. Chem., Int. Ed. Engl. 26: 458
13. Ziegler T, Hörsch B, Effenberger F (1990) Synthesis 575
14. Becker W, Pfeil E (1966) J. Am. Chem. Soc. 88: 4299
15. Becker W, Pfeil E (1966) Biochem. Z. 346: 301
16. Effenberger F, Hörsch B, Weingart F, Ziegler T, Kühner S (1991) Tetrahedron Lett. 2605
17. Niedermeyer U, Kula M-R (1990) Angew. Chem., Int. Ed. Engl. 29: 386
18. Effenberger F, Hörsch B, Förster S, Ziegler T (1990) Tetrahedron Lett. 1249
19. Seely MK, Criddle RS, Conn EE (1966) J. Biol. Chem. 241: 4457
20. Bourquelot E, Danjou E (1905) J. Pharm. Chim. 22: 219
21. Kuroki GW, Conn EE (1989) Proc. Natl. Acad. Sci. USA 86: 6978
22. Wehtje E, Adlercreutz P, Mattiasson B (1990) Biotechnol. Bioeng. 36: 39
23. Ognyanov VI, Datcheva VK, Kyler KS (1991) J. Am. Chem. Soc. 113: 6992
24. Hill RL, Teipel JW (1971) Fumarase and Crotonase, In: The Enzymes, Boyer P (ed) vol 5, p 539, Academic Press, New York

25. Hanson KR, Havir EA (1972) The Enzymic Elimination of Ammonia, In: The Enzymes, Boyer P (ed) vol. 7, p. 75, Academic Press, New York
26. Botting NP, Akhtar M, Cohen MA, Gani D (1987) J. Chem. Soc., Chem. Commun. 1371
27. Nuiry II, Hermes JD, Weiss PM, Chen C, Cook PF (1984) Biochemistry 23: 5168
28. Woods SA, Miles JS, Roberts RE, Guest JR (1986) Biochem. J. 237: 547
29. Findeis MA, Whitesides GM (1987) J. Org. Chem. 52: 2838
30. Fronza G, Fuganti C, Grasselli P, Poli G, Servi S (1988) J. Org. Chem. 53: 6154
31. Fronza G, Fuganti C, Grasselli P, Poli G, Servi S (1990) Biocatalysis 3: 51
32. Abraham W-R, Arfmann H-A (1989) Appl. Microbiol. Biotechnol. 32: 295
33. Chibata I, Tosa T, Sato T (1976) Methods Enzymol. 44: 739
34. Akhtar M, Botting NP, Cohen MA, Gani D (1987) Tetrahedron 43: 5899
35. Akhtar M, Cohen MA, Gani D (1986) J. Chem. Soc., Chem. Commun. 1290

2.6 Glycosyl-Transfer Reactions

Oligosaccharides and polysaccharides are important classes of naturally occurring compounds [1]. They play a vital rôle in (i) intracellular migration and secretion of glycoproteins, (ii) cell-cell interactions, (iii) oncogenesis and (iv) interaction of cell-surfaces with pathogens [2-4]. The ready availability of such oligosaccharides of well-defined structure is critical for the investigation of possible applications as well as fundamental studies. Isolation of these materials from natural sources is a complex task and is not economical on a large scale due to the low concentration of the structures of interest in the complex mixtures of carbohydrates obtained from natural sources. Chemical synthesis of complex oligosaccharides is perhaps the greatest challenge facing synthetic organic chemistry since it requires many protection and deprotection steps which result in low yields. Moreover, stereospecific chemical synthesis of oligosaccharides, in particular of the important α-sialylated structures, is difficult. In this context, biocatalysts are attractive as they allow the regio- and stereospecific synthesis of a target structure with a minimum of reaction and purification steps [5-9].

There are two groups of enzymes which can be used for the synthesis of oligosaccharides. *Glycosyl transferases* are the biocatalysts which are responsible for the build-up of oligosaccharides in vivo. They require that the sugar monomer which is to be linked to the growing oligosaccharide chain is activated by phosphorylation prior to the condensation step. The activating group on the anomeric centre (LG, see Scheme 2.179) is either a nucleoside phosphate (usually a *di*phosphate in the Leloir-pathway [10]) or a simple phosphate (in non-Leloir pathway enzymes [11]). Glycosyl transferases are highly specific with respect to their substrate(s) and the nature of the glycosidic bond to be formed.

On the other hand, *glycosidases*, belonging to the class of hydrolytic enzymes (therefore also termed 'glycohydrolases'), have a catabolic function in vivo as they cleave glycosidic linkages to form mono- or oligosaccharides. Consequently, they are, in general, less specific when compared to glycosyl transferases. Both of these types of enzymes can be used for the synthesis of oligosaccharides and related compounds.

2.6.1 Glycosyl Transferases

Three fundamental steps constitute the Leloir-pathway: activation, transfer and modification [12]. These steps represent biological solutions to the

2.6 Glycosyl-Transfer Reactions

problems also faced by chemists, i.e. chemical activation of sugars, regio- and stereospecific formation of glycosidic linkages and elaboration of the products. In the first step, a sugar is phosphorylated by a kinase to give a sugar-1-phosphate. This activated sugar subsequently reacts with a nucleoside triphosphate (usually uridine triphosphate, UTP) under catalysis of a nucleoside transferase and forms a chemically activated nucleoside diphosphate sugar (NDP, see Scheme 2.179). These key nucleoside diphosphate sugars constitute the 'donors' in the subsequent condensation with the 'acceptors' (a mono- or oligosaccharide, a protein or a lipid). The latter step is catalysed by a glycosyl transferase. To ensure proper functioning of the cell, a large number of glycosyl transferases seem to be neccessary. Each NDP-sugar requires a distinct group of glycosyl transferases; more than one hundred glycosyl transferases have been identified to date and each one appears to catalyse the formation of a unique glycosidic linkage [13]. It is therefore not surprising that the exact details on the specificity of the various glycosyl transferases are not known.

Scheme 2.179. Synthesis of oligosaccharides by glycosyl transferases.

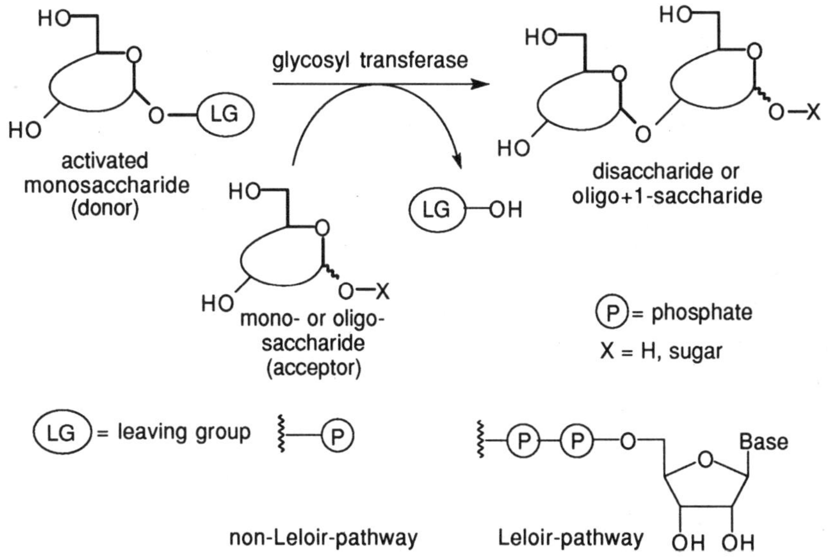

Chemists have begun to apply enzymes of the Leloir-pathway to the synthesis of oligosaccharides. Two requirements are critical for the success of this approach, namely the availability of the sugar nucleoside phosphates at practical costs and the availability of the glycosyl transferases. Although the first issue is being resolved for most of the common NDP-sugars utilizing phosphorylating enzymes (see Section 2.1.4), the supply glycosyl transferases remains a problem. Only few of these enzymes are commercially available; isolation is difficult, because the unstable membrane-bound proteins are present only in low concentrations [14]. Genetic engineering will have a major impact in this area. Enzymatic methods [15] of preparing the NDP-sugars have several advantages over the chemical methods [16]. For example, the NDP-sugar may be generated in situ (e.g. by epimerisation of UDP-Glc to UDP-Gal) making it possible to drive unfavourable equilibria in the required direction. Purification steps may be eliminated because the by-products of enzyme-catalysed methods do not interfere with further enzymatic steps.

The point of interest to synthetic chemists is the range of acceptors and donors that can be used in glycosyl transferase catalysed reactions. Fortunately, the specificity of glycosyl transferases is high but not absolute. UDP-Galactosyl (UDP-Gal) transferase is the best-studied transferase in terms of specificity for the acceptor sugar. It has been demonstrated that this enzyme catalyses the transfer of UDP-Gal to a remarkable range of acceptor substrates of the carbohydrate-type [17-21]. Other glycosyl transferases, although less well-studied than UDP-Gal transferase, also appear to accept a range of acceptor substrates [22-25].

Table 2.5. Glycoside synthesis using galactosyl transferase (donor = UDP-Gal).

Acceptor	Product
Glc-OH	β-Gal-(1→4)-Glc-OH
GlcNAc-OH	β-Gal-(1→4)-GlcNAc-OH
β-GlcNAc-(1→4)-Gal-OH	β-Gal-(1→4)-β-GlcNAc-(1→4)-Gal-OH
β-GlcNAc-(1→6)-Gal-OH	β-Gal-(1→4)-β-GlcNAc-(1→6)-Gal-OH
β-GlcNAc-(1→3)-Gal-OH	β-Gal-(1→4)-β-GlcNAc-(1→3)-Gal-OH

The use of the multi-enzyme systems, which arise due to the need to prepare the activated UDP-sugar in situ, is exemplified with the synthesis of N-acetyllactosamine [26] (see Scheme 2.180). Glucose-6-phosphate is isomerized to its 1-phosphate by phosphoglucomutase. Transfer of the activating group (UDP) from UTP is catalysed by UDP-glucose pyrophosphorylase liberating pyrophosphate, which is destroyed by inorganic pyrophosphatase. Then, the centre at carbon 4 is epimerized by UDP-galactose epimerase in order to drive the process away from the equilibrium. Finally, UDP-galactose is linked to N-acetylglucosamine using galactosyl transferase to yield N-acetyllactosamine. The liberated UDP is recycled back to the respective triphosphate by pyruvate kinase at the expense of phosphoenol pyruvate. The overall yield of this sequence was in the range of 70% and it was performed on a scale greater than 10g.

Scheme 2.180. Synthesis of *N*-acetyllactosamine using a six-enzyme system.

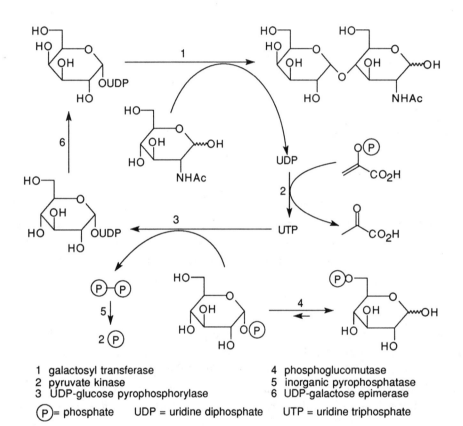

1 galactosyl transferase
2 pyruvate kinase
3 UDP-glucose pyrophosphorylase
4 phosphoglucomutase
5 inorganic pyrophosphatase
6 UDP-galactose epimerase

(P) = phosphate   UDP = uridine diphosphate   UTP = uridine triphosphate

Alternatively, oligosaccharides have also been prepared by non-Leloir routes. In this case the activated donor is a sugar-1-phosphate. For example, trehalose, one of the major storage carbohydrates in plants, fungi and insects, was synthesized from glucose and its glucose-1-phosphate using trehalose phosphorylase as the catalyst [27].

Scheme 2.181. Synthesis of trehalose via the non-Leloir pathway.

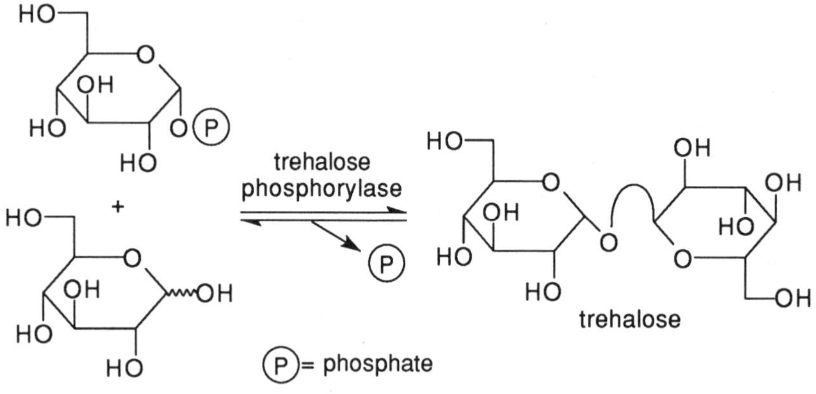

2.6.2 Glycosidases

The glycosidases (also termed glycohydrolases) cleave glycosidic bonds in the catabolism of carbohydrates. Since they are 'true' hydrolases they are independent of any cofactor. Two groups of them exist: *exo*-glycosidases only cleave terminal sugar residues, whereas *endo*-glycosidases are also able to split a carbohydrate chain in the middle. In general, the glycosidases show high (but not absolute) specificity for both the glycosyl moiety and the nature of the glycosidic linkage, but little if any specificity for the aglycon component (R, see Scheme 2.183) [28]. It has long been recognized that the nucleophile (NuH, which is water in the 'normal' hydrolytic pathway) can be replaced by other nucleophiles, such as another sugar or a primary or secondary non-natural alcohol. This allows the degradative nature of glycosyl hydrolysis to be turned into the more useful synthetic direction [29]. Major advantages of glycosidase-catalysed glycosyl transfer are that there is minimal (or zero) need for protection and that the stereochemistry at the newly formed anomeric centre can be completely controlled. However, regiocontrol with respect to the acceptor remains a problem, particularly when it is a mono- or oligo-saccharide carrying multiple hydroxy groups.

The observation that hydrolysis of glycosidic linkages leads to net retention of configuration at the anomeric centre, led to the suggestion of two possible mechanisms [30, 31].

- A double-step nucleophilic displacement sequence, with a covalently bound enzyme-glycoside intermediate, or

2.6 Glycosyl-Transfer Reactions

- an acid-catalysed cleavage involving a carbocation intermediate (analogous to the acid-catalysed cleavage of glycosidic bonds). In this case, binding of the carbocation must be arranged in such a way that attack of the incoming nucleophile is only possible from one side.

For a β-glucosidase from *Agrobacterium* the former mechanism involving a double nucleophilic displacement sequence was proven recently [32]. Thus, the carboxylate moiety of a glutamic acid residue attacks the glycoside donor forming a covalent enzyme-substrate intermediate. The latter is subsequently cleaved by another nucleophilic displacement involving the acceptor nucleophile, affording a net retention of configuration at the anomeric centre.

Scheme 2.182. Mechanism of β-glucosidase from *Agrobacterium*.

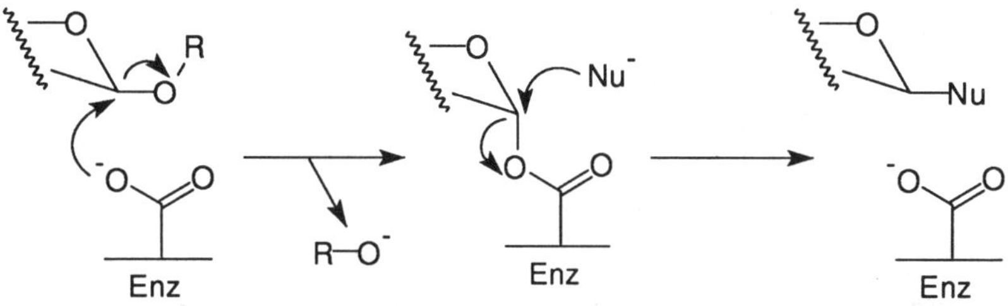

Glycosidases can be used for the synthesis of glycosides in two modes. The first (thermodynamic) approach is the direct reversal of glycoside hydrolysis by shifting the equilibrium of the reaction from hydrolysis to synthesis. This procedure uses a free monosaccharide as substrate and has been referred to as 'direct glycosylation' or 'reverse hydrolysis' [33-36]. Since the equilibrium constant for this reaction lies strongly in favour of hydrolysis, high concentrations of both the monosaccharide and the nucleophilic component (carbohydrate or alcohol) must be used. Yields in these reactions are generally low and reaction mixtures comprising of up to 75% by weight of thick syrups are not amenable to scale-up. Other methods to improve such systems make use of aqueous-organic two-phase systems [37, 38] and polyethylene glycol modified glycosidases [39]. Interestingly, the former approach was reported as long ago as 1913 [40]. Alternatively, the glycoside so formed can be removed from the reaction medium by selective adsorption [41].

Scheme 2.183. Glycoside synthesis using glycosidases (glycohydrolases).

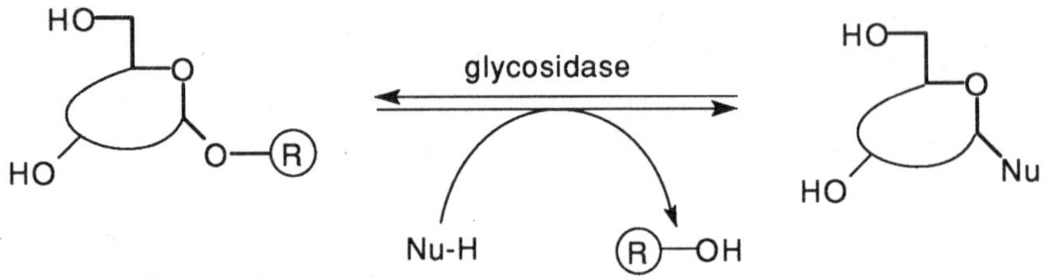

Nu = nucleophile: primary or secondary alcohol, sugar
Direct glycosylation: R = H
Transglycosylation: R = F, glycosyl, Ph, p-Ph-NO$_2$

The second route - the kinetic approach - utilizes a preformed activated glycoside, which is hydrolysed by an appropriate glycosidase; this process is referred to as 'transglycosylation'. The enzyme-glycoside intermediate is then trapped by a nucleophile other than water to yield a new glycoside. Several species have been used as the activated donor glycoside, including disaccharides, glycosyl fluorides [42, 43], -azides and aryl (usually p-nitrophenyl) glycosides. Transglycosylation gives higher yields and is generally the method of choice [44-46]. Since the product of the reaction is also a substrate for the enzyme, the success of this procedure as a preparative method depends on the rate of hydrolysis of the product being slower than that of the glycosyl donor. In practice this condition can be attained readily.

Table 2.6. Transglycosylation catalysed by glycosidases.

Enzyme	Glycoside	Nucleophile	Product(s)
α-galactosidase	α-Gal-O-p-Ph-NO$_2$	α-Gal-O-allyl	α-Gal-(1→3)-α-Gal-O-allyl
α-galactosidase	α-Gal-O-p-Ph-NO$_2$	α-Gal-O-Me	α-Gal-(1→3)-α-Gal-O-Me
			α-Gal-(1→6)-α-Gal-O-Me[a]
α-galactosidase	α-Gal-O-p-Ph-NO$_2$	β-Gal-O-Me	α-Gal-(1→3)-β-Gal-O-Me[a]
			α-Gal-(1→6)-β-Gal-O-Me
β-galactosidase	β-Gal-O-o-Ph-NO$_2$	α-Gal-O-Me	β-Gal-(1→6)-α-Gal-O-Me
β-galactosidase	β-Gal-O-o-Ph-NO$_2$	β-Gal-O-Me	β-Gal-(1→6)-β-Gal-O-Me[a]
			β-Gal-(1→3)-β-Gal-O-Me

[a] Minor product.

2.6 Glycosyl-Transfer Reactions

Table 2.7. Asymmetric glycosylation of cyclic *meso*-diols by β-galactosidase.

Glycoside	Nucleophile	Product	d.e. [%]
β-Gal-O-Ph	*cis*-cyclopentane-1,2-diol	mono-β-Gal cyclopentanediol	89
β-Gal-O-Ph	*cis*-4-cyclopentene-1,3-diol	mono-β-Gal cyclopentenediol	50
β-Gal-O-Ph	*cis*-cyclohexane-1,2-diol	mono-β-Gal cyclohexanediol	90
lactose	*cis*-cyclohexa-3,5-diene-1,2-diol	mono-β-Gal cyclohexadienediol	80
β-Gal-O-Ph	norbornene-diol	mono-β-Gal norbornene-diol	75

Table 2.8. Asymmetric glycosylation of cyclic *meso*-diols by β-galactosidase.

Glycoside	Nucleophile	Product	d.e. [%]
β-Gal-(1→4)-Glc	solketal (rac)	β-Gal-O-solketal	<10
β-Gal-(1→4)-Glc	glycidol (rac)	β-Gal-O-glycidyl	40
α-Glc-(1→4)-Glc	*trans*-cyclohexane-1,2-diol (rac)	mono-α-Glc cyclohexanediol	100

The primary advantage of using glycosidases in comparison to glycosyl transferases is that expensive activated sugar nucleosides are not required. The major drawbacks, however, are low yields and the frequent formation of product mixtures due to the limited selectivity of glycosidases and in particular due to the formation of undesired 1,6-linkages. The regio- and stereoselectivity of transglycosylation reactions is influenced by a number of parameters such as reaction temperature [47], concentration of organic cosolvent, the reactivity of the activated donor [48], the anomeric configuration of the acceptor glycoside and the nature of the aglycon [49, 50].

Besides the synthesis of natural glycosides, a considerable number of non-natural alcohols have been employed as nucleophiles in glycoside-transfer reactions [51, 52]. With some exceptions, the selectivity was not pronounced. The types of transformation include the asymmetrisation of *meso*-diols and the kinetic resolution of racemic primary and secondary alcohols.

As shown in Table 2.7, cyclic *meso*-1,2- or 1,3-diols have been transformed into the corresponding monoglycosides in good to excellent diastereoselectivity using β-galactosidase which is readily available from the dairy industry [53, 54].

When racemic primary and secondary alcohols were used as nucleophiles, a kinetic resolution was observed in some cases. As may be seen from Table 2.8, the selectivity depends on the structure of the aglycon component [55-57]. However, in light of more recent investigations [58], the high diastereoselectivity of some of the reactions previously reported [59] seem to be erroneous.

References

1. Kennedy JF, White CA (1983) Bioactive Carbohydrates, Ellis Horwood, West Sussex, GB
2. Ginsburg V, Robbins PW (eds) (1984) Biology of Carbohydrates, Wiley, New York
3. Karlsson K-A (1989) Ann. Rev. Biochem. 58: 309
4. Hakomori S (1984) Ann. Rev. Immunol. 2: 103
5. Nilsson KGI (1988) Trends Biotechnol. 6: 256
6. Toone EJ, Simon ES, Bednarski MD, Whitesides GM (1989) Tetrahedron 45: 5365
7. Okamoto K, Goto T (1990) Tetrahedron 46: 5835
8. Gigg J, Gigg R (1990) Topics Curr. Chem. 154: 77
9. Edelman J (1956) Adv. Enzymol. 17: 189
10. Leloir LF (1971) Science 172: 1299
11. Sharon N (1975) Complex Carbohydrates, Addison-Wesley, Reading, MA
12. Schachter H, Roseman S (1980) In The Biochemistry of Glycoproteins and Proteoglycans, p 85, Lennarz WJ (ed) Plenum Press, New York
13. Beyer TA, Sadler JE, Rearick JI, Paulson JC, Hill RL (1981) Adv. Enzymol. 52: 23
14. Sadler JE, Beyer TA, Oppenheimer CL, Paulson JC, Prieels J-P, Rearick JI, Hill RL (1982) Methods Enzymol. 83: 458
15. Ginsburg V (1964) Adv. Enzymol. 26: 35
16. Kochetkov NK, Shibaev VN (1973) Adv. Carbohydr. Chem. Biochem. 28: 307
17. Palcic MM, Srivastava OP, Hindsgaul O (1987) Carbohydr. Res. 159: 315

18. Berliner LJ, Robinson RD (1982) Biochemistry 21: 6340
19. Lambright DG, Lee TK, Wong SS (1985) Biochemistry 24: 910
20. Beyer TA, Sadler JE, Rearick JI, Paulson JC, Hill RL (1981) Adv. Enzymol. 52: 23
21. Augé C, David S, Mathieu C, Gautheron C (1984) Tetrahedron Lett. 1467
22. Srivastava OP, Hindsgaul O, Shoreibah M, Pierce M (1988) Carbohydr. Res. 179: 137
23. Palcic MM, Venot AP, Ratcliffe RM, Hindsgaul O (1989) Carbohydr. Res. 190: 1
24. Nunez HA, Barker R (1980) Biochemistry 19: 489
25. David S, Augé C (1987) Pure Appl. Chem. 59: 1501
26. Wong C-H, Haynie SL, Whitesides GM (1982) J. Org. Chem. 47: 5416
27. Haynie SL, Whitesides GM (1990) Appl. Biochem. Biotechnol. 23: 155
28. Wallenfels K, Weil R (1972) In The Enzymes, Boyer PD (ed) vol VII, p 618, Academic Press, New York
29. Pan SC (1970) Biochemistry 9: 1833
30. Post CB, Karplus M (1986) J. Am. Chem. Soc. 108: 1317
31. Withers SG, Street IP (1988) J. Am. Chem. Soc. 110: 8551
32. Withers SG, Warren RAJ, Street IP, Rupitz K, Kempton JB, Aebersold R (1990) J. Am. Chem. Soc. 112: 5887
33. Veibel S (1936) Enzymologia 1: 124
34. Li Y-T (1967) J. Biol. Chem. 242: 5474
35. Johansson E, Hedbys L, Mosbach K, Larsson P-O, Gunnarson A, Svensson S (1989) Enzyme Microb. Technol. 11: 347
36. Likolov ZL, Meagher MM, Reilly PJ (1989) Biotechnol. Bioeng. 34: 694
37. Rastall RA, Bartlett TJ, Adlard MW, Bucke C (1990) The Production of Hetero-Oligosaccharides using Glycosidases, in Opportunities in Biotransformations, Copping LG, Martin RE, Pickett JA, Bucke C, Bunch AW (eds) p 47, Elsevier, London
38. Vulfson EN, Patel R, Beecher JE, Andrews AT, Law BA (1990) Enzyme Microb. Technol. 12: 950
39. Beecher JE, Andrews AT, Vulfson EN (1990) Enzyme Microb. Technol. 12: 955
40. Bourquelot E, Bridel M (1913) Ann. Chim. Phys. 29: 145
41. Fujimoto H, Nishida H, Ajisaka K (1988) Agric. Biol. Chem. 52: 1345
42. Gold AM, Osber MP (1971) Biochem. Biophys. Res. Commun. 42: 469
43. Drueckhammer DG, Wong C-H (1985) J. Org. Chem. 50: 5912
44. Nilsson KGI (1990) Asymmetric Synthesis of Complex Oligosaccharides, in Opportunities in Biotransformations, Copping LG, Martin RE, Pickett JA, Bucke C, Bunch AW (eds) p 131, Elsevier, London
45. Nilsson KGI (1988) Carbohydr. Res. 180: 53
46. Nilsson KGI (1990) Carbohydr. Res. 204: 79
47. Nilsson KGI (1988) Ann. N. Y. Acad. Sci. USA 542: 383
48. Nilsson KGI (1987) A Comparison of the Enzyme-catalysed Formation of Peptides and Oligosaccharides in Various Hydroorganic Solutions Using the Nonequilibrium Approach, In Biocatalysis in Organic Media, Laane C, Tramper J, Lilly MD (eds) p 369, Elsevier, Amsterdam
49. Nilsson KGI (1987) Carbohydr. Res. 167: 95
50. Nilsson KGI (1989) Carbohydr. Res. 188: 9
51. Huber RE, Gaunt MT, Hurlburt KL (1984) Arch. Biochem. Biophys. 234: 151
52. Ooi Y, Mitsuo N, Satoh T (1985) Chem. Pharm. Bull. 33: 5547
53. Gais H-J, Zeissler A, Maidonis P (1988) Tetrahedron Lett. 5743
54. Crout DHG, MacManus DA, Critchley P (1991) J. Chem. Soc., Chem. Commun. 376
55. Boos W (1982) Methods Enzymol. 89: 59
56. Björkling F, Godtfredsen SE (1988) Tetrahedron 44: 2957
57. Mitsuo N, Takeichi H, Satoh T (1984) Chem. Pharm. Bull. 32: 1183
58. Crout DHG, MacManus DA, Critchley P (1990) J. Chem. Soc., Perkin Trans I, 1865
59. Boos W, Lehmann J, Wallenfels K (1968) Carbohydr. Res. 7: 381

2.7 Halogenation and Dehalogenation Reactions

Halogen-containing compounds are not only produced by man, but also by Nature [1]. A brominated indole derivative - Tyrian purple dye - was isolated from the mollusc *Murex brandaris* by the Phoenicians. Since that time, numerous halogenated natural products of various structural types have been isolated from sources such as bacteria, fungi, algae, higher plants, marine molluscs, insects and mammals [2]. Whereas fluorinated and iodinated species are rather rare, chloro- and bromo-derivatives are found more often. The former are predominantly produced by terrestrial species [3] and the latter in marine organisms [4]. Although the natural function of halogenating enzymes is not yet known, they do seem to be involved in the defence mechanism of their hosts. For instance, some algae produce halometabolites, which makes them inedible to would-be predators [5]. In contrast to hydrolytic or redox enzymes, which were intensively investigated as early as around the turn of the century, halogen-converting biocatalysts have been a subject of research for about the last thirty years [6-10].

2.7.1 Halogenation

Although an impressive number of halometabolites have been identified, only one type of halogenating enzyme has been isolated to date, i.e. haloperoxidases [11, 12]. This type of redox enzyme is widely distributed in nature and is capable of carrying out a multitide of reactions following the general equation shown below, where X stands for halide (Cl^-, Br^- and I^-, but not F^-). The individual enzymes are called chloro-, bromo- and iodoperoxidase; the mane reflects the smallest halide ion that they can oxidize. Bearing in mind their unique position as halogenating enzymes and the large variety of structurally different halometabolites produced by them, the majority of haloperoxidases are characterized by a low product selectivity and wide substrate tolerance. As a consequence, any asymmetric induction observed in haloperoxidase-catalysed reactions is usually very low.

$$\text{substrate} + H_2O_2 + X^- + H^+ \xrightarrow{\text{haloperoxidase}} \text{halogenated product} + 2\,H_2O$$

The most intensively studied haloperoxidases are the chloroperoxidase from the mold *Caldariomyces fumago* [13] and the bromoperoxidases from algae [14] and bacteria such as *Pseudomonas aureofaciens* [15], *Ps. pyrrocinia*

[16] and *Streptomyces* sp. [17]. The only iodoperoxidase of preparative use is isolated from horseradish root [18]. Recently, a haloperoxidase isolated from milk has been reported to be useful for the transformation of halohydrins [19].

Halogenation of Alkenes

Haloperoxidases have been shown to transform alkenes by a formal addition of hypohalous acid to produce α,β-halohydrins. The reaction mechanism is presumed to proceed via a halonium intermediate [20, 21], similarly to the chemical formation of halohydrins.

Scheme 2.184. Haloperoxidase-catalysed transformation of alkenes.

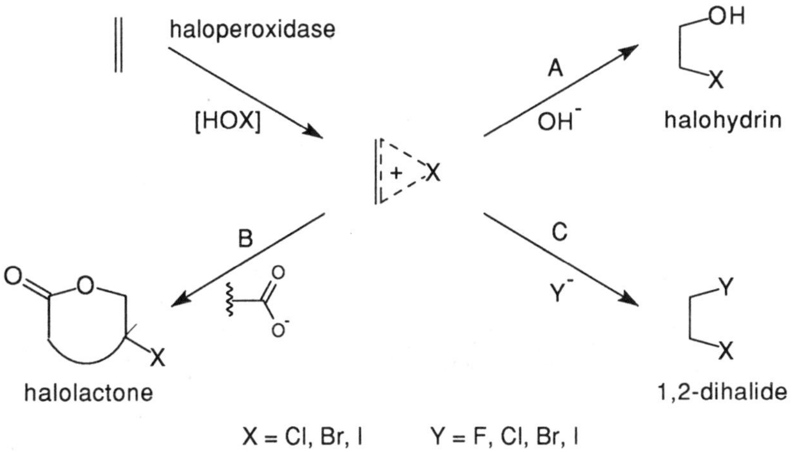

Functional groups present in the alkene can lead to products other than the expected halohydrin by competing with hydroxyl anion (pathway A) for the halonium intermediate. Unsaturated carboxylic acids, for instance, are transformed into the corresponding halolactones due to the nucleophilicity of the carboxylate group (pathway B) affording a halolactonisation [22, 23]. Similarly, if the concentration of halide ion in the reaction mixture is increased, 1,2-dihalides are formed (pathway C) [24]. Although this latter transformation may primarily lead to an undesired side-reaction, it offers the unique possibility to introduce fluorine, which is not oxidized by haloperoxidases, into the substrate via an enzyme-catalysed process. Furthermore, migration of functional groups such as halogen [25] and loss of carbon-containing units such as acetate and formaldehyde may occur, particularly when an oxygen substituent is attached to the C=C bond [26, 27].

All types of carbon-carbon double bonds - isolated (e.g. propene), conjugated (e.g. butadiene) and cumulative (e.g. allene) - are reactive [28]. The size of the substrate seems to be of little importance since steroids [29] and sterically demanding bicyclic alkenes [30] are accepted equally well.

Scheme 2.185. Formation of halohydrins from alkenes.

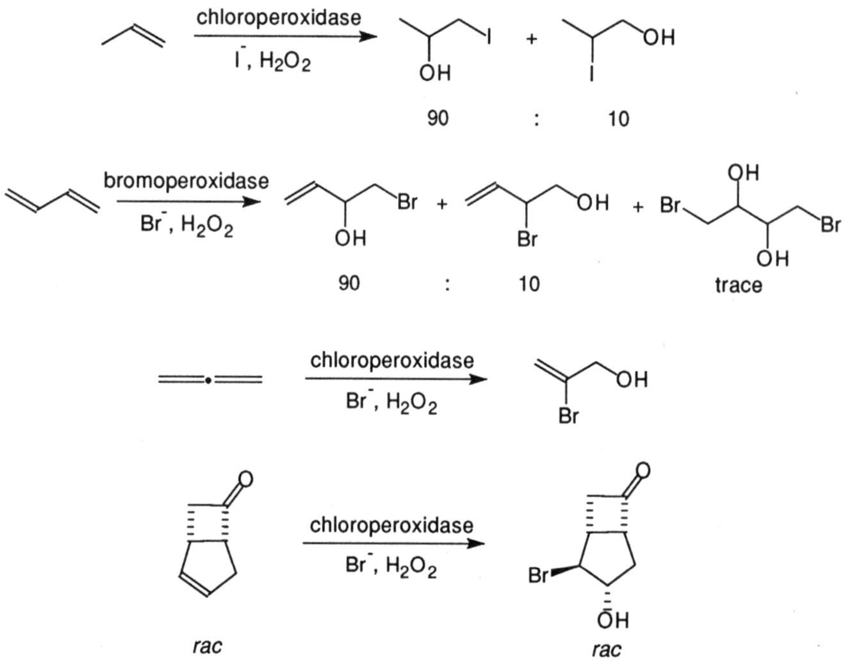

Halogenation of Alkynes

With alkyne substrates, haloperoxidase catalysed reactions yield α-haloketones (Scheme 2.186) [31]. As with alkenes, the product distribution in the reaction with alkynes is dependent on the halide ion concentration. Both homogeneous and heterogeneous dihalides can be formed, dependent upon whether a single halide species or a mixture of halide ions are present. Highly strained cyclopropanes undergo a ring fission to yield α,γ-halohydrins. However, this reaction has not yet been studied in detail.

Scheme 2.186. Haloperoxidase-catalysed reactions of alkynes and cyclopropanes.

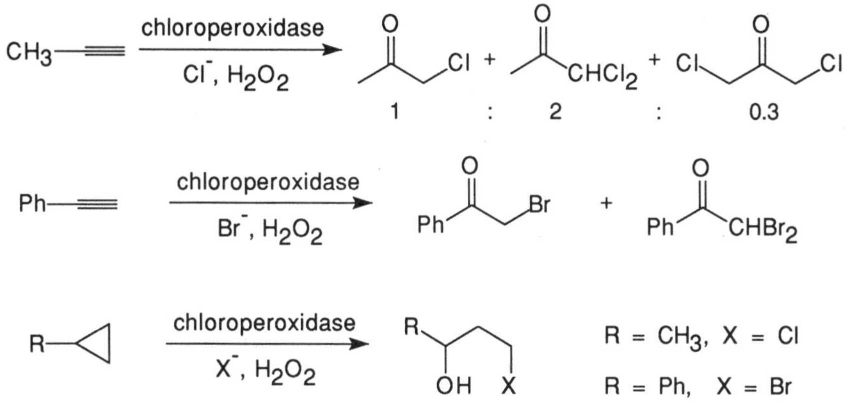

Halogenation of Aromatic Compounds

A wide range of electron-rich aromatic and heteroaromatic compounds are readily halogenated by haloperoxidases (Scheme 2.187) [32-34]. Particularly well accepted are phenols [35, 36] and anilines [37] as well as their respective O- and N-alkyl derivatives. The colour change of phenolic dyes such as phenol red or fluorescein upon halogenation serves as a useful and simple assay for haloperoxidases [38]. Since haloperoxidases are also peroxidases, they also can catalyse halide-independent peroxidation reactions of aromatics (see Section 2.3.2.2). Thus, dimerisation, polymerisation, oxygen insertion and de-alkylation reactions are encountered as undesired side-reactions, particularly whenever the halide ion is omitted or depleted from the reaction mixture.

Halogenation of C-H Groups

Halogenation of C-H groups is only possible if they are sufficiently activated by adjacent electron-withdrawing substituents, for example carbonyl groups. Since the reactivity seems to be a function of the enol content of the substrate, simple ketones like 2-heptanone are unreactive [39]. 1,3-Diketones, however, are readily halogenated to give the corresponding 2-mono- or 2,2-dihalo derivatives (Scheme 2.188) [40]. As with the formation of halohydrins from alkenes, the reactivity of the substrate is independent of its size. For example, monocyclic compounds such as barbituric acid [41] and polycyclic steroids [42] are equally well accepted. β-Ketoacids are also halogenated, but the spontaneous decarboxylation of the intermediate α-halo-β-ketoacid affords the correponding α-haloketones [43].

Scheme 2.187. Halogenation of aromatic compounds.

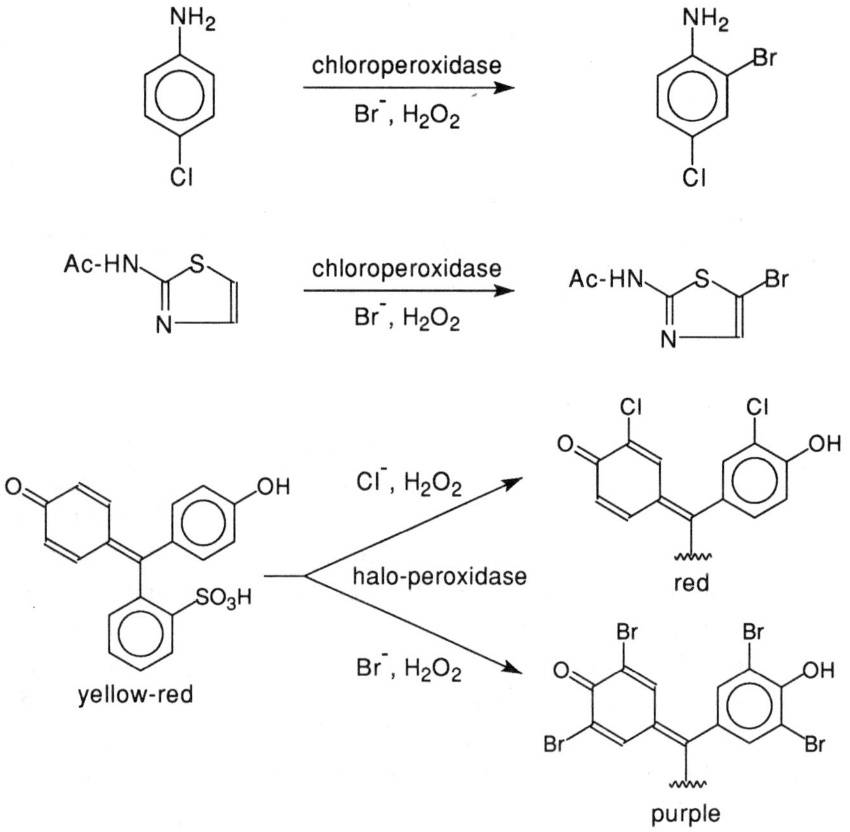

Scheme 2.188. Halogenation of C-H groups.

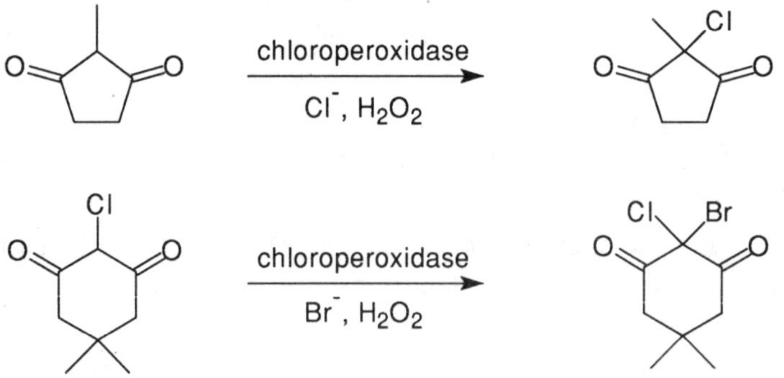

2.7 Halogenation and Dehalogenation Reactions

Halogenation of N- and S-Atoms

Amines are halogenated to form unstable haloamines, which readily deaminate or decarboxylate, liberating the halogen [44]. This pathway constitutes a part of the natural mammalian defence system against microorganisms, parasites and, perhaps, tumor cells. It is of no synthetic use.

In an analogous fashion, thiols are oxidized to yield the corresponding sulfenyl halides. These highly reactive species are prone to undergo nucleophilic attack by hydroxyl ion or by excess thiol [45, 46]. As a result, sulphoxides or disulphides are formed, respectively.

In view of the results described above, it seems that enzymatic halogenation reactions involving haloperoxidase enzymes do not show any significant advantage over the usual chemical reactions due to their lack of regio- and stereoselectivity. A benefit, however, lies in the mild reaction conditions employed.

2.7.2 Dehalogenation

The concentrations of halo-organic compounds in the ecosphere remained reasonably constant due to the establishment of an equilibrium between biosynthesis and biodegradation. Due to man´s recent activity, a large number of halogen-containing compounds - most of which are recalcitrant - are liberated either by intent (e.g. insecticides), or because of poor practice (lead-scavengers in gasoline) or through abuse (dumping of waste) into the ecosystem. These halogenated compounds would rapidly pollute the earth if there were no microbial dehalogenation pathways [47, 48]. Five major pathways for enzymatic degradation of halogenated compounds have been discovered (Table 2.9) [49-52].

Table 2.9. Major biodegradation pathways of halogenated compounds.

Reaction Type	Starting Material		Product(s)
Reductive dehalogenation	C-X	→	C-H + X-
Oxidative degradation	H-C-X	→	C=O + HX
Dehydrohalogenation	H-C-C-X	→	C=C + HX
Epoxide formation	HO-C-C-X	→	epoxide + HX
Hydrolysis	C-X + H_2O	→	C-OH + HX

Redox enzymes are responsible for the replacement of the halogen by a hydrogen atom (reductive dehalogenation [53, 54]) and for oxidative degradation [55]. Elimination of hydrogen halide leads to the formation of an alkene [56], which is further degraded by oxidation. Since all of these pathways proceed either with a loss of a functional group or through a removal of a chirality center, they are of little use for the biocatalytic synthesis of organic compounds. On the other hand, the enzyme-catalysed formation of an epoxide from a halohydrin and the hydrolytic replacement of a halide by a hydroxyl functionality may take place in a stereospecific manner and are therefore of synthetic interest.

Halohydrin Epoxidase

To date, only *Flavobacterium* sp. has been identified as a source of an enzyme capable of catalysing the formation of epoxides from α,β-halohydrins - halohydrin epoxidase [57]. Chloro-, bromo- and iodohydrins are suitable substrates, with the bromohydrins usually being most readily accepted [58]. From mixtures of positional isomers of 1,2-halohydrins (e.g. 1-bromo-2-propanol and 2-bromo-1-propanol) the terminal halogen-containing isomer reacts more rapidly than its positional counterpart. 1,3-Halohydrins and fluorohydrins are not accepted as substrates. The stereospecificity of halohydrin epoxidases is related to that of haloperoxidases, i.e. whenever asymmetric induction is observed, it is usually low to moderate.

Scheme 2.189. Formation of epoxides by halohydrin epoxidases.

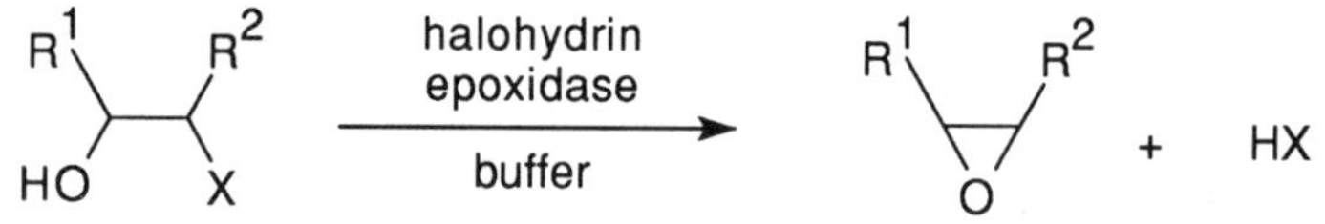

R^1	R^2	X	Relative Activity [%]
H	H	F	0
H	H	Cl	18
H	H	Br	60
H	H	I	24
CH$_3$	H	Br	100
H	CH$_3$	Br	5

Halido-Hydrolase

Hydrolytic dehalogenation catalysed by halidohydrolases (or 'dehalogenases') proceeds by nucleophilic substitution of the halogen atom

2.7 Halogenation and Dehalogenation Reactions

with a hydroxyl ion [59]. Neither co-factors nor metal ions are required for the enzymatic activiy. Due to the SN_2-type of the reaction, the configuration of the reacting centre is inverted [60].

Two main types of enzymes have been characterized which are classified according to their preferred type of substrate: 2-Haloacid dehalogenase is able to accept all kinds of short-chain 2-haloacids [61, 62], while haloacetate dehalogenase exclusively acts on haloacetates [63, 64]. Both of them are inactive on non-activated isolated halides. Interestingly, the reactivity of organic halides (fluoride through iodide derivatives) in dehalogenase reactions depends on the source of enzyme. In some cases it increases from iodine- to fluorine-derivatives, which is in sharp contrast to the comparable chemical reactivity with nucleophiles such as hydroxyl ion [65]. The most intriguing aspect of dehalogenases is their enantiospecificity [66]. Depending on the growth conditions, the microbial production of (R)- or (S)-specific enzymes may be induced [67, 68].

(S)-2-Chloropropionic acid is a key chiral synthon required for the synthesis of a range of important herbicides. Several attempts to resolve racemic 2-chloropropionic acid via enzymatic methods using 'classic' hydrolases have been reported to proceed with varying degrees of selectivity [69]. An elegant approach makes use of an (R)-specific dehalogenase enzyme from *Pseudomonas putida* [70]. Thus from a racemic mixture of chloropropionic acid the (R)-enantiomer is converted into (S)-lactate via inversion of configuration leaving the (S)-chloropropionate behind.

Scheme 2.190. Resolution of 2-chloropropanoic acid by dehalogenase.

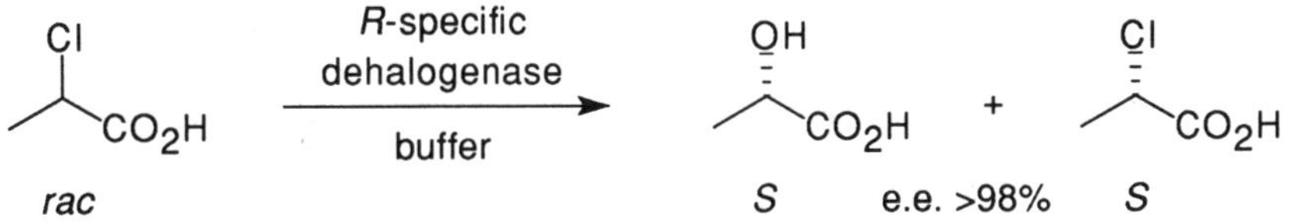

References

1. Petty MA (1961) Bacteriol. Rev. 25: 111
2. Siuda JF, De Barnardis JF (1973) Lloydia 36: 107
3. Fowden L (1968) Proc. Royal Soc. London B 171: 5
4. Fenical W (1979) Recent Adv. Phytochem. 13: 219
5. Faulkner DJ (1984) Nat. Prod. Rep. 1: 251
6. Morris DR, Hager LP (1966) J. Biol. Chem. 241: 1763
7. van Pée K-H (1990) Kontakte (Merck) 41
8. Neidleman SL, Geigert J (1986) Biohalogenation, Ellis Horwood, Chichester
9. Franssen MCR, van der Plas HC (1992) Adv. Appl. Microbiol. 37: 41
10. Neidleman SL (1980) Hydrocarbon Proc. 60: 135

11. Hewson WD, Hager LP (1979) The Porphyrins 7: 295
12. Neidleman SL, Geigert J (1987) Endeavour 11: 5
13. Shaw PD, Hager LP (1961) J. Biol. Chem. 236: 1626
14. Itoh N, Izumi Y, Yamada H (1985) J. Biol. Chem. 261: 5194
15. van Pée K-H, Lingens F (1985) J. Bacteriol. 161: 1171
16. Wiesner W, van Pée K-H, Lingens F (1985) Hoppe-Seylers Z. Physiol. Chem. 366: 1085
17. van Pée K-H, Lingens F (1985) J. Gen. Microbiol. 131: 1911
18. Wagner AP, Psarrou E, Wagner LP (1983) Anal. Biochem. 129: 326
19. Fukuzawa A, Aye M, Murai A (1990) Chem. Lett. 1579
20. Thomas JA, Morris DR, Hager LP (1970) J. Biol. Chem. 245: 3135
21. Yamada H, Itoh N, Izumi Y (1985) J. Biol. Chem. 260: 11962
22. Turk J, Henderson WR, Klebanoff SJ, Hubbard WC (1983) Biochim. Biophys. Acta 751: 189
23. Boeynaems JM, Watson JT, Oates JA, Hubbard WC (1981) Lipids 16: 323
24. Geigert J, Neidleman SL, Dalietos DJ, DeWitt SK (1983) Appl. Environ. Microbiol. 45: 1575
25. Lee TD, Geigert J, Dalietos DJ, Hirano DS (1983) Biochem. Biophys. Res. Commun. 110: 880
26. Neidleman SL, Oberc MA (1968) J. Bacteriol. 95: 2424
27. Levine SD, Neidleman SL, Oberc MA (1968) Tetrahedron 24: 2979
28. Geigert J, Neidleman SL, Dalietos DJ, DeWitt SK (1983) Appl. Environm. Microbiol. 45: 366
29. Neidleman SL, Levin SD (1968) Tetrahedron Lett. 4057
30. Ramakrishnan K, Oppenhuizen ME, Saunders S, Fisher J (1983) Biochemistry 22: 3271
31. Geigert J, Neidleman SL, Dalietos DJ (1983) J. Biol. Chem. 258: 2273
32. Neidleman SL, Cohen AI, Dean L (1969) Biotechnol. Bioeng. 2: 1227
33. van Pée K-H, Lingens F (1984) FEBS Lett. 173: 5
34. Itoh N, Izumi Y, Yamada H (1987) Biochemistry 26: 282
35. Jerina D, Guroff G, Daly J (1968) Arch. Biochem. Biophys. 124: 612
36. Matkovics B, Rakonczay Z, Rajki SE, Balaspiri L (1971) Steroidologia 2: 77
37. Corbett MD, Chipko BR, Batchelor AO (1980) Biochem. J. 187: 893
38. Loo TL, Burger JW, Adamson RH (1964) Proc. Soc. Exp. Biol. Med. 114: 60
39. Beissner RS, Guilford WJ, Coates RM, Hager LP (1981) Biochemistry 20: 3724
40. Libby RD, Thomas JA, Kaiser LW, Hager LP (1982) J. Biol. Chem. 257: 5030
41. Franssen MCR, van der Plas HC (1984) Recl. Trav. Chim. Pays-Bas 103: 99
42. Neidleman SL, Diassi PA, Junta B, Palmere RM, Pan SC (1966) Tetrahedron Lett. 5337
43. Theiler R, Cook JC, Hager LP, Siuda JF (1978) Science 202: 1094
44. Grisham MB, Jefferson MM, Metton DF, Thomas EL (1984) J. Biol. Chem. 259: 10404
45. Silverstein RM, Hager LP (1974) Biochemistry 13: 5069
46. Tsan M-F (1982) J. Cell. Physiol. 111: 49
47. Lal R, Saxena DM (1982) Microbiol. Rev. 46: 95
48. Alexander M (1977) Introduction to Soil Microbiology, p. 438, Wiley, Chichester
49. Ghisalba O (1983) Experientia 39: 1247
50. Rothmel RK, Chakrabarty AM (1990) Pure Appl. Chem. 62: 769
51. Müller R, Lingens F (1986) Angew. Chem., Int. Ed. Engl. 25: 778
52. Vogel TM, Criddle CS, McCarthy PL (1987) Environ. Sci. Technol. 21: 722
53. Castro CE, Wade RS, Belser NO (1985) Biochemistry 24: 204
54. Chacko CI, Lockwood JL, Zabik M (1966) Science 154: 893
55. Markus A, Klages V, Krauss S, Lingens F (1984) J. Bacteriol. 160: 618
56. Yoshida M, Fujita T, Kurihara N, Nakajima M (1985) Pest. Biochem. Biophysiol. 23: 1
57. Castro CE, Bartnicki EW (1968) Biochemistry 7: 3213
58. Geigert J, Neidleman SL, Liu T-N, DeWitt SK, Panschar BM, Dalietos DJ, Siegel ER (1983) Appl. Environ. Microbiol. 45: 1148

59. Leigh JA, Skinner AJ, Cooper RA (1988) FEMS Microbiol. Lett. 49: 353
60. For an example exhibiting a retention of conversion see: Weightman AJ, Weightman AL, Slater JH (1982) J. Gen. Microbiol. 128: 1755
61. Motosugi K, Esaki N, Soda K (1982) Agric. Biol. Chem. 46: 837
62. Allison N, Skinner AJ, Cooper RA (1983) J. Gen. Microbiol. 129: 1283
63. Kawasaki H, Miyoshi K, Tonomura K (1981) Agric. Biol. Chem. 45: 543
64. Walker JRL, Lien BC (1981) Soil Biol. Biochem. 13: 231
65. Kawasaki H, Tone N, Tonomura K (1981) Agric. Biol. Chem. 45: 35
66. Onda M, Motosugi K, Nakajima H (1990) Agric. Biol. Chem. 54: 3031
67. Tsang JSH, Sallis PJ, Bull AT, Hardman DJ (1988) Arch. Microbiol. 150: 441
68. Little M, Williams PA (1971) Eur. J. Biochem. 21: 99
69. Cambou B, Klibanov AM (1984) Appl. Biochem. Biotechnol. 9: 255
70. Taylor SC (1990) *S*-2-Chloropropanoic Acid by Biotransformation, in: Opportunities in Biotransformations, Copping LG, Martin RE, Pickett JA, Bucke C, Bunch AW (eds) p 170, Elsevier, London

3 Special Techniques

Most biocatalysts can be used in a straightforward manner by regarding them as chemical chiral catalysts and by applying standard methodology but, in addition, some special techniques have been developed in order to broaden their range of applications. In particular, using biocatalysts in non-aqueous media rather than in water can lead to the gain of some significant advantages as long as some specific guidelines are followed. Furthermore, 'fixation' of the enzyme by immobilisation may be necessary, and the use of membrane technology may be advantageous as well. For both of the latter topics only the most simple techniques which can be adopted in an average organic laboratory are discussed.

3.1 Enzymes in Organic Solvents

Water is a poor solvent for nearly all reactions in preparative organic chemistry because most organic compounds are insoluble in this medium. Furthermore, the removal of water is tedious and expensive due to its high boiling point and high heat of vaporisation. Side-reactions such as hydrolysis, racemisation, polymerisation and decomposition are often facilitated in the presence of water. These limitations have been circumvented long ago by the introduction of organic solvents for the majority of organic-chemical processes. On the other hand, conventional biocatalysis has mainly been performed in aqueous solutions due to the perceived notion that enzymes are most active in water and that organic solvents only serve to destroy their catalytic power. However, this commonly held opinion is certainly too simplistic, due to the fact that in Nature many enzymes or multienzyme complexes function in hydrophobic environments, for instance in the presence of, or bound onto, a membrane [1]. Therefore it should not be surprising that enzymes can be catalytically active in the presence of organic solvents [2-8]. The rôle of water in functioning biological systems is contradictory. Thus the enzyme depends on water for the majority of the non-covalent interactions that help to maintain its catalytically active conformation [9] but water also participates in most of the reactions which lead to denaturation. Thus, it may be anticipated that replacing *some* (but not *all*) of the water with an organic solvent would retain the enzymatic activity. Equally, completely anhydrous solvents are incapable of supporting enzymatic activity; some water is always necessary for catalysis. The crucial answer to the question concerning how much water is required to retain catalytic activity is enzyme-dependent. For example, α-chymotrypsin needs only 50 molecules of water per enzyme molecule to remain catalytically active [10]; this is much less than is needed to form a monolayer of water around the enzyme. Other enzymes, like subtilisin and various lipases are similar in their need for trace quantities of water [11]. In other cases, however, much more water is needed. Polyphenol oxidase, for instance, requires the presence of about 3.5×10^7 molecules of water [12].

The water present in a biological system can be separated into two physically distinct categories [13]. Whereas the majority of the water (>98%) serves as a true solvent ('bulk water'), a small fraction of it is tightly bound to the enzyme's surface ('bound water'). The physical state of bound water as monitored by differences in melting point and spectroscopical properties is clearly distinct from the bulk water and it should be regarded as crucial for the enzyme structure rather than as adventitious residual solvent. If one extends these facts to the consideration of enzymatic catalysis in organic media, it should be possible to replace the bulk water by an organic solvent without significant alteration of the enzyme's environment.

Biocatalytic transformations performed in organic media offer the following advantages:
- The overall yields of processes performed in organic media are usually better due to the omission of an extractive step during work-up. The recovery of product(s) is facilitated by the use of low-boiling organic solvents.
- Non-polar substrates are transformed at better rates due to their increased solubility [14].
- Since an organic medium is a hostile environment for living cells, microbial contamination is negligable.
- Deactivation and/or inhibition of the enzyme caused by lipophilic substrates and/or products is minimized due to their solubility in the organic medium since their local concentration at the enzyme's surface is lower.
- Many side-reactions [such as hydrolysis of labile groups (e.g. acid anhydrides [15]), polymerisation of quinones [16], racemisation of

cyanohydrins [17] or acyl-migration [18] are water-dependent and are therefore largely suppressed in an organic medium.
- Immobilization of enzymes is often unneccessary because they may be recovered by simple filtration after the reaction due to their insolubility in organic solvents. Nevertheless, if it is desired, experimentally simple adsorption onto the surface of a macroscopic carrier such as diatomaceous earth (Celite), silica or glass beads is possible. Desorption from the carrier into the medium - 'leaking' - cannot occur in a lipophilic environment.
- Since many of the reactions which are responsible for the denaturation of enzymes (see Section 1.4) are hydrolytic reactions and therefore require water, it is obvious that enzymes should be more stable in an environment of low water-content [19]. For instance, porcine pancreatic lipase is active for many hours at 100 °C in a 99% organic medium but it is rapidly denatured at this temperature when placed in water [20].
- During the partial unfolding and refolding of the enzyme during the formation of the enzyme-substrate complex ('induced-fit'), numerous hydrogen bonds are broken. This process is greatly facilitated in an aqueous medium, which ensures that the broken bonds are rapidly replaced by hydrogen bonds to the surrounding water. Thus, it serves as a 'molecular lubricant' [21]. This process is impeded in an organic solvent and, as a consequence, enzymes appear to be more rigid. Thus it is often possible to control some of the enzyme's properties such as the substrate-specificity [22-25], the regioselectivity [26] and the enantioselectivity [27-30] by variation of the solvent.
- The most important advantage, however, is the possibility to shift thermodynamic equilibria to favour synthesis over hydrolysis. Thus, by using hydrolase enzymes (mainly lipases and proteases) esters [31-33], polyesters [34, 35], lactones [36, 37], amides [38, 39] and peptides [40] can be *synthesized* in a chemo-, regio- and enantioselective manner.

The organic solvents which have commonly been used for enzyme-catalysed reactions can be classified into three different categories.

Enzyme Dissolved in a Monophasic Aqueous-Organic Solution.
The enzyme and the substrate and/or product are dissolved in a monophasic solution consisting of water and a water-miscible organic co-solvent, such as dimethyl sulphoxide, dimethyl formamide, tetrahydrofuran, dioxane, acetone or one of the lower alcohols. Systems of this type are mainly used for the transformation of lipophilic substrates, which are sparingly soluble in an

3.1 Enzymes in Organic Solvents

aqueous system alone and which would therefore react at low reaction rates. In some cases, selectivities of esterases and proteases may be enhanced by using water-miscible co-solvents (see Section 2.1.3.1). As a rule, most water-miscible solvents can be applied in concentrations up to 10% of the total volume but in some enzyme/solvent combinations 50-70% of co-solvent may be used. If the proportion of the solvent system exceeds this value, the essential bound water is stripped from the enzyme´s surface leading to deactivation. Only rarely do enzymes remain catalytically active in water-miscible organic solvents with an extremely low water content; these cases are limited to unusually stable enzymes such as subtilisin and some lipases [41-43]. Water-miscible organic solvents have also been successfully used to decrease the freezing temperature of aqueous systems when biocatalytic reactions were conducted at temperatures below 0 °C ('cryoenzymology') [44, 45].

Enzyme Dissolved in a Biphasic Aqueous-Organic Solution

Reaction systems consisting of two macroscopic phases [46, 47], namely the aqueous phase containing the dissolved enzyme, and a second phase of a non-polar organic solvent (preferably lipophilic and of high molecular weight) such as hydrocarbons, ethers or chlorinated hydrocarbons, may be advantageous to achieve a spatial separation of the biocatalyst from the organic phase. The biocatalyst is in a favourable aqueous environment and not in direct contact with the organic solvent, where most of the substrate is located. The limited concentrations of organic material in the aqueous phase may circumvent inhibition phenomena. Furthermore, the removal of product from the enzyme surface drives the reaction towards completion. Due to the fact that in such biphasic systems the enzymatic reaction proceeds only in the aqueous phase, a mass-transfer of the reactant(s) to and product(s) from the catalyst and between the two phases is necessary [48]. It is obvious that shaking or stirring represents a crucial parameter in such systems.

The number of phase-distributions encountered in a given reaction depends on the number of reactants and products (A, B, C, D) which are involved in the transformation (Table 3.1). Each distribution, measured as the partition coefficient, is dependent on the solubilities of substrate(s) and product(s) in the two phases and this represents a potential rate-limiting factor.

Therefore, in biphasic systems the partition coefficient (a thermodynamic dimension) and the mass-transfer coefficient (a kinetic dimension) may dominate the k_{cat} of the enzyme. As a consequence, the overall reaction rate is determined by the physical properties of the system (such as solubilities and stirring) and not by the enzyme´s catalytic power. In other words, the enzyme could work faster, but is unable to get enough substrate. Enhanced agitation (stirring, shaking) would improve the mass-transfer but on the other hand it may lead to deactivation of the enzyme. Nevertheless, water-organic solvent two-phase systems have been successfully used to transform highly lipophilic substrates such as steroids [49], fats [50] and alkenes [51].

Table 3.1. Partition coefficients involved in biphasic reactions.

Type of Reaction	Number of Partition Coefficients
A → B	3
A + B → C	4
A → B + C	4
A + B → C + D	7
Any type[a]	1

[a] For monophasic systems.

Enzyme Suspended in a Monophasic Organic Solution

Replacing all of the bulk water (which counts for >98%) by a water-immiscible organic solvent leads to a suspension of the solid enzyme in a monophasic organic solution [52]. Although the biocatalyst seems to be 'dry', it must have the necessary residual bound water to remain catalytically active. Most of the research on such systems (which proved that they are extremely reliable, versatile and easy to use) has been performed over the past decade but it is striking that the first biotransformation of this kind was already reported in 1900! [53]. Due to the importance of this technique and its simplicity, all of the examples discussed below have been performed using solid 'dry' enzymes in organic solvents having a water content of <2%.

One particularly important aspect is the effect of the pH of the reaction medium. In organic solutions that lack a distinct aqueous phase, pH cannot be measured easily [54]. On the other hand, the ionization state of the enzyme which is a function of the pH, determines its conformation and hence its properties such as activity and selectivity. Since the ionization state of charged groups of a protein does not change, when placed in an organic solvent, it is important to employ solid enzymes that have been recovered by lyophilization or precipitation from a buffer at their pH optimum [55].

The physical state of the enzyme may be crystalline, lyophilized or precipitated. Adsorption of enzymes onto a macroscopic surface generates a better distribution of the bioctalyst and gives significantly enhanced reaction rates, in some cases up to one order of magnitude [56]. Any inorganic material such as Celite, silica gel or an organic non-ionic support (e.g. XAD-8) may be used as the carrier.

In order to provide a measure for the hydrophobicity and therefore the 'compatibility' of an organic solvent with high enzyme activity, many parameters such as the Hildebrandt solubility parameter (δ), the dielectric constant (ε), and the dipole moment (μ), have been proposed [57]. The most reliable results were obtained by using the logarithm of the partition coefficient (log P) of a given solvent between 1-octanol and water (Table 3.2) [58].

Table 3.2. Compatibility of solvents with enzymatic activity.

log P	Water-Miscibility	Effects on Enzyme Activity
-2.5 to 0	completely miscible	may be used to solubilize lipophilic substrates in concentrations up to 50% v/v without affecting the enzyme
0 to 2	partially miscible	limited use due to rapid enzyme deactivation
2 to 4	low miscibility	causes weak enzyme distortion, may be used with caution, since activity is often unpredictable
>4	immiscible	causes no enzyme distortion and ensures high retention of activity

Table 3.3. Log P values for common organic solvents.

Solvent	log P	Solvent	log P
dimethylsulphoxide	-1.3	pentyl acetate	2.2
N,N-dimethylformamide	-1.0	toluene	2.5
ethanol	-0.24	octanol	2.9
acetone	-0.23	dibutyl ether	2.9
tetrahydrofuran	0.49	carbon tetrachloride	3.0
ethyl acetate	0.68	cyclohexane	3.2
propyl acetate	1.2	hexane	3.5
butyl acetate	1.7	octane	4.5
chloroform	2.0	dodecane	6.6

Note that if the log P value is not available in the literature (Table 3.3), it can be calculated from hydrophobic fragmental constants [59]. As may be deduced from the log P values of some selected common organic solvents, water-miscible hydrophilic solvents such as DMF, DMSO, acetone and lower alcohols, are usually incompatible, whereas water immiscible lipophilic solvents such as alkanes, ethers, aromatics and haloalkanes retain an enzyme´s high catalytic activity due to the fact that they do not strip off the bound water from the enzyme´s surface. To further ensure this state of affairs, solvents should be saturated with water. Only in certain cases, in which polar substrates such as polyhydroxy compounds are involved, can water-miscible solvents such as dioxane, tetrahydrofuran, 3-methyl-3-pentanol or pyridine be considered. However, in these solvents, most enzymes are deactivated and only exceptionally stable enzymes (for instance subtilisin) can be used.

The remarkable catalytic activity of solid proteins in organic solvents can be explained by their special properties [60]. In contrast to densely packed crystals of organic compounds of comparatively low molecular weight, solid proteins represent delicate structures. There is sufficient mobility within the single enzyme unit to permit minor conformational changes conconant with formation of the enzyme-substrate complex [61]. The total surface of solid enzymes is in the range of $1-3 \times 10^6$ m^2 per kg, which is close to that of silica or activated carbon, and about 1/3 to 2/3 of the total volume is hollow, often filled with solvent. Thus, if sufficient agitation is provided, the substrate is not only transformed by the active sites exposed to the surface of the crystal but also at those partly buried inside.

Instead of a lipophilic organic solvent, supercritical gases such as carbon dioxide (T_{crit} 31 °C and p_{crit} 73 bar) which exhibit solubility properties similar to that of hexane, can be used as solvent or co-solvent for the enzymatic transformation of lipophilic organic compounds [62, 63]. Enzymes are as stable in this medium as in lipophilic organic solvents and the use of supercritical gases is not restricted to a particular class of enzyme, but not surprisingly, the use of hydrolases is dominant. For instance, esterification [64] transesterification [65, 66], alcoholysis [67] and hydrolysis [68] are known as well as hydroxylation [69] and dehydrogenation reactions [70]. The most striking advantages of this type of solvent are a lack of toxicity, easy removal and the low viscosity which is intermediate between those of gases and those of liquids. This latter property ensures a high diffusion about one to two orders higher than in common solvents. Furthermore, small variations in temperature or pressure may result in large solubility changes near the critical point, which make it possible to control reaction rates. Some disadvantages should be mentioned. The high-pressure equipment represents a considerable initial investment and the depressurization step may cause enzyme deactivation [71]. The main use of supercritical gases as solvents resides in the production of 'natural' compounds used for cosmetics and food supplies.

The following basic rules should be considered for the application of solid ('dry') enzymes in organic media having a low water content:

- Hydrophobic solvents are more compatible than hydrophilic ones (log P of the organic solvent should be greater than 3 or 4),
- the water layer bound to the enzyme must be maintained and this is accomplished by using water-saturated organic solvents.
- The 'micro-pH' must be that of the pH-optimum of the enzyme in water, a prerequisite that is fulfilled if the protein was isolated from an aqueous solution at the appropriate pH and
- stirring, shaking or sonication is necessary in order to maximize diffusion of substrate to the catalysts´s surface.

3.1.1 Ester Synthesis

Esterification

In every synthetic reaction where a net amount of water is formed (such as an ester-synthesis from an alcohol and a carboxylic acid [72-74]) physicochemical problems are often observed. Due to the fact that the lipophilic solvent (log P >4) is unable to accommodate the water which is gradually produced during the course of the reaction, it is collected at the hydrophilic enzyme surface. The water gradually forms a discrete aqueous phase which encompasses the enzyme. Thus, substrate and enzyme are gradually separated from each other by an interface. As a consequence, the reaction stops before reaching the desired extent of conversion. In order to solve this problem, two techniques have been developed. Firstly, removal of water from the system e.g. by evaporation [75] or chemical drying [76, 77] via addition of molecular sieves or water scavenging inorganic salts and secondly, avoiding the formation of water by employing an acyl-transfer step rather than an esterification reaction.

For example, 6-O-acyl derivatives of alkyl glucopyranosides, useful as biodegradable nonionic surfactants, were synthesized from fatty acids and the corresponding 6-O-acyl glucopyranosides in the presence of a thermostable lipase (Scheme 3.1) [78]. In order to drive the reaction towards completion, the water produced during the reaction was evaporated (70 °C, 0.01 bar).

Both of the methods for the removal of water have inherent disadvantages. Evaporation of water from the reaction mixture can only be efficient if the alcohol and acid reactants have a low volatility (high boiling point). On the other hand, recovery of enzymes from organic solvents in the presence of solid inorganic water-scavengers is troublesome.

Scheme 3.1. Biocatalytic synthesis of glucose esters.

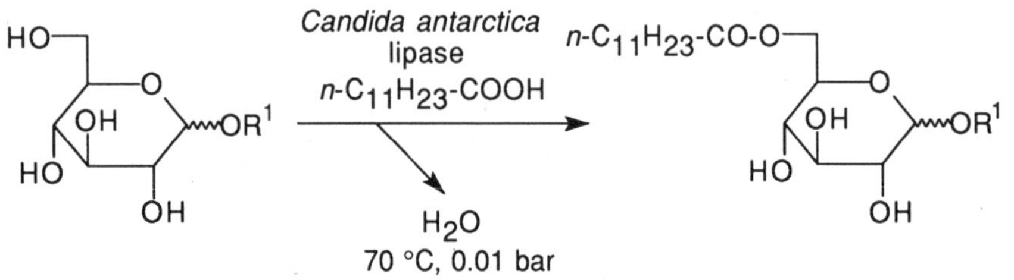

R^1	Yield of Monoester [%]	Yield of Diester [%]
H	<5	0
Me	53	4
Et	93	5
i-Pr	93	4
n-Pr	96	17
n-Bu	94	22

Acyl-Transfer

Trans- or inter-esterifications, which do not release water in the course of the reaction, are easier to perform [79]. Furthermore, the water-content of the reaction medium, which is a crucial parameter for retaining the enzyme's activity, is kept at a constant level. In acyl transfer reactions any traces of chemically available 'bulk' water which may be present in the reaction medium, are quickly consumed at the expense of acyl donor, which is usually used in excess. The 'bound' water, which is required to retain the enzyme's activity and which is tightly bound onto the enzyme's surface, is not removed.

In contrast to hydrolytic reactions, where the nucleophile, water, is always in excess (55 moles per liter) the concentration of the nucleophile in acyl transfer reactions is always limited. As a result, trans- and interesterification reactions involving 'normal' esters are generally *reversible* in contrast to the *irreversible* nature of a hydrolytic reaction. This leads to a slow reaction rate and can have some important consequences on the selectivity of the reaction (see Section 2.1.1).

In order to avoid the undesired depletion of the optical purity of (mainly) the remaining substrate during an enzymatic resolution under reversible reaction conditions, two tricks can be applied to shift the equilibrium of the reaction. The presence of an excess of acyl donor may be expensive and not always compatible to retain a high enzyme activity, but it may be helpful in some cases. A better solution however, is the use of special acyl donors which ensure a more or less irreversible type of reaction.

The reversibility of transesterification reactions is caused by the comparable nucleophilicity of the attacking nucleophile and the leaving group of the acyl donor, both of which compete for the acyl-enzyme intermediate in the forward- and the reverse reaction. If the nucleophilicity of the leaving group is depleted, for instance by the introduction of electron withdrawing substituents, the reaction is shifted to the right, i.e. towards completion. This concept has been verified with the introduction of 'activated' esters [80], such as 2-haloethyl, cyanomethyl and oxime esters. Although acyl-transfer using activated esters is still reversible in principle, the equilibrium of the reaction is shifted so far to the product side, that for preparative purposes it can be regarded as irreversible. To indicate this, the term 'quasi-irreversible' was proposed recently [81].

Scheme 3.2. Enzymatic acylation of (±)-2-octanol.

$$RS\text{-2-octanol} + \text{acyl donor} \underset{Et_2O}{\overset{PPL}{\rightleftharpoons}} R\text{-2-octyl ester} + S\text{-2-octanol}$$

Acyl Donor	Initial Rate [%]
ethyl acetate	.3
methyl butanoate	5
2-chloroethyl acetate	1
ethyl cyanoacetate	6
trichloroethyl trichloroacetate	7
methyl bromoacetate	14
tributyrin	34
trichloroethyl butanoate	58
trichloroethyl heptanoate	100

The relative rate of the enantioselective acylation of (±)-2-octanol catalysed by porcine pancreatic lipase (PPL) was one to two orders of magnitude faster when 'activated' esters were used as acyl donors instead of 'non-activated' methyl or ethyl alkanoates.

The following parameters should be considered before the 'activated' ester is chosen: cyanomethyl esters have been used only rarely due to toxicity problems (formaldehyde cyanohydrin is liberated). 2-Haloethyl esters have been applied more widely. Trifluoroethyl esters are the acyl donors of choice when reactions are performed on a laboratory scale. For larger batches trichloroethyl esters are more economic but the removal of trichloroethanol during work-up can be troublesome due to its high boiling point (151 °C). 2-Chloroethyl esters are cheap and the resulting 2-chloroethanol is easier to remove (bp 130 °C), but their degree of activation is limited.

Scheme 3.3. Enzymatic acyl-transfer using activated esters.

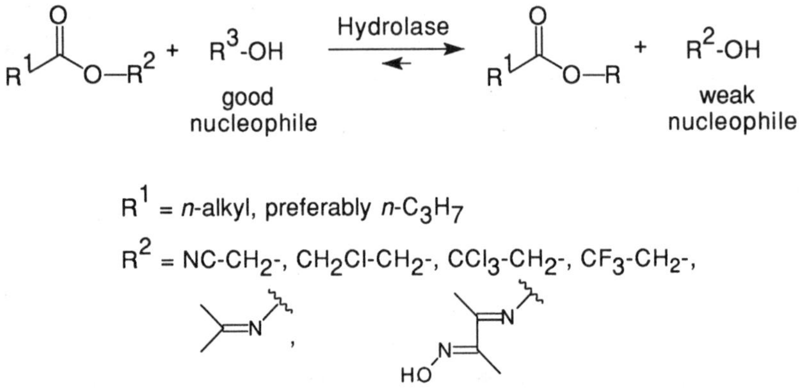

$$R^1-CO-O-R^2 + R^3-OH \xrightarrow{\text{Hydrolase}} R^1-CO-O-R^3 + R^2-OH$$

good nucleophile → weak nucleophile

R^1 = n-alkyl, preferably n-C_3H_7

R^2 = NC-CH_2-, CH_2Cl-CH_2-, CCl_3-CH_2-, CF_3-CH_2-, >C=N-, HO-N=C<

More recently, oxime esters have been proposed as acyl donors for 'irreversible' acyl-transfer reactions [82]. During the reaction a weakly nucleophilic oxime is liberated which is unable to compete with the substrate alcohol for the acyl-enzyme intermediate. While it was shown that the reaction is still reversible, the reverse reaction may be neglected from a preparative point of view. However, co-substrate inhibition and problems in separating the non-volatile oxime from the substrate alcohol during work-up may be encountered.

In contrast to the above mentioned acyl-donors which shift the equilibrium of the reaction to the product side by liberating a weakly nucleophilic co-product alcohol, two concepts have recently been proposed for making the reaction completely irreversible.

Enol esters such as vinyl or isopropenyl esters liberate unstable enols as co-products, which instantaneously tautomerise to give the stable aldehydes or ketones [83, 84]. Thus, the reaction becomes completely irreversible and this ensures that all the benefits with regard to a rapid reaction rate and a high

3.1 Enzymes in Organic Solvents

selectivity are accrued. Acyl transfer using enol esters has been shown to be about only 10 times slower than hydrolysis and about 10 to 100 times faster than acyl transfer reactions using 'activated' esters. In contrast, when non-activated esters such as ethyl acetate were used, reaction rates of about 10^{-2} to 10^{-4} of that of the hydrolytic reaction are observed [85].

Scheme 3.4. Enzymatic acylation using enol esters and acid anhydrides.

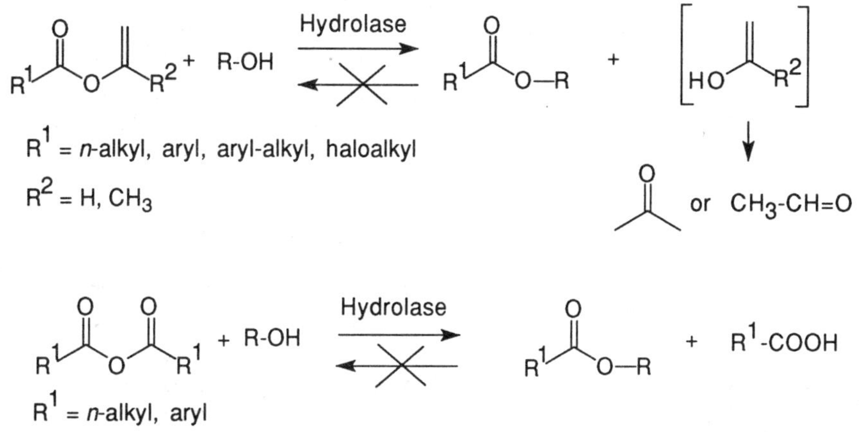

R^1 = n-alkyl, aryl, aryl-alkyl, haloalkyl
R^2 = H, CH_3

or $CH_3-CH=O$

R^1 = n-alkyl, aryl

Due to steric reasons, vinyl esters give better reaction rates than isopropenyl esters and the former are therefore used most widely, but their use is not without drawbacks. Acetaldehyde, which is liberated during the reaction, is known to act as an alkylating agent by forming Schiff's bases in a Maillard-type of reaction [86] from, for example, the terminal amino residues of lysine [87]. Thus, a positive charge is removed from the enzyme´s surface during the course of this reaction, which may cause enzyme deactivation. The extent of this depends on the nature of the enzyme. For instance, as *Pseudomonas* sp. lipases (PSL) seems to be almost completely inert in this respect, the lipase from *Candida cylindracea* (CCL) has been shown to be more sensitive.

Covalent immobilization of CCL onto an epoxy-activated macroscopic carrier leads to selective monoalkylation of the lysine amino residues which are involved in the deactivation reaction with retention of the positive charge. In contrast to the native enzyme, the immobilized enzyme is inert towards the formation of Schiff´s-bases, which results not only in a greatly stabilized activity but also in a significant enhancement in selectivity [88]. The addition of molecular sieve to the medium in order to trap acetaldehyde seems to have some benefit [89, 90].

Another useful method of achieving a complete irreversible acyl transfer reaction is the use of *acid anhydrides* [91]. The selectivities achieved are usually high and the reaction rates are about the same as with enol esters. One of the advantages of this technique is that no aldehydic byproducts are formed and the enzyme is not acylated under the conditions employed, making its re-use possible.

Scheme 3.5. Selectivity enhancement via addition of base.

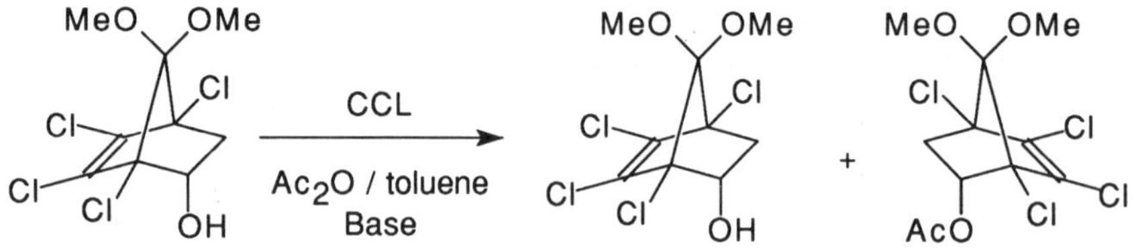

Base	Reaction Rate	Selectivity (E)
none	good	18
KHCO$_3$	low	210
2,6-lutidine	good	210

For some enzymes such as PSL the liberated acid does not present any problems, but others like CCL are more sensitive and require more protection. For instance when acetic anhydride is used, the liberated acetic acid may lead to a decrease of the pH in the micro-environment of the enzyme, thus leading to possible depletion of activity and selectivity. The CCL-catalysed resolution of the bicyclic tetrachloro-alcohol shown in Scheme 3.5 using acetic anhydride as acyl donor initially proceeded only with moderate selectivity (E = 18). Addition of a weak inorganic or (preferably) organic base such as 2,6-lutidine which functions as an acid scavenger, led to a more than ten-fold increase in selectivity [92]. A similar acid-quenching effect could be observed by immobilisation of CCL onto diatomaceous earth.

Besides the often used acyl donors mentioned above, others which would also ensure an irreversible type of reaction have been investigated [93]. Bearing in mind that most of the problems of irreversible enzymatic acyl-transfer arise from the formation of unavoidable by-products, emphasis has been put on finding acyl donors that possess cyclic structures, which would not liberate any byproducts at all. Unfortunately, with candidates such as lactones, lactams, cyclic anhydrides and enol lactones the drawbacks dominated their merits.

3.1 Enzymes in Organic Solvents

Enzyme-catalysed acyl transfer can be applied to a number of different synthetic problems. The majority of applications that have been reported involve the asymmetrisation of prochiral and *meso*-diols or the kinetic resolution of racemic primary and secondary alcohols. Since, as a rule, an enzyme's preference for a specific enantiomer is not reversed when water is replaced by an organic solvent, it is always the same enantiomer, which is accepted. Due to the fact that hydrolysis and esterification represent reactions into opposite directions, products of opposite configuration are obtained via hydrolysis or esterification. In other words, in case the (*R*)-enantiomer of an ester is *hydrolysed* at a faster rate than its (*S*)-counterpart [yielding an (*R*)-alcohol and an (*S*)-ester], *esterification* of the racemic alcohol will lead to the formation of an (*R*)-ester and an (*S*)-alcohol.

Representative examples for stereo- and regioselective transformations (e.g. separation of *E/Z*-mixtures, and the regioselective protection of polyhydroxy compounds) are given below.

Stereoisomeric mixtures of the allylic terpene alcohols geraniol and nerol, which are used as additives to flavour and fragrance preparations, were separated by selective acylation with an acid anhydride using porcine pancreatic lipase (PPL) as catalyst [94]. Depending on the acyl donor employed, the slightly less hindered geraniol was more quickly acylated to give geranyl acetate leaving nerol unreacted. Anhydride proved to be unsuitable giving a low yield and poor selectivity, but longer-chain acid anhydrides were used successfully.

Scheme 3.6. *E/Z*-Stereoselective enzymatic acylation of terpene alcohols.

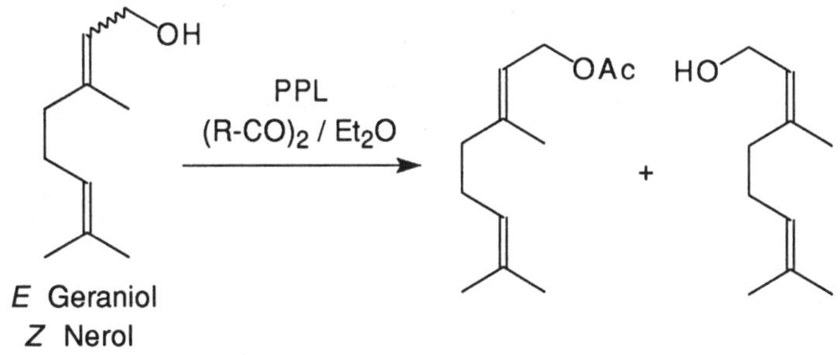

R	Geranyl Acetate [%]	Nerol [%]	Selectivity [k_E/k_Z]
n-C$_3$H$_7$-	85	16	11
n-C$_5$H$_{11}$-	66	7	13
n-C$_7$H$_{15}$-	72	7	15

Chiral 1,3-propanediol derivatives are considered to be useful building blocks for the preparation of enantiomerically pure bioactive compounds such as phospholipids [95], platelet activating factor (PAF), PAF-antagonists [96] and renin inhibitors [97]. A simple access to these synthons starts from 2-substituted 1,3-propanediols, which in turn are obtained from malonic acid derivatives. Depending on the substituent in position 2 and on the lipase used, (R)- or (S)-monoesters were obtained in excellent optical purities [98-100]. PPL was equally effective as a biocatalyst as was *Pseudomonas* sp. lipase (PSL). The last three entries demonstrate an enhancement in selectivity which may be obtained when the reaction temperature is lowered [101].

Scheme 3.7. Asymmetric acylation of 2-substituted propane-1,3-diols.

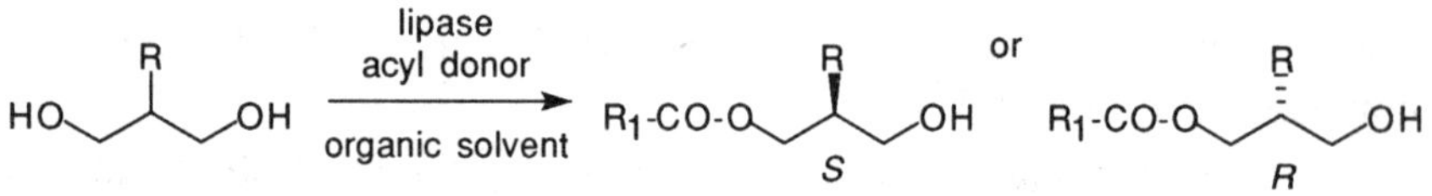

R	Lipase	Acyl Donor	R^1	Solvent	Configuration	e.e. [%]
Me	PSL	vinyl acetate	Me	CHCl$_3$	S	>98
CH$_2$-Ph	PSL	vinyl acetate	Me	none	R	>94
CH$_2$-1-naphtyl	PSL	vinyl acetate	Me	none	R	86
O-CH$_2$-Ph	PSL	vinyl stearate	n-C$_{17}$H$_{35}$	i-Pr$_2$O	S	92
O-CH$_2$-Ph	PSL	i-propenyl acetate	Me	CHCl$_3$	S	96
NH-Z	PPL	vinyl pentanoate	n-C$_4$H$_9$	THF	R	97
O-CH$_2$-Ph	PSL	vinyl acetate	Me	none	S	90[a]
O-CH$_2$-Ph	PSL	vinyl acetate	Me	none	S	92[b]
O-CH$_2$-Ph	PSL	vinyl acetate	Me	none	S	94[c]

[a] Performed at 25 °C, [b] at 17 °C and [c] at 8 °C.

Cyclic *meso-cis*-diols were asymmetrically acylated quite efficiently to give the respective chiral monoester by a PSL [102]. Whereas a slow reaction rate was observed when ethyl acetate was used as acyl donor, the reaction was about ten times faster when vinyl acetate was employed.

Primary alcohols may be resolved with moderate to good selectivities by *Pseudomonas* sp. lipase (PSL) using vinyl acetate [103] or acetic anhydride as the acyl donor (Scheme 3.8). Whereas the selectivities achieved were moderate with alkyl and aryl substituents, the introduction of a sulfur atom in R^2 helped considerably (Scheme 3.8). In this way, chiral isoprenoid synthons having a C$_5$-backbone were obtained in >98% enantiomeric excess.

Scheme 3.8. Enantioselective acylation of primary alcohols.

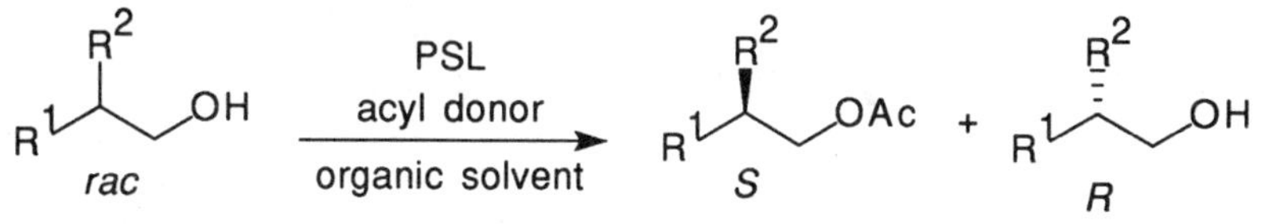

Acyl Donor	Solvent	R^1	R^2	e.e. of		Selectivity
				Ester [%]	Alcohol [%]	(E)
Ac_2O	benzene	Me	Ph	8	28	12
Ac_2O	benzene	Et	n-C_4H_9	17	36	2
vinyl acetate	$CHCl_3$	Me	$(CH_2)_2SPh$	>98	>98	100
vinyl acetate	$CHCl_3$	Me	$(CH_2)_2SO_2$-Ph	>98	>98	100

Numerous acyclic secondary alcohols have been separated into their enantiomers using a lipase-catalysed acyl-transfer reaction [104]. As long as the substituents are different in size and one of them carries an aromatic group, excellent selectivities were obtained with *Pseudomonas* sp. lipases.

The Katsuki-Sharpless epoxidation of allylic alcohols constitutes one of the most important developments in asymmetric synthesis during the last decade [105]. Concepts based upon asymmetric epoxidations also include some highly selective kinetic resolutions [106], but there are some limitations in this approach. For instance, they are not generally applicable to substrates other than allyllic alcohols and the products have to be purified from significant amounts of catalyst residues. A recent study shows that secondary alcohols containing an alkene or alkyne moiety can be conveniently resolved using an irreversible acyl-transfer approach (Scheme 3.9) [107]. The optical purities generally were excellent and the procedures may be readily scaled up using a standard organic laboratory equipment.

Enantiomerically pure *trans*-cycloalkane-1,2-diols are of interest for the synthesis of optically active crown-ethers [108] or as chiral auxiliaries for the preparation of bidentate ligands [109]. A convenient method for their preparation is shown in Scheme 3.10. PSL catalysed enantioselective acylation produces varying amounts of diester, monoester and remaining non-accepted diol in excellent optical purities [110]; the ratio of products is dependent on the ring size. Although compounds of such a type have been previously obtained via enzymatic hydrolysis of the corresponding diesters, the product monoesters are notoriously unstable in an aqueous medium due to the ease of acyl group migration [111-113]. Using acyl-transfer reactions, which are performed in an organic solvent and in the absence of water, acyl migration is not observed.

Scheme 3.9. Resolution of allylic and acetylenic alcohols.

R¹	R²	e.e. Acetate [%]	e.e. Alcohol [%]	Selectivity (E)
Me	Ph-(C=CH$_2$)-	>95	>95	>20
CH$_2$=CH-	Ph-	>95	46	>10
CH$_2$=CH-	(E)-Ph-CH=CH-	>95	>95	>20
CH$_2$=CH-	Ph-C≡C-	>95	>95	>20
HC≡C-	(E)-Ph-CH=CH-	>95	68	>20
Me	(E,E)-Me-(CH=CH)$_2$-	69	79	10-20
Me	n-Bu-C≡C-	>95	87	>20
Me	Ph-C≡C-	>95	>95	>20
Et	n-Bu-C≡C-	>95	82	>20
Me	Me$_3$Si-C≡C-	>95	>95	>20

Scheme 3.10. Resolution of *trans*-cycloalkanediols.

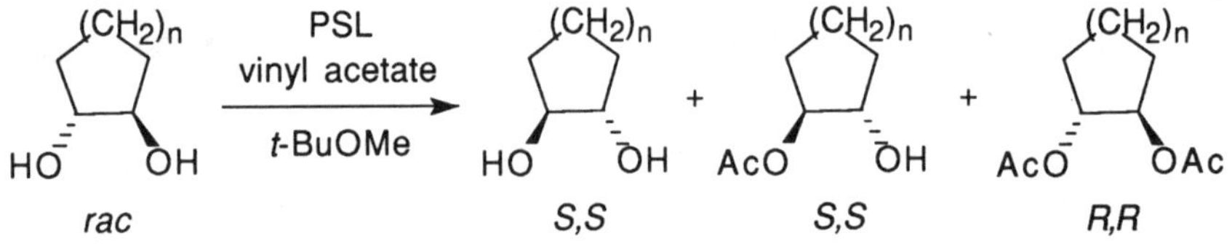

n	Diol		Monoacetate		Diacetate	
	Yield [%]	e.e. [%]	Yield [%]	e.e. [%]	Yield [%]	e.e. [%]
1	traces	n.d.	42	>95	44	>98
2	31	83	16	66	39	>98
3	24	>95	20	70	42	>98

An alternative method for the preparation of optically active epoxides makes use of a lipase-catalysed resolution of halohydrins (Scheme 3.11). *Pseudomonas* sp. lipase catalysed acylation of racemic halohydrins affords a readily separable mixture of (R)-halohydrin and the corresponding (S)-ester in good to excellent optical purities [114, 115]. Alkaline treatment of the latter leads to the formation of epoxides with no loss of optical purity. A semiquantitative comparison of the reaction rate obtained with different acyl donors using substrates of this type revealed that they were in an order of ethyl acetate << trichloroethyl acetate < isopropenyl acetate < vinyl butanoate ~ vinyl octanoate ~ vinyl acetate [116].

Scheme 3.11. Resolution of halohydrins.

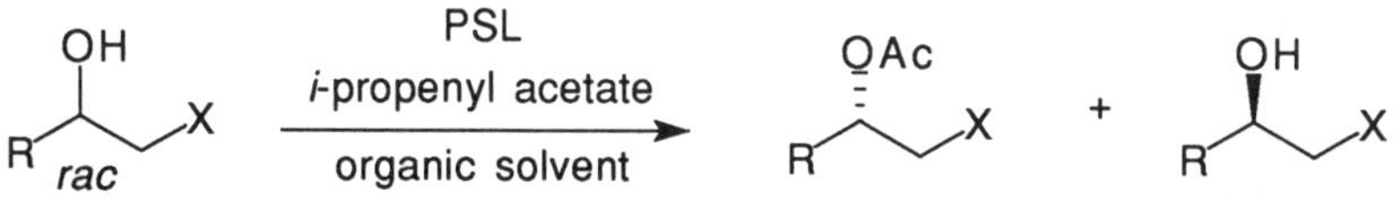

R	X	Solvent	e.e. of		Selectivity
			Acetate [%]	Alcohol [%]	(E)
Ph	Cl	i-Pr$_2$O	92	97	100
2-naphtyl	Br	i-Pr$_2$O	95	80	95
4-Br-C$_6$H$_4$-	Br	i-Pr$_2$O	95	94	140
4-MeO-C$_6$H$_4$-	Br	i-Pr$_2$O	93	87	80
3,4-(MeO)$_2$-C$_6$H$_3$-	Cl	i-Pr$_2$O	97	87	>180
TsO-CH$_2$-[a]	Cl	hexane	96	70	>100
TsO-CH$_2$-	n-C$_3$H$_7$-	hexane	86	66	26
TsO-CH$_2$-	n-C$_9$H$_{19}$-	hexane	87	37	21

[a] Ts = 4-Me-C$_6$H$_4$-SO$_2$-

The remarkable synthetic potential of enzyme-catalysed irreversible acyl-transfer, performed in nearly anhydrous organic solvents, can be demonstrated particularly well by the transformation of alcoholic substrates which are prone to decomposition reactions in an aqueous medium and thus cannot be transformed via enzyme-catalysed hydrolysis reactions.

For instance, organometallic compounds such as hydrolytically labile chromium-tricarbonyl complexes, which are of interest as chiral auxiliary reagents for asymmetric synthesis [117], were easily resolved by PSL (Scheme 3.12) [118, 119]. A remarkable enhancement in selectivity was obtained when the acyl moiety of the vinyl ester used as acyl donor was varied.

Scheme 3.12. Resolution of metalloorganic hydroxy compounds.

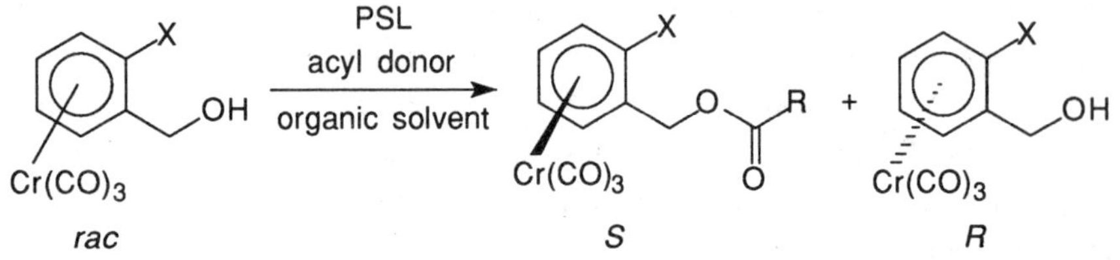

X	Acyl Donor	R	Solvent	e.e. of Ester [%]	e.e. of Alcohol [%]	Selectivity (E)
Me	i-propenyl acetate	Me	none	98	100	>200
OMe	i-propenyl acetate	Me	none	97	95	>200
SiMe3	i-propenyl acetate	Me	none	84	85	30
Me	vinyl benzoate	Ph	toluene	87	96	38
Me	vinyl acetate	Me	toluene	88	61	39
Me	vinyl octanoate	n-C$_7$H$_{15}$	toluene	90	96	67
Me	vinyl palmitate	n-C$_{15}$H$_{31}$	toluene	>98	>98	>200

Chiral 1-ferrocenylethanol, which cannot be well resolved via enzymatic hydrolysis due to the lability of 1-ferrocenyl acetate [120], was obtained in optical purities of greater than 90% using the same PSL-vinyl acetate system [121].

Optically pure cyanohydrins are required for the synthesis of synthetic pyrethroids, which are more environmentally acceptable agents for agricultural pest control than the classic highly chlorinated phenol derivatives [122]. They are important intermediates for the synthesis of chiral α-hydroxyacids, α-hydroxyaldehydes [123] and aminoalcohols [124]. By asymmetric hydrolysis of their respective acetates by microbial lipases [125], only the remaining non-accepted substrate enantiomer can be obtained in high optical purity because the cyanohydrin so-formed is spontaneously racemised since it is in equilibrium with the corresponding aldehyde and hydrocyanic acid at pH values >4 (see Section 2.1.3.2). In the absence of water, however, the cyanohydrins are stable and can be isolated in high optical and chemical yields (Scheme 3.13) [126, 127]. The enhanced stability of cyanohydrins in organic solvents has been shown to be advantageous for the oxynitrilase-catalysed formation of cyanohydrins [128].

3.1 Enzymes in Organic Solvents

Scheme 3.13. Enantioselective acylation of cyanohydrins.

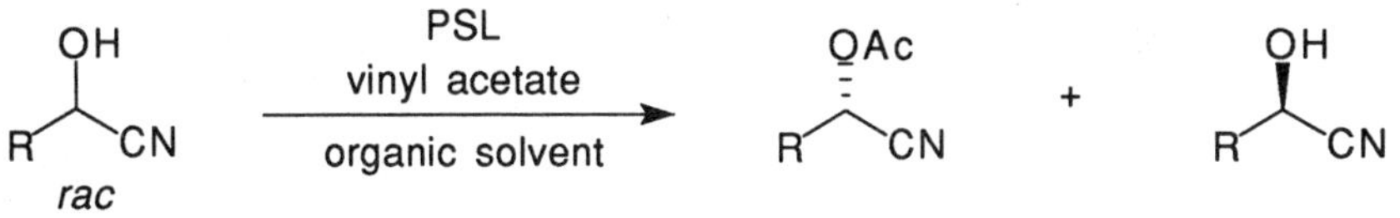

R	Solvent	e.e. of Acetate [%]	e.e. of Cyanohydrin [%]	Selectivity (E)
Ph-$(CH_2)_2$-	none	90[a]	98[b]	24
1-naphtyl-O-CH_2-	none	91[a]	96[b]	28
Ph-CH_2-O-CH_2-	none	55[c]	95[c]	12
n-C_3H_7-	CH_2Cl_2	55	33	5
Ph-	CH_2Cl_2	68	80	13
Ph-$(CH_2)_2$-	CH_2Cl_2	95	86	100
Benzo[1,3]dioxol-5-yl-	CH_2Cl_2	92	98	110
4-HO-C_6H_4-	CH_2Cl_2	79	50	14

[a] Conversion 22-25 %, [b] conversion 56-59%, [c] conversion 63%.

Chiral hydroxyesters can be obtained via enzymatic hydrolysis of their acyloxy derivatives, but a disadvantage which is commonly encountered in such resolutions is an undesired side-reaction involving the carboxyl moiety which leads to the formation of hydroxyacids as byproducts. Thus, low yields are often reported [129]. If the resolution is carried out as an acyl-transfer reaction, the unwanted side-reaction is completely suppressed because the hydroxyl functionality is the only nucleophile in the substrate molecule and no hydrolysis can take place on the carboxyl moiety due to the absence of water [130]. This concept was successfully applied for the resolution of γ-hydroxy-α,β-unsaturated esters which are used for the synthesis of statin analogues (Scheme 3.14) [131, 132]. An interesting reversal of the stereochemical preference of PSL was observed upon variation of the acyl chain of the substrate. With straight-chain agents, the (R)-enantiomer was acylated but when sterically more demanding branched derivatives of the acyl donor were chosen, the (S)-counterpart was preferred.

Scheme 3.14. Resolution of γ-hydroxy-α,β-unsaturated esters.

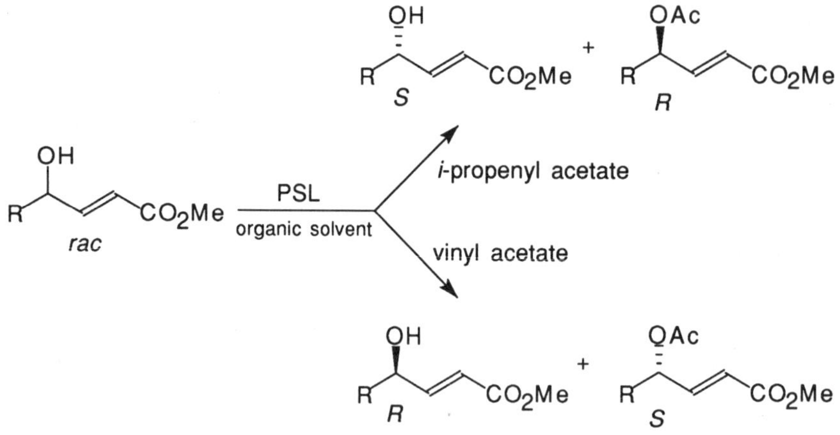

R	Acyl Donor	Acetoxy ester e.e. [%]	Configuration	Hydroxy ester e.e. [%]	Configuration	Selectivity (E)
Me	A	91	R	>95	S	>30
Et	A	>95	R	>95	S	>150
n-Pr	A	74	R	>95	S	>20
i-Pr	B	19	S	14	R	1.6
i-Pr-CH$_2$-	B	28	S	37	R	2.5
C$_6$H$_{11}$-CH$_2$-	B	77	S	54	R	13
Me$_2$ThexSiO-(CH$_2$)$_2$-	B	>95	S	72	R	>150

Acyl donor: A i-propenyl acetate, B vinyl acetate.

This strategy has been successfully applied to the preparation of optically active α-methylene-β-hydroxy esters and -ketones [133], which cannot be resolved using the Sharpless-epoxidation technique because of the deactivating influence of the electron-withdrawing alkene-unit [134]. Similarly, optically active cyclopentanoids carrying a terminal carboxylate group useful for prostaglandin synthesis were obtained without the occurrence of undesired side-reactions [135].

Racemic hydroperoxides may be resolved in organic solvents via lipase-catalysed acyl-transfer (Scheme 3.15). Although the so-formed acetylated (R)-peroxy-species is unstable and readily decomposes to form the corresponding ketone via elimination of acetic acid, the remaining (S)-hydroperoxide was isolated in varying optical purity [136]. This concept was recently applied to a hydroperoxy derivative of an unsaturated fatty acid ester [137].

Scheme 3.15. Resolution of hydroperoxides.

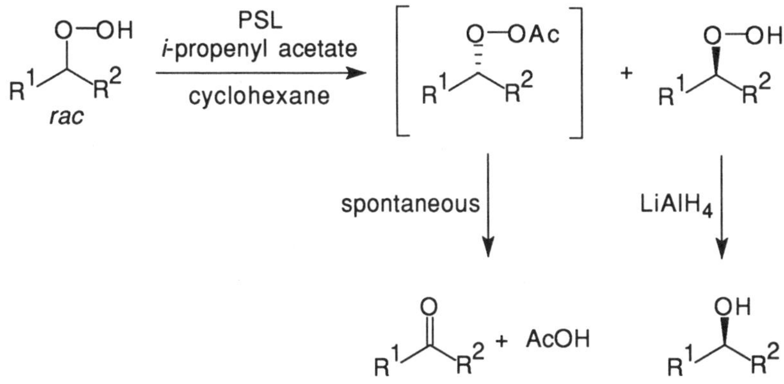

R¹	R²	e.e. Hydroperoxide [%]	Selectivity (E)
Me	Ph	100	>20
Me	2-naphtyl	58	2.3
Et	Ph	62	3.7
Me	n-Pr	10	1.2

Regioselective Protection of Polyhydroxy Compounds

Selective protection and deprotection of compounds containing multiple hydroxyl groups such as carbohydrates and steroids is a current problem in organic synthesis [138]. Often a series of steps is required to achieve the desired combination of protected and free hydroxyl functionalities using standard methodology. Enzymatic acyl-transfer reactions in organic solvents have been proven to be extremely powerful for such transformations by making use of the regioselectivity of hydrolytic enzymes. Whereas the acylation of steroids in lipophilic organic solvents having a desired log P of greater than 3 or 4 is comparatively facile, carbohydrate derivatives are only scarcely soluble in these circumstances. Thus, more polar solvents such as dioxane, THF, 3-methyl-3-pentanol, DMF, or even pyridine have to be used. As a consequence, only the most stable enzymes such as PPL or subtilisin can be used.

Paralleling the difference in reactivity between primary and secondary hydroxyl groups, the former can be selectively acylated by using PPL in THF [139] or pyridine [140] as the solvent (Scheme 3.16). Unlike most of the other hydrolases, the protease subtilisin is stable enough to remain active even in anhydrous DMF [141]. In general, activated esters such as trihaloethyl esters have been used as acyl donors.

Scheme 3.16. Regioselective protection of primary hydroxy groups of carbohydrates.

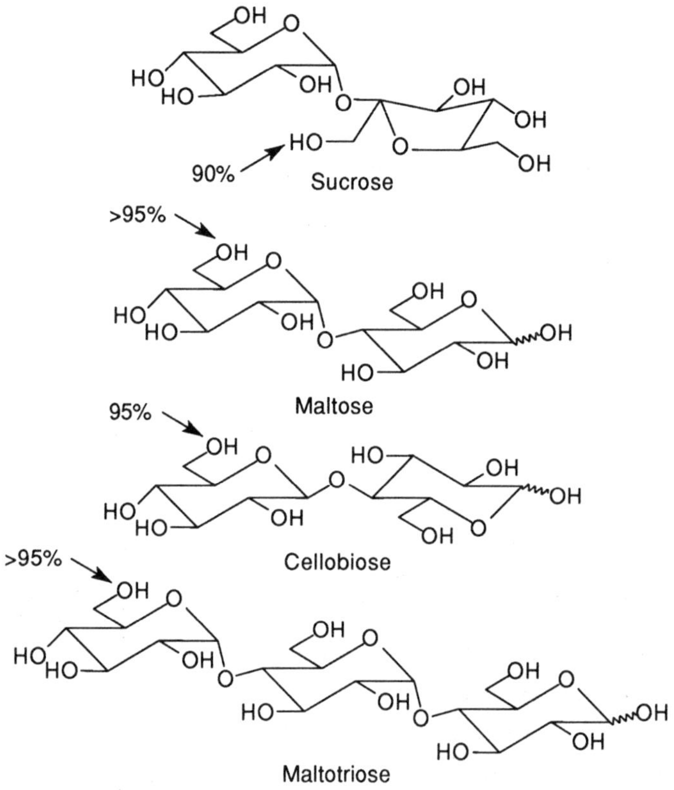

A greater challenge, however, is the regioselective discrimination of secondary hydroxyl groups due to the close similarity of their reactivity. This has been accomplished with steroids using anhydrous acetone [142] or benzene as solvent [143]. Due to the more lipophilic character of the solvents used, several lipases have also been employed in addition to subtilisin. For the regioselective acylation of secondary hydroxyl groups of sugar derivatives with blocked primary hydroxyl groups, pyridine [144] or mixed solvent systems (CH$_2$Cl$_2$/THF/acetone) have been used [145]. The potential of different lipases for the regioselective acylation of secondary hydroxyl groups is illustrated below. A highly selective discrimination between the two secondary hydroxyl

3.1 Enzymes in Organic Solvents

groups of a 1,4-anhydro-D-arabinitol derivative was achieved by using either a lipase from *Rhizopus javanicus* or from *Humicola lanuginosa* [146].

Scheme 3.17. Regioselective protection of secondary hydroxyl groups.

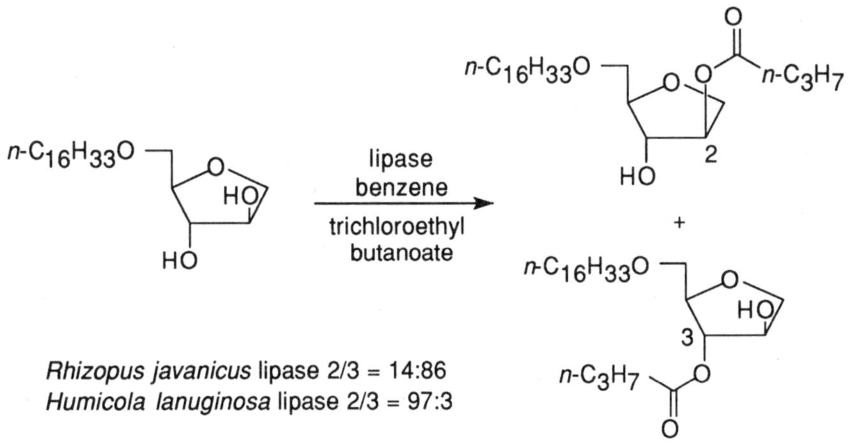

Rhizopus javanicus lipase 2/3 = 14:86
Humicola lanuginosa lipase 2/3 = 97:3

Scheme 3.18. Regioselective acylation of castanospermine.

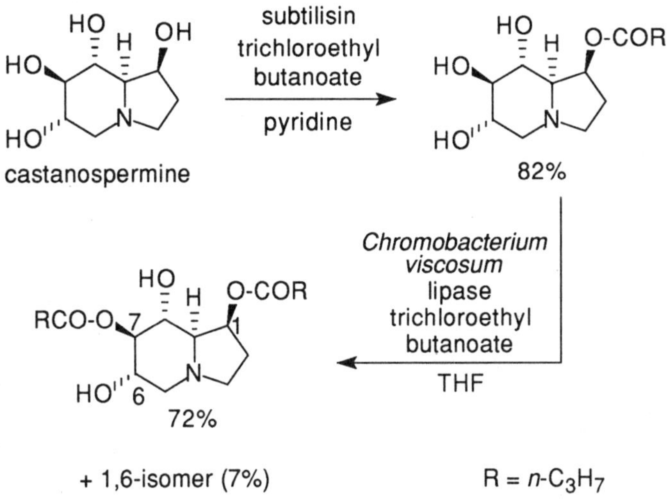

+ 1,6-isomer (7%)   R = n-C$_3$H$_7$

The plant alkaloid castanospermine is a potent glucosidase-inhibitor and is currently being considered as an antineoplastic agent and also for the treatment of AIDS [147]. Some of the corresponding *O*-acyl derivatives, which have been reported as being more active than castanospermine itself, were obtained by an enzyme-catalysed regioselective acylation reaction (Scheme 3.18) [148]. Thus, the 1-OH group was selectively acylated using subtilisin as catalyst and *Chromobacterium viscosum* lipase was employed to esterify the hydroxyl group in position 7. Only minor amounts of the 1,6-isomer were detected.

3.1.2 Lactone Synthesis

Bearing in mind the enzyme catalysed esterification and transesterification reactions, it is not surprising that lactones may be obtained from hydroxyacids or -esters via intramolecular cyclisation reactions [149]. Under chemical catalysis, the course of the reaction is relatively simple. The formation of either lactones or open-chain oligomers mainly depends on the ring-size of the product. Lactones with less than five and more than seven atoms in the ring are not favoured and linear condensation products are formed predominantly; in contrast, five-membered lactones are easily formed. The formation of six membered structures is often accompanied by the formation of straight-chain oligomers. The corresponding cyclic dimers - diolides - are usually not obtained by chemical methods.

Scheme 3.19. Lipase-catalysed formation of lactones, diolides and oligomers.

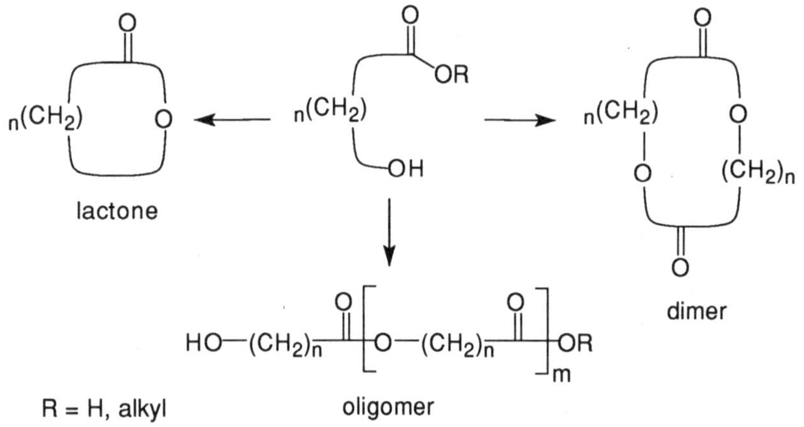

In contrast, the success or failure of lipase-catalysed lactonisation depends on several parameters. The product profile depends not only on the length of the hydroxy acid, but also on the type of lipase, the solvent, the dilution, and even on the temperature [150, 151]. When racemic or prochiral hydroxyacids

3.1 Enzymes in Organic Solvents

are employed as substrates, a kinetic resolution [152] or asymmetrisation may be accomplished with high selectivities [153]. It is obvious that this is particularly easy with γ-hydroxy derivatives. The most important synthetic aspect of enzymatic lactone formation, however, is the possibility to direct the condensation reaction towards the formation of macrocyclic lactones and dilactones - i.e. macrolides and macrodiolides respectively (Scheme 3.19) [154, 155]. This strategy was employed as the key step in the synthesis of the naturally occurring antifungal agent (-)-pyrenophorin, a 16-membered macrocyclic diolide [156].

3.1.3 Amide Synthesis

The aminolysis of esters leads to the formation of amides. Enzymatic catalysis of this reaction may be advantageous for the synthesis of amides, such as propargylamides, which cannot be obtained by using chemical catalysis [157]. The analogous hydrazinolysis of esters leading to the formation of hydrazides under mild reaction conditions, may be performed using enzymatic catalysis in a similar manner [158-160].

Scheme 3.20. Resolution of α-methylbenzylamine via ester-aminolysis.

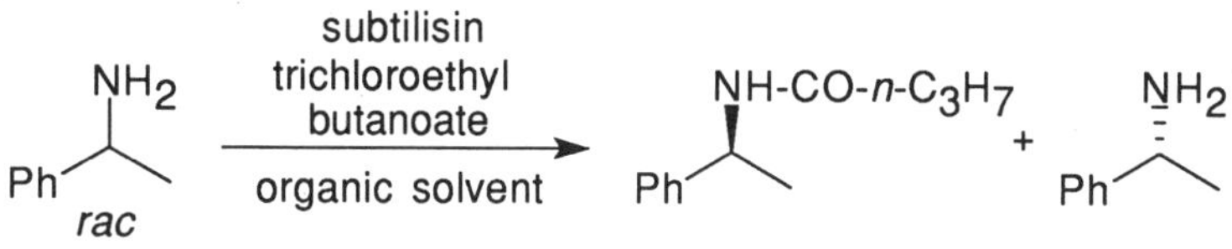

Solvent	Selectivity [v_S/v_R]
octane	1.4
n-But$_2$O	1.8
pyridine	2.5
tetrahydrofuran	3.5
2-methyl-2-butanol	4.1
3-methyl-3-pentanol	7.7

More important, however, is the enantioselective formation of amides, where a centre of chirality may be located either on the ester [161, 162], the amine [163] or on both entities. The latter reaction leads to the formation of diastereomeric amides [164]. One example for the resolution of an amine via enantiospecific ester-aminolysis is given in Scheme 3.20 [165]. α-Methylbenzylamine was resolved using trifluoroethyl butanoate and subtilisin as catalyst. The influence of the organic solvent on the selectivity of the reaction, expressed as the ratio of the specific rate constants v_S and v_R, was significant. Replacement of lipophilic solvents by more polar ones led to a more than five-fold increase in selectivity.

3.1.4 Peptide Synthesis

Peptides display a diverse range of biological activity. They may be used as sweeteners and toxins, antibiotics and chemotactics, as well as growth factors. They play an important rôle in hormone release as either stimulators or inhibitors. The most recent application is their use as immunogens for the generation of specific antisera. At present, the most frequently used methods of peptide synthesis are purely chemical in nature, generally proceeding through a sequence of four steps. First, all the functionalities of the educts, which are not to participate in the reaction, must be selectively protected. Then, the carboxyl group must be activated in a second step to enable the formation of the peptide bond. Finally, the protective groups have to be removed in toto, if the synthesis is completed, or α-amine- or α-carboxyl groups have to be selectively liberated, if the synthesis is to be continued. Two of the major problems associated with chemical peptide synthesis are a danger of racemisation - particularly during the activation step - and the tedious purification of the final product from isomeric peptides with a closely related sequence. To circumvent these problems, peptide synthesis is increasingly carried out by making use of the specificities of enzymes, in particular proteases [166-172]. This field has been intentely investigated during the last decade, although the first report on an enzyme-catalysed peptide synthesis dates back to 1938 [173]. The pros and cons of conventional versus enzymatic peptide synthesis may be summarized in Table 3.4.

Table 3.4. Pros and cons of chemical and enzymatic peptide synthesis.

	Chemical	Enzymatic
stereoselectivity	low	high
regioselectivity	low	high
amino acid range	broad	limited
protective group requirements	high	low
purity requirements of starting materials	high	moderate
by-products	some	negligible
danger of racemisation	some	none

3.1 Enzymes in Organic Solvents

Scheme 3.21. Principles of enzymatic peptide synthesis.

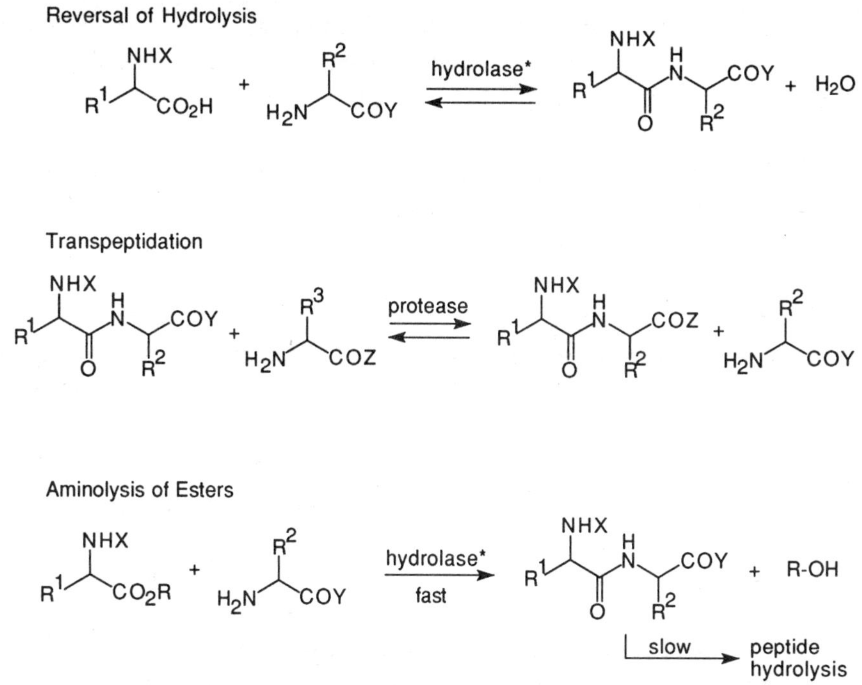

Reversal of Hydrolysis

$$R^1\text{-CH(NHX)-CO}_2\text{H} + \text{H}_2\text{N-CH}(R^2)\text{-COY} \xrightleftharpoons{\text{hydrolase*}} R^1\text{-CH(NHX)-CO-NH-CH}(R^2)\text{-COY} + \text{H}_2\text{O}$$

Transpeptidation

$$R^1\text{-CH(NHX)-CO-NH-CH}(R^2)\text{-COY} + \text{H}_2\text{N-CH}(R^3)\text{-COZ} \xrightleftharpoons{\text{protease}} R^1\text{-CH(NHX)-CO-NH-CH}(R^3)\text{-COZ} + \text{H}_2\text{N-CH}(R^2)\text{-COY}$$

Aminolysis of Esters

$$R^1\text{-CH(NHX)-CO}_2R + \text{H}_2\text{N-CH}(R^2)\text{-COY} \xrightarrow[\text{fast}]{\text{hydrolase*}} R^1\text{-CH(NHX)-CO-NH-CH}(R^2)\text{-COY} + R\text{-OH}$$

$\xrightarrow{\text{slow}}$ peptide hydrolysis

X = N-terminal blocking group (e.g. Ph-CH$_2$-O-CO-, *t*-Bu-O-CH-)
Y, Z = C-terminal blocking group (e.g. *t*-Bu-, Ph-CH$_2$-, Ph-NH-NH-)
R = leaving group (e.g. Me, Et)
* esterase, lipase or protease

Enzyme-catalysed peptide synthesis may be conducted via three basic ways: *Reversal of hydrolysis, transpeptidation* (which is used to a lesser extent) and *aminolysis of esters* (Scheme 3.21).

Both of the former methods represent a reversible type of reaction and are thermodynamically controlled.

Under physiological conditions, the equilibrium position in protease-catalysed reactions is far on the side of proteolysis. In order to create a driving force in the direction towards peptide synthesis, the following constraints may be applied.
- One of the reactants may be used in excess.
- Removal of product via formation of an insoluble product [174], by specific complex formation [175], or by extraction of the product into an organic phase by using a water-immiscible organic cosolvent.

- Lowering the water-activity (concentration) of the system by addition of water-miscible organic cosolvents. In this respect, polyhydroxy compounds such as glycerol or butane-1,4-diol have been shown to conserve enzyme activity better than the solvents which are more commonly employed, such as DMF, DMSO, ethanol, acetone or acetonitrile [176].

The third method - aminolysis of esters - involves a kinetically controlled irreversible type of reaction, in which two nucleophiles (water and an amine) are competing for the acyl-enzyme intermediate. Thus it is not surprising that besides proteases, other hydrolases which are capable of forming an acyl-enzyme intermediate (mainly lipases such as PPL [177], PLE and CCL [178, 179]) may be effectively used for this reaction. On the other hand, this method is not applicable to metallo- and carboxyproteases. Since the peptide formed during aminolysis of an ester is hydrolytically cleaved in a subsequent slow reaction, these reactions have to be terminated before the equilibrium is reached. Of course this second reaction may be neglected when non-proteolytic hydrolases are used.

Because all enzymes exhibit substrate selectivity to some extent, the availability of a library of enzymes which can cover all possible types of peptide bonds is of crucial importance. Although the existing range of proteases is far from complete, it provides a reasonably coverage (Table 3.5). The most striking 'shortage' involves proline derivatives. The most commonly used proteases and their crude selectivities are listed below, where X stands for an unspecified amino acid or peptide residue.

The dipeptide Asp-Phe-OMe (Aspartame) is used in large amounts as low calorie sweetener. One of the most economical strategies for its synthesis involves an enzymatic step (Scheme 3.22). Z-Protected aspartic acid is linked with Phe-OMe in a thermodynamically controlled condensation catalysed by thermolysin. Removal of the product via formation of an insoluble salt was used as driving force to shift the equilibrium of the reaction towards the synthetic direction [180].

The potential of enzymatic peptide synthesis is illustrated with the chemo-enzymatic synthesis of an enkephalin derivative - (D-Ala$_2$, D-Leu$_5$)-enkephalin amide (Scheme 3.23) [181]. The enkephalins commonly undergo rapid enzymatic degradation in vivo, and therefore exert only limited biological activities. However, by substitution of a D-amino acid into the sequence, 'chirally muted' derivatives are obtained, which can elicit long-lasting pharmacological effects [182]. The Tyr-D-Ala and Gly-Phe subunits were obtained via an ester-aminolysis and a condensation reaction using α-

3.1 Enzymes in Organic Solvents

chymotrypsin and thermolysin, respectively. The latter dipeptide was extended by a D-Leu unit using the same methodology. After converting the C-terminal hydrazide protective group into an azide moiety and removal of the N-terminal benzyloxycarbonyl-(Z)-group by hydrogenolysis, the fragments were coupled by conventional methodology.

Table 3.5. Selectivities of proteases.

Protease	Type	Specificity
Achromobacter protease	serine protease	-Lys-X
α-chymotrypsin, subtilisins	serine protease	-Trp(Tyr,Phe,Leu,Met)-X
carboxypeptidase Y	serine protease	non-specific
elastase	serine protease	-Ala(Ser,Met,Phe)-X
trypsin, *Streptomyces griseus* protease	serine protease	-Arg(Lys)-X
Staphylococcus aureus V8 protease	serine protease	-Glu(Asp)-X
papain, ficin	thiol protease	-Phe(Val,Leu)-X
clostripain, cathepsin B	thiol protease	-Arg-X
cathepsin C	thiol protease	H-X-Phe(Tyr,Arg)-X
thermolysin, *B. subtilis* protease	metalloprotease	-Phe(Gly,Leu)-Leu(Phe)
Myxobacter protease II	metalloprotease	X-Lys
pepsin, cathepsin D	carboxyl protease	-Phe(Tyr,Leu)-Trp(Phe,Tyr)

Scheme 3.22. Enzymatic synthesis of Aspartame.

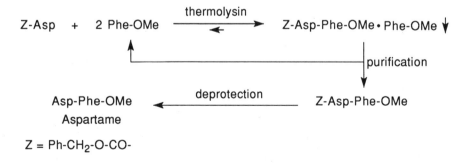

Z = Ph-CH$_2$-O-CO-

Scheme 3.23. Chemo-enzymatic synthesis of an enkephalin-derivative.

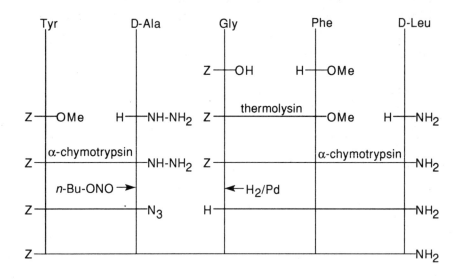

Scheme 3.24. Enzymatic conversion of porcine into human insulin.

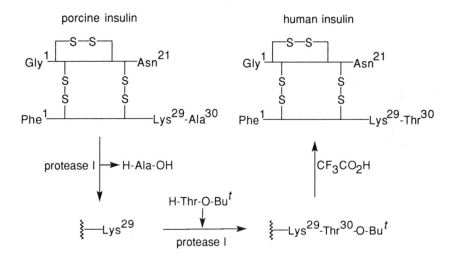

Probably the most prominent example of an enzymatic peptide synthesis is the transformation of porcine insulin into its human counterpart. As millions of diabetics suffer from insulin deficiency, and due to the fact that the demands cannot be satisfied by exploiting natural sources for obvious reasons, numerous attempts have been undertaken to convert porcine insulin into human insulin via

3.1 Enzymes in Organic Solvents

exchange of a terminal alanine by a threonine residue [183-185]. A protease from *Achromobacter lyticus* which is completely specific for peptide bonds formed by a lysine residue [186] is used to hydrolyse the terminal Ala^{30} residue from porcine insulin (Scheme 3.24). The same enzyme is then used to catalyse the condensation reaction with threonine, protected as its *tert*-butyl ester. The *tert*-butyl group was removed by acid treatment to yield human insulin [187].

3.1.5 Peracid Synthesis

In contrast to the highly acidic conditions usually applied for in situ generation of peroxycarboxylic acids, they may be generated under virtually neutral conditions in a suitable organic solvent directly from the parent carboxylic acid and hydrogen peroxide via lipase catalysis (Scheme 3.25).

Scheme 3.25. Lipase-catalysed peracid formation and catalytic epoxidation.

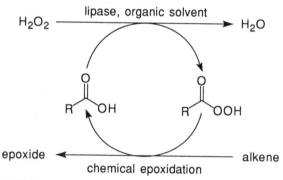

Alkene	Yield Epoxide [%]
1-octene	~5
3-ethylpent-2-ene	~80
tetramethylethylene	>95
cyclohexene	>95

The mechanism involves a perhydrolysis of the acyl-enzyme intermediate by the nucleophilic hydrogen peroxide. The peroxy acids formed can be used in situ for the epoxidation of alkenes, liberating the fatty acid which re-enters the cyclic process [188]. It should be noted that the epoxidation reaction takes place without enzymatic catalysis, therefore no selectivities can be expected. The main advantages of this method are the mild conditions employed and a higher safety margin due to the fact that only catalytic concentrations of peracid are involved. Medium-chain alkanoic acids (C_8 to C_{16}) and a biphasic aqueous-organic solvent system containing toluene or hexane give the best yields. Among various lipases tested, an immobilized lipase from *Candida antarctica* was shown to be superior to CCL and PSL.

3.1.6 Redox Reactions

In contrast to hydrolases, redox enzymes such as dehydrogenases and oxidases have been used more scarcely in organic solvents because they require cofactors, e.g. nicotinamide adenine dinucleotide species.

Scheme 3.26. Asymmetric reduction and oxidation using HLADH.

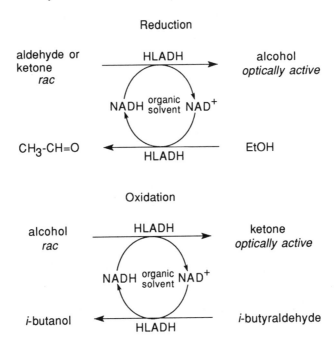

rac-Substrate	Product	e.e. Product [%]
2-phenylpropionaldehyde	(-)-2-phenylpropanol	95
2-chlorocyclohexanone	(+)-*trans*-2-chloro-cyclohexanol	98
trans-3-methylcyclohexanol	(S)-3-methylcyclohexanone	100
cis-2-methylcyclopentanol	(S)-2-methylcyclopentanone	96

3.1 Enzymes in Organic Solvents

The latter are highly polar compounds and are therefore completely insoluble in a lipophilic medium. The employment of such systems seems to be severely hampered by ineffective interaction of two suspended insoluble species - i.e. the enzyme and the cofactor. These limitations can be circumvented by co-precipitation of enzyme and cofactor onto a macroscopic carrier [189] provided that a minimum amount of water is present. Thus, the cofactor is able to freely enter and exit the active site of the enzyme but it cannot disaggregate from the carrier into the milieu because it is trapped together with the enzyme in the hydration layer on the carrier surface. One particular advantage of this technique is an enhanced cofactor stability which results in significantly higher turnover numbers.

The direction of the reduction catalysed by horse liver alcohol dehydrogenase can be reversed if the recycling system is appropriately modified (Scheme 3.26). Reduction of aldehydes/ketones and oxidation of alcohols is effected by NADH- or NAD$^+$-recycling, using ethanol or *iso*-butyraldehyde respectively.

Scheme 3.27. Polyphenol oxidase catalysed regioselective hydroxylation of phenols.

R = H-, Me-, MeO-, HO$_2$C-(CH$_2$)$_2$-, HO-CH$_2$-, HO-(CH$_2$)$_2$-.

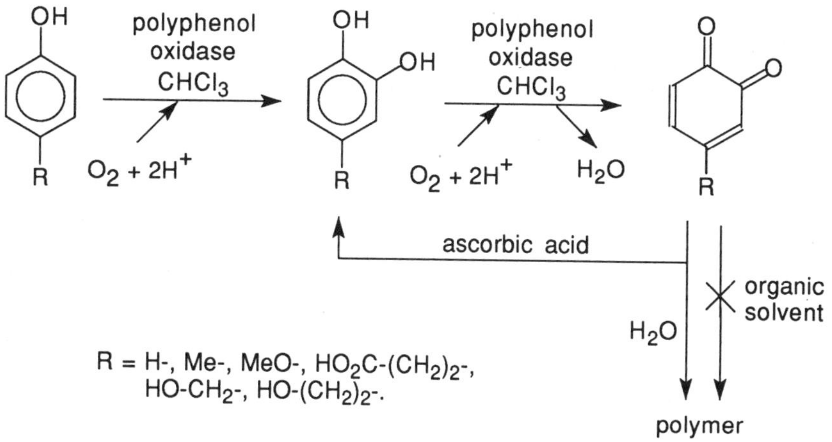

Polyphenol oxidase - an oxygenase - catalyses the hydroxylation of phenols to catechols and subsequent dehydrogenation to *o*-quinones [190]. The preparative use of this enzyme for the regioselective hydroxylation of phenols is impeded by the instability of *o*-quinones in aqueous media, which causes rapid polymerisation to form polyaromatic pigments and subsequent enzyme deactivation [191]. Since water is an essential component of the polymerisation reaction, the *o*-quinones formed are stable when the enzymatic reaction is performed in an organic solvent (Scheme 3.27). Subsequent nonenzymatic chemical reduction of the *o*-quinones by e.g. ascorbic acid to form catechols leads to a net regioselective hydroxylation of phenols [192]. Depending on the substituent R in *para*-position (*o*-, and *m*-cresols were unreactive and electron-withdrawing and bulky substituents decreased the reactivity), cresols were obtained in good yields. The preparative use of this method was demonstrated by the conversion of *N*-acetyl-L-tyrosin ethyl ester into the corresponding DOPA-derivative. An incidential observation that the related enzyme horseradish peroxidase remains active in nearly anhydrous organic solvent was reported in the late 1960s [193].

3.2 Immobilization

In practice, two significant drawbacks are often encountered in enzyme-catalysed reactions.
- Many enzymes are not sufficiently stable under the operational conditions and they may lose catalytic activity due to autooxidation, self-digestion and/or denaturation by the solvent, the solutes or due to mechanical shear forces.
- Since enzymes are water-soluble molecules, their repeated use, which is important to ensure an economic process [194], is problematic due to the fact that they are difficult to separate from substrates and products.
- The productivity of industrial processes, measured as the space-time yield is often low due to the limited tolerance of enzymes to high concentrations of substrate(s).

These problems may be overcome by immobilization of the enzyme [195-200]. This technique involves either the attachment of an enzyme to a solid support (coupling onto a carrier) or linkage of the enzyme molecules to each other (cross-linking). Alternatively, the biocatalyst may be confined to a restricted area from which it cannot leave but where it remains catalytically active (entrapment into a solid matrix or a membrane-restricted compartment). As a consequence, homogeneous catalysis using a native enzyme turns into heterogeneous catalysis when immobilized biocatalysts are employed. Depending on the immobilization technique, the properties of the biocatalyst such as stability, selectivity [201], K_m-value, pH- and temperature characteristics, may be significantly altered [202], sometimes for the better, sometimes for the worse. At present, predictions about the effects of immobilization are very difficult to make. The following immobilization techniques are known (Figure 3.1) [203].

Adsorption

Adsorption of a biocatalyst onto a water-insoluble macroscopic carrier is the easiest and oldest method of immobilization. It may be equally well applied to isolated enzymes as well as to whole viable cells. For example, adsorption of whole cells of *Acetobacter* to wood chips for the fermentation of vinegar from ethanol was first used in 1815! Adsorbing forces are of different types (Van der Waals, ionic interactions and hydrogen bonding) and are relatively weak. The appealing feature of immobilization by adsorption is the simplicity of the procedure. Additionally, losses in enzyme activity are usually low, but desorption (leakage) from the carrier is caused by minor changes in the reaction parameters, such as a variation of substrate concentration, temperature or pH.

Numerous inorganic and organic materials have been used as carriers: activated charcoal [204], aluminium oxide [205], diatomaceous earth (Celite) [206], cellulose [207] and synthetic resins [208]. In contrast to the majority of enzymes, which preferably adsorb to materials having a polar surface, lipases are better adsorbed to lipophilic carriers due to their peculiar physicochemical character (see Section 2.1.3.2) [209-211]. Adsorption is the method of choice when enzymes are used in lipophilic organic solvents, where desorption cannot occur.

Figure 3.1. Principles of immobilization techniques.

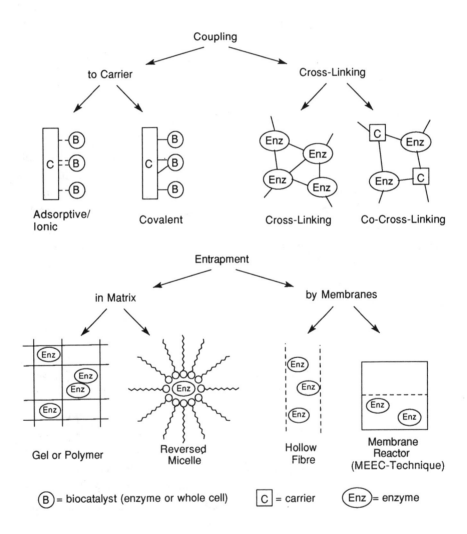

Ionic Binding

Ion exchange resins readily adsorb proteins and have been widely employed for enzyme immobilization. Both cation exchange resins such as carboxymethyl cellulose or Amberlite IRA [212], and anion exchange resins, e.g. diethyl-aminoethylcellulose (DEAE-cellulose) [213] or sephadex [214], are used industrially. Although being stronger in nature compared to the forces involved in simple physical adsorption, ionic binding forces are particularly susceptible to the presence of other ions. As a consequence, proper maintainance of ion concentrations and pH is important for continued immobilization by ionic binding and for prevention of desorption of the enzyme. When the biocatalytic activity is exhausted, the carrier may be reused by re-loading it with fresh biocatalyst.

Covalent Attachment

Covalent binding of an enzyme to a macroscopic carrier leads to the formation of stable linkages thus inhibiting leakage completely. A disadvantage of this method is that rather harsh conditions are often employed since the biocatalyst undergoes a chemical reaction. Consequently, some loss of activity due to conformational changes of the enzyme is always observed. As a rule of thumb, each bond attached to an enzyme decreases its native activity by about one fifth. Consequently, residual activities of 60 to 80% (compared to that of the native enzyme) are normal. The functional groups of the enzyme which are commonly involved in covalent binding are the nucleophilic species: i.e. mainly α- and ε-amino groups, but also carboxy-, sulfhydryl-, hydroxyl- and phenolic functions. In general, covalent immobilization involves two steps, i.e. (i) activation of the carrier with a reactive spacer group and (ii) enzyme attachment. Since viable cells usually do not survive the drastic reaction conditions required for the formation of covalent bonds, this type of immobilization is only recommended for isolated enzymes.

Porous glass is a popular inorganic carrier for covalent immobilization [215]. Activation is achieved by silylation of the hydroxy groups using aminoalkylethoxy- or aminoalkyl-chlorosilanes as shown in Scheme 3.28. In the next step, the aminoalkyl groups attached to the glass surface are either transformed into reactive isothiocyanates or into Schiff´s bases by treatment with thiophosgene or glutardialdehyde, respectively. Both of the latter species are able to covalently bind an enzyme through its amino groups.

Carriers based on natural polymers of the polysaccharide type (such as cellulose [216], dextran [217], starch [218], chitin or agarose [219]) can be useful alternatives to inorganic material due to their well-defined pore size.

Activation is achieved by reaction of adjacent hydroxyl functions with cyanogen bromide leading to the formation of reactive imidocarbonates (Scheme 3.29). Again, coupling of the enzyme involves its amino groups. To avoid the use of hazardous cyanogen bromide, some of these pre-activated polysaccharide-type carriers may be obtained from commercial sources.

Scheme 3.28. Covalent immobilization of enzymes to inorganic carriers.

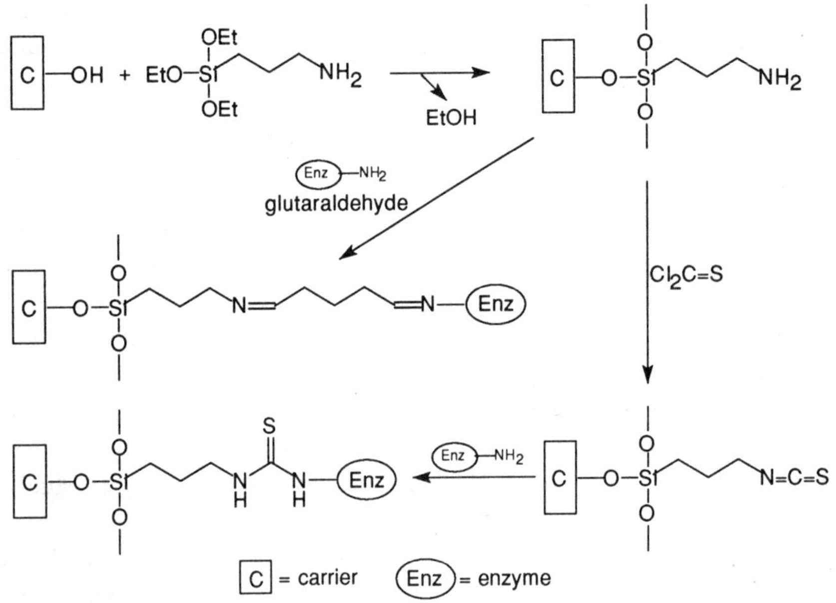

Scheme 3.29. Covalent immobilization of enzymes onto natural polymers.

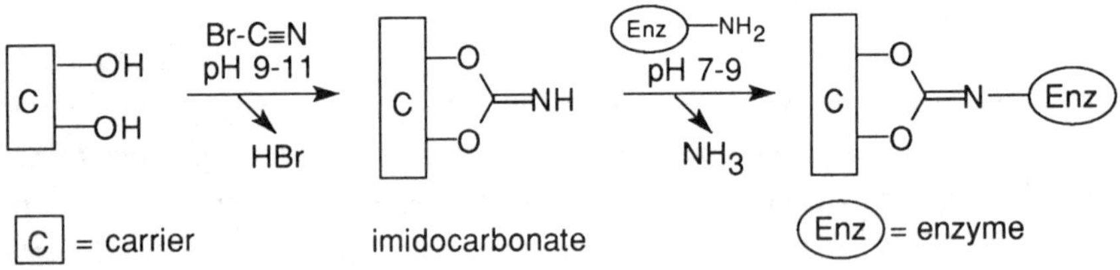

More recently, synthetic co-polymers (e.g. based on polyvinyl acetate) have become popular [220-222]. Partial hydrolysis of some of the acetates liberates hydroxyl functions, which are activated with epichlorohydrin [223]. A number of such epoxy-activated resins are commercially available (e.g. VA-Epoxy

Biosynth, Eupergit). Enzyme attachment occurs via nucleophilic opening of the epoxide groups by amino groups of the enzyme, under mild conditions, with the formation of a stable bond. In contrast to the majority of the above mentioned covalent immobilization reactions which remove a positive charge from the enzyme´s surface (due to the formation of Schiff´s bases or acylation involving an amino group), this method preserves the charge distribution of the enzyme since it constitutes an alkylation process.

Scheme 3.30. Covalent immobilization of enzymes onto synthetic polymers.

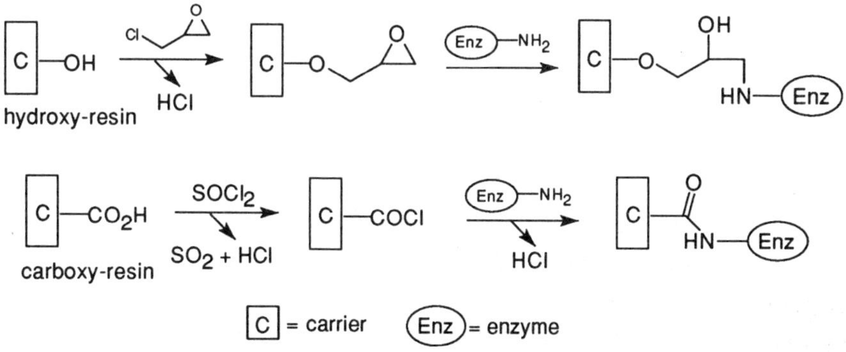

Alternatively, cation exchange resins can be activated by transforming their carboxyl groups into acid chlorides, which form stable amide bonds with the amino groups of an enzyme.

Cross-Linking

Attachment of enzymes to each other by covalent bonds is termed 'cross-linking'. By this means, insoluble high-molecular aggregates are obtained. The enzyme molecules may be cross-linked either with themselves or may be co-cross-linked with other inactive 'filler' proteins such as albumins. The most widely used bifunctional reagents used for this type of immobilization are α,ω-glutardialdehyde [224, 225], dimethyl adipimidate, dimethyl suberimidate and hexamethylenediisocyanate or -isothiocyanate. The advantage of this method is its simplicity, but it is not without drawbacks. The soft aggregates are often of gelatine-like nature, which prevents their use in packed bed reactors. Furthermore, the activities achieved are often limited due to diffusional problems, since many of the biocatalyst molecules are buried inside the complex structure which impedes their access to the substrate. The reactive groups involved in the cross-linking of an enzyme are not only free amino functions but also sulfhydryl- and hydroxyl groups.

Scheme 3.31. Cross-linking of enzymes by glutardialdehyde.

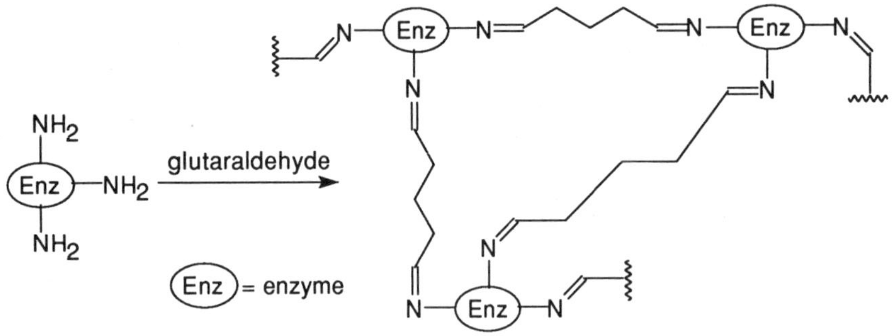

Entrapment into Gels

Biocatalysts may be physically encaged in a macroscopic matrix. To ensure catalytic activity, it is necessary that substrate and product molecules can pass into and out of the macroscopic structure unhindered. Depending on the type of reactor, the aggregate may be used in almost any shape (such as beads, fibres, foils or cylinders) thau is most convenient. Due to the lack of covalent binding, entrapment is a mild immobilization method which is also applicable to the immobilization of viable cells.

Figure 3.2. Gel-entrapment of viable cells.

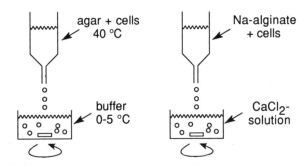

Entrapment into a biological matrix such as agar gels [226], alginate gels [227] or κ-carragenan [228] is frequently used for viable cells. As depicted in Figure 3.2, the gel-formation may be initiated either by variation of the

3.2 Immobilization

temperature or by changing the ionotropic environment of the system. For instance, an agar gel is easily obtained by dropping a mixture of cells suspended in a warm (40 °C) solution of agar into well-stirred ice-cold buffer. Alternatively, calcium-alginate or κ-carrageenan gels are prepared in a similar fashion by addition of a sodium alginate solution to a $CaCl_2$- or KCl-solution, respectively. The main drawbacks of such biological matrices are their instability towards changes in temperature or the ionic environment and their low mechanical stability.

A tight network which is able to contain isolated enzyme molecules (which are obviously much smaller than whole cells) may be obtained by polymerisation of synthetic monomers such as polyacrylamide in the presence of the enzyme [229-231]. It is obvious that the harsh conditions required for the polymerisation makes this method inapplicable to whole cells.

Scheme 3.32. Entrapment of enzymes by polymerisation.

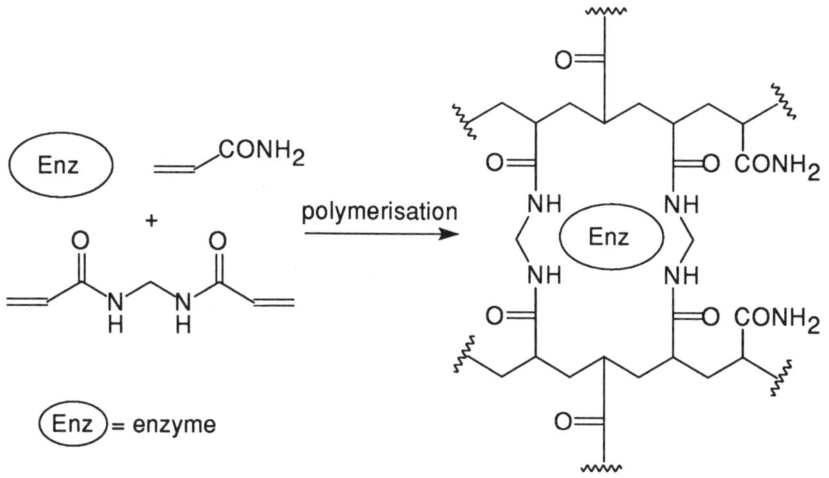

Entrapment into Membrane-Compartments

Enzymes may be enclosed in a restricted compartment bordered by a membrane. Although this does not lead to an 'immobilization' per se, it provides a restricted space for the enzyme which is then separated from the rest of the reaction vessel. Small substrate and/or product molecules can freely diffuse through the membrane, but the biocatalyst can not. The separation of a reaction volume into compartments by membranes is a close imitation of 'biological immobilization' within a living cell. Many enzymes are membrane-bound in order to provide a safe micro-environment for them within the cell.

Two general methods exist for the entrapment of enzymes into membrane-restricted compartments.

Figure 3.3. Entrapment of enzymes in reversed micelles and vesicles.

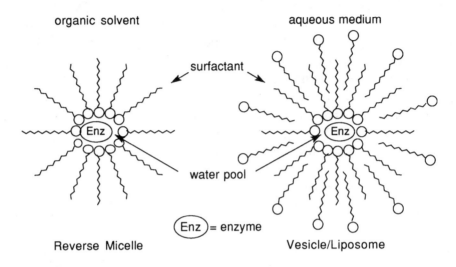

Mixtures of certain compositions containing water, an organic solvent and a detergent, give transparent solutions in which the organic solvent is the continuous phase [232]. The water is present in tiny droplets with a diameter of 6-40 nm [233], which are surrounded by the surfactant, such as Aerosol OT (sodium di-2-ethylhexyl sulphosuccinate). The whole structure is embedded in the organic solvent and represents a micelle which is turned inside out. It is therefore termed a 'reverse micelle' (see Figure 3.3). When water constitutes the bulk phase, micelles may be formed by a symmetrical double layer of surfactant. The latter structure constitutes a vesicle (liposome). The water trapped inside these micro-environments has several chemical and physical properties that deviate from 'normal' water, such as restricted molecular motion, decreased hydrogen bonding, increased viscosity and a depressed freezing point [234, 235]. Enzymes can be accommodated in these water-pools and can stay catalytically active [236, 237]. The exchange of material from one micelle to another occurs by means of collisions and is a very fast process. The catalytic activity of enzymes may be altered through entrapment in micelles. For instance, the activity [238, 239] and the temperature stability [240] is often enhanced; in some cases the specificity is changed [241, 242]. From a preparative point of view, it is important that, in these systems, compounds which are sparingly soluble in water (e.g. steroids) can be converted in much higher rates than would have been possible in aqueous media [243]. The disadvantages encountered when using micelles on a preparative scale are mainly operational problems during workup, due to the presence of a considerable amount of surfactant.

A practical alternative to the use of sensitive biological membranes is the use of synthetic membranes [244] based on polyamide or polyethersulfone (Figure 3.4). They have long been employed for the separation of enzymes by ultrafiltration, which makes use of the large difference in size between high-molecular biocatalysts and small substrate/products molecules. Synthetic membranes of defined pore-size covering the range between 500 to 300 000 Dalton are commercially available at reasonable cost. The biocatalyst is detained in the reaction compartment by the membrane, but small substrate/product molecules can freely diffuse through the barrier. This principle allows biocatalytic reactions to be performed in highly desirable continuous processes. Synthetic membranes are available in various shapes such as foils or hollow fibres.

A simplified form of a membrane reactor which does not require any special equipment may be obtained by using an enzyme solution enclosed in a dialysis bag (Figure 3.4). This simple technique termed 'membrane-enclosed-enzymatic-catalysis' (MEEC) seems to be applicable to most types of enzymes except lipases [245, 246]. It consists of a dialysis bag containing the enzyme solution, mounted on a gently rotating magnetic stirring bar.

Figure 3.4. Principle of a membrane reactor and the MEEC-technique.

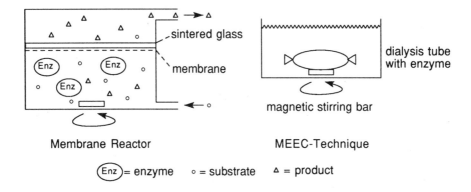

Immobilization of Cofactors

All of the above mentioned immobilization techniques can readily be used for enzymes which are independent of cofactors and for those in which the cofactors are tightly bound (e.g. flavines, see Section 1.4.3). Some enzymes depend on cofactors, which readily dissociate into the medium (e.g. NAD(P)H or ATP), and in these cases the cofactor has to be immobilized as well, to ensure a proper functioning of its recycling system (see Sections 2.1.4 and 2.2.1). Two solutions to this problem have been put forward.

Thus the cofactor may be bound to the surface of a cross-linked enzyme or it may be attached to a macroscopic carrier. In either case it is essential that the spacer arm is long enough so that the cofactor can freely swing back and forth between both enzymes - Enz^1 to perform the reaction and Enz^2 to perform the regeneration. These requirements are very difficult to meet in practice.

Figure 3.5. Co-immobilisation of cofactors.

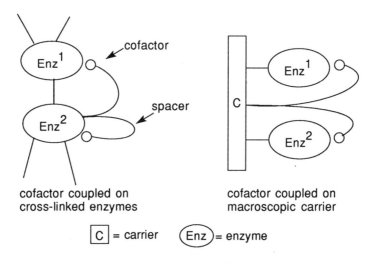

In membrane reactors, a more promising approach has been developed in which the molecular weight of the cofactor is artificially increased by attachment of large groups such as polyethylene glycol (MW 20 000) [247-249], polyethylenimine [250] or dextran moieties [251]. Although the cofactor is freely dissolved, it cannot pass the membrane barrier due to its high-molecular weight ballast.

3.3 Modified and Artificial Enzymes

An amazingly wide variety of non-natural substrates may be converted by means of non-modified native enzymes often with high specificities. In some instances, however, the 'natural' properties of the biocatalyst such as activity, selectivity or stability are insufficient. To extend the preparative applicability of enzymes, a number of techniques have been developed aimed to improve the biocatalysts by chemical modification. Numerous methods for the the site-directed chemical modification of enzymes by group-specific reagents have become established techniques but it has been mainly used for the elucidation of enzyme structure and mechanism [252, 253]. The types of chemically modified enzymes which may have some potential for the biotransformation of non-natural compounds are discussed in this section. Enzyme modification by 'natural' methods (genetic engineering) is excluded since it requires special expertise and it cannot be regarded as a 'standard technique' in this context.

3.3.1 Polyethylene Glycol Modified Enzymes

Enzymes acting in nearly anhydrous organic solvents always give rise to heterogeneous systems (see Section 3.1). To achieve a desirable homogeneous catalysis, they can be chemically modified in order to make them soluble in lipophilic organic solvents. This can be readily achieved by covalent attachment of the amphipathic polymer polyethylene glycol (PEG) to the surface of enzymes [254]. The pros and cons of PEG-modified enzymes are as follows:
- They dissolve in organic solvents such as benzene, toluene or chlorinated hydrocarbons (chloroform, 1,1,1-trichloroethane, trichloroethylene [255]).
- Their properties such as stability, activity [256, 257] and specificity [258, 259] may be altered.
- Due to their solubility in various organic solvents, spectroscopic studies on the conformation of enzymes are simplified [260].

The most widely used modifier is monomethyl PEG having a molecular weight of about 5000. Linkage of the polymer chains onto the enzyme's surface may be achieved by several methods, all of which involve reaction at the ε-amino groups of lysine residues. The latter are preferably located on the surface of the enzyme. The most prominent 'linkers' are shown in Scheme 3.33. For instance, nucleophilic displacement of two of the chlorine atoms of cyanuric chloride by monomethyl PEG yields a very popular modifier, which is attached via an alkylation reaction. Alternatively, a synthetic PEG-containing copolymer, bearing reactive succinic anhydride groups as linkers, can be used to form stable amide bonds [261]. Urethane derivatives are obtained when an activated carbonate ester of monomethyl PEG is used as the modifier. In this case, the reactive *p*-nitrophenyl group is displaced by amino groups of the enzyme. Finally, α-methyl-ω-carboxy-PEG may be linked to an enzyme via succinimide activation of the carboxyl moiety [262].

Scheme 3.33. Modification of enzymes by polyethylene glycol.

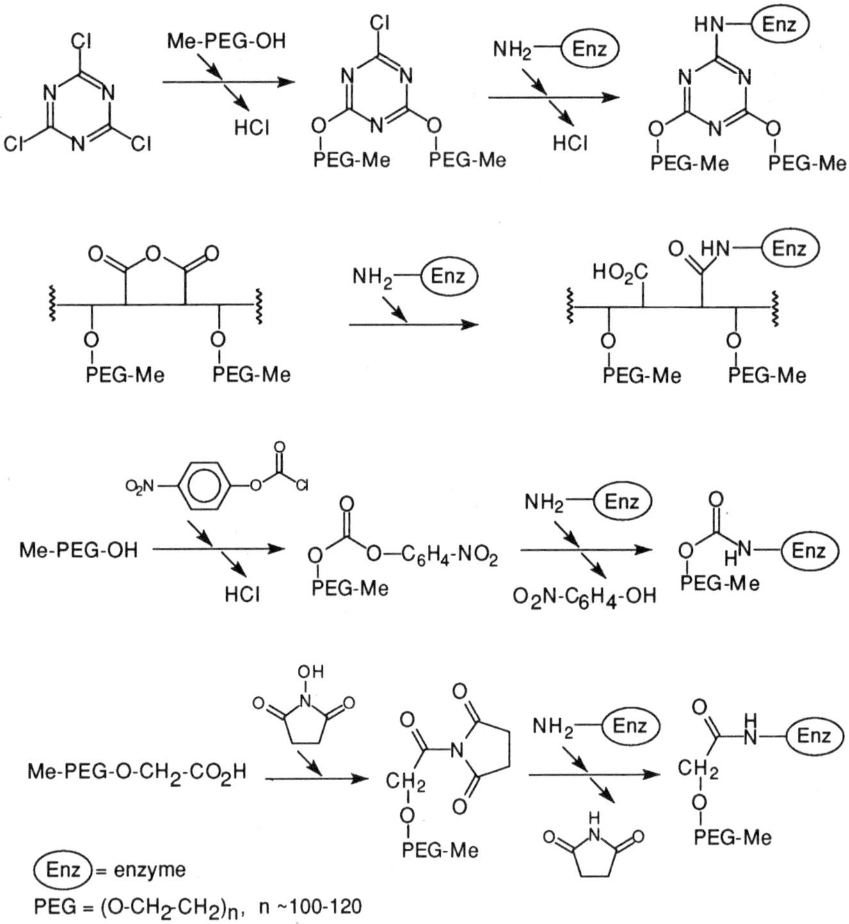

Enz = enzyme
PEG = $(O\text{-}CH_2\text{-}CH_2)_n$, n ~100-120

This technique seems to work for all classes of enzymes, including hydrolases (lipases, proteases), redox enzymes (dehydrogenases, oxidases [263]) and transferases (glucosidases [264]). The residual activities are usually high (50-80%), and for most enzymes about five to ten PEG chains per enzyme

3.3 Modified and Artificial Enzymes

molecule is sufficient to render them soluble in organic solvents. Care has to be taken to avoid extensive modification which leads to deactivation.

PEG-modified enzymes may be recovered from a benzene or toluene solution by precipitation upon the addition of petroleum ether or hexane [265]. A special type of PEG-modification, which allows the simple recovery of the enzyme from solution by using magnetic forces, is shown below [266]. When α,ω-dicarboxyl-PEG is exposed to a mixture of ferric ions and hydrogen peroxide, ferromagnetic magnetite particles are formed (Fe_3O_4) which are tightly adsorbed to the carboxylate. The remaining 'free end' of the magnetic modifier is then covalently linked to the enzyme via succinimide activation. When a magnetic field of moderate strength (5000-6000 Oe ≅ 53-75 Am^{-1}) is applied to the reaction vessel, the modified enzyme is removed from the solvent by being pulled to the walls of the container.

Scheme 3.34. Modification of enzymes by magnetite-polyethylene glycol.

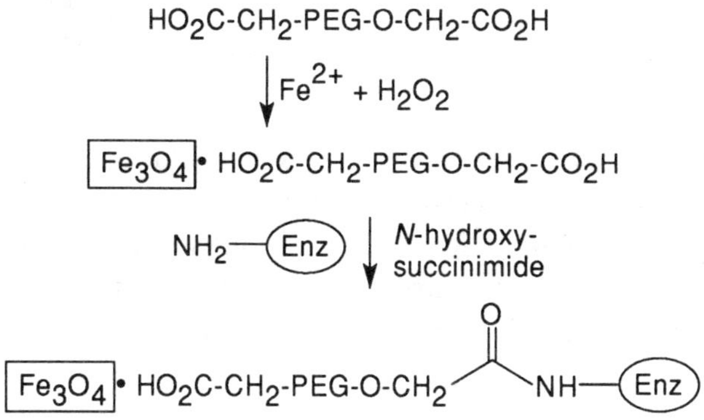

$$HO_2C-CH_2-PEG-O-CH_2-CO_2H$$

$$\downarrow Fe^{2+} + H_2O_2$$

$$\boxed{Fe_3O_4} \bullet HO_2C-CH_2-PEG-O-CH_2-CO_2H$$

$$NH_2-\text{Enz} \quad \downarrow \text{N-hydroxy-succinimide}$$

$$\boxed{Fe_3O_4} \bullet HO_2C-CH_2-PEG-O-CH_2-C(=O)-NH-\text{Enz}$$

The majority of applications using PEG-modified enzymes have involved the synthesis of esters [267, 268], polyesters [269], amides [270] and peptides [271, 272].

3.3.2 Semisynthetic Enzymes

Besides varying the physicochemical properties of enzymes (such as their solubility), attempts have been undertaken to fundamentally modify the properties of an enzyme. Chemical modification of the reactive group in the active site (the chemical operator) leads to 'semisynthetic' enzymes, which often do not have much in common with their natural ancestors [273]. For example, the nucleophilic serine hydroxy group of subtilisin may be transformed to its sulphur- or selenium analogue by a two-step procedure [274-276]. Thus, a serine hydrolase is converted into a cysteine hydrolase. The so-called thiol- and seleno-subtilisins exhibit dramatically altered specificities in comparison with native subtilisin. For instance, the ratio of the rate constants for aminolysis vs. hydrolysis is enhanced from 30 (native subtilisin) to 21 000 for seleno-subtilisin [277].

Scheme 3.35. Synthesis of thiol- and seleno-subtilisin.

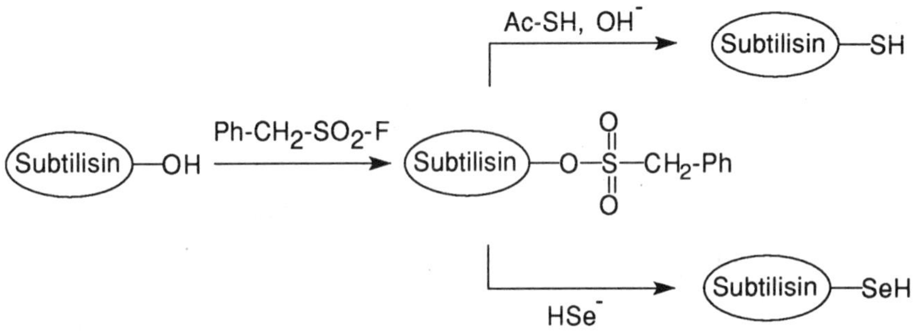

It should be noted, however, that the exchange of the reactive chemical operator did not lead to a 'better' (in terms of activity), but to a 'different' enzyme (when regarding its specificity). The often exhibited low activity of modified enzymes may be explained by the fact that the whole environment in the active site plays a rôle in the catalytic mechanism, not just the chemical operator.

Scheme 3.36. Synthesis of a semisynthetic redox enzyme from papain.

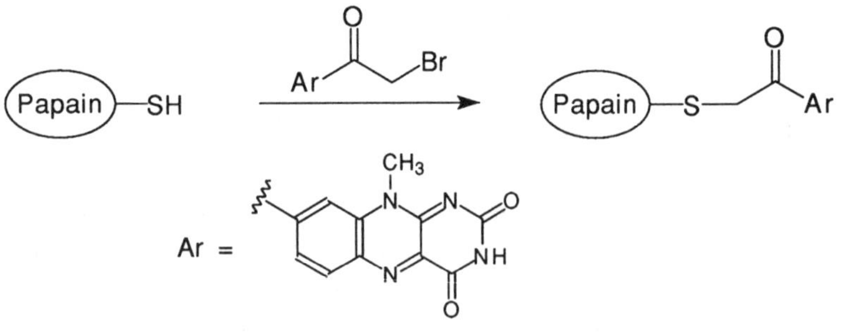

The attachment of a redox-cofactor derivative to the nucleophilic thiol group in the active site of papain causes an even more dramatic change (Scheme 3.36). By this means, a hydrolase is transformed into a redox enzyme [278, 279]. For instance, linkage of an isoalloxazine derivative to the active site of papain gave rise to a semisynthetic redox enzyme - flavopapain [280-282].

At present, studies on semisynthetic enzymes are mainly directed towards getting more insight into the catalytic mechanism of enzymes rather than creating synthetic biocatalysts for preparative use. Future studies will tell whether semi-synthetic enzymes will be also of general use for organic synthesis.

3.3.3 Catalytic Antibodies

Obviously enzymes cannot be rationally designed and then be prepared by a de novo synthesis. As a consequence, some synthetically useful transformations such as the Diels-Alder reaction are excluded from enzymatic catalysis. Furthermore, a search for biocatalysts possessing opposite stereochemical properties is usually a tedious empirical undertaking which is often unsuccessful. This gap may be filled by the development of synthetic enzymes, the catalytic antibodies [283-285]. To indicate their derivation, they are commonly called 'abzymes'.

The immune system produces a vast repertoire - in the range of 10^8 to 10^{10} [286] - of exquisitely specific proteins (antibodies) that protect vertebrates from 'foreign' invaders such as pathogenic bacteria and viruses, parasites and cancer cells. Antibodies have become invaluable tools in the detection, isolation and analysis of biological materials and are the key elements in many diagnostic procedures. Their potential, however, is not limited to biology and medicine, as they possess a potential for biocatalysis as well. Because they can also be elicited to a large array of synthetic molecules using hybridoma technology, they offer a unique approach for generating tailor-made enzyme-like catalysts. In a very simplified version they can be regarded as enzymes possessing an active site, but no chemical operator.

Antibodies are large proteins consisting of four peptide chains which have a molecular weight of about 150 000 daltons. There are two identical heavy (H, 50 000 Da) and light chains (L, 25 000 Da) which are cross-linked by disulfide bonds. The light chains are divided into two domains, the variable (V) and the constant (C) regions (V_L, C_L), while the heavy chains consist of V_H, C_H1, C_H2 and C_H3 domains. Both the variable V_H and V_L domains, which are located, to the first approximation, in the first 110 amino acids of the heavy and light chains are highly polymorphic and change with each antigen (hapten). Consequently, binding occurs in this region. On the other hand, the constant regions represent the framework of the antibody and are relatively invariant.

Figure 3.6. Simplified structure of a catalytic antibody.

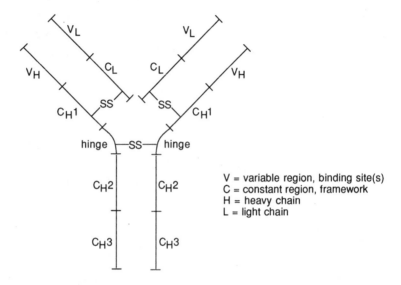

V = variable region, binding site(s)
C = constant region, framework
H = heavy chain
L = light chain

The general approach which has been used to generate catalytic antibodies is based on the complementary of an antibody to its corresponding hapten, i.e. the ligand against which the antibody is raised. It may be compared with the 'negative' in terms of a photographic picture. This fact has been exploited to generate artificial cavities (corresponding to the active site of an enzyme) that are complementary to the rate-determining transition state of the reaction which is to be catalysed. Thus, entropy requirements which are involved in the correct orientation of the reaction partners can be met. If necessary, these combining sites may be equipped with an appropriately positioned catalytic amino-acid side chain or a cofactor, which may be introduced either directly by site-directed mutagenesis or indirectly by chemical modification. Antibodies bind molecules ranging in size from about 6 to 34 Å with association constants of 10^4 to 10^{14} M^{-1}. As with enzymes, binding occurs by Van-der-Waals, hydrophobic, electrostatic and hydrogen-bonding interactions.

The general protocol to obtain catalytic antibodies is as follows. After determination of the transition state of the reaction to be catalysed, a stable transition state mimic is synthesized. This 'model' is then used as an antigen to

3.3 Modified and Artificial Enzymes

elicit antibodies. If required, the latter can then be genetically or chemically modified by attachment of a reactive group [287], or a cofactor, in order to catalyse redox reactions [288]. Furthermore, like enzymes, they can be immobilized and used in organic solvents [289].

Scheme 3.37. Antibody-catalysed ester hydrolysis and lactonisation.

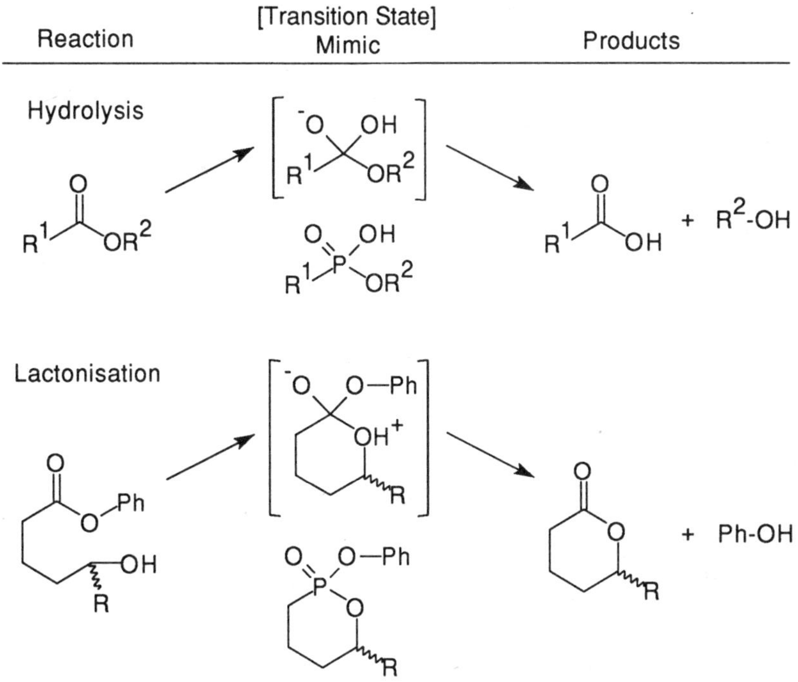

As shown in Scheme 3.37, catalytic antibodies have been produced to catalyse an impressive variety of chemical reactions. Stable phosphonic esters (or lactones, respectively) have been used as transition state mimicks for the hydrolysis of esters and amides [290-292], acyl transfer [293] and lactonisation [294] reactions, which all proceed via a tetrahedral carbanionic intermediate. In some instances, the reactions proved to be enantiospecific [295]. Elimination reactions [296] and photochemical reactions [297] can also be catalysed.

More importantly from a preparative viewpoint, pericyclic reactions, such as the Claisen-rearrangement [298, 299] or the Diels-Alder reaction [300], have been catalysed by antibodies, which were raised against bicyclic and tricyclic transition state mimicks (Scheme 3.28). The latter reactions cannot be catalysed using enzymes. The cleavage of an ether [301] and *cis-trans* isomerisation of alkenes has been reported more recently [302].

Scheme 3.38. Antibody-catalysed pericyclic reactions.

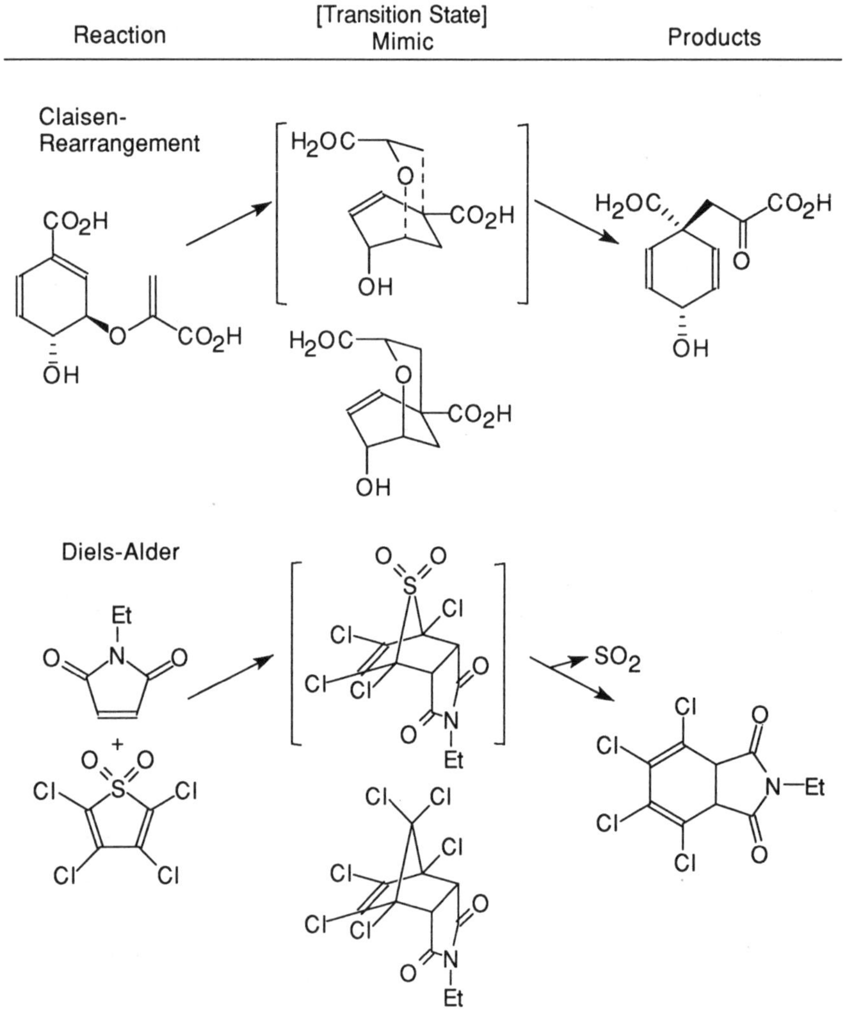

The catalytic power of abzymes is measured as the rate acceleration compared to the uncatalysed reaction. Although the early catalytic antibodies had only modest rate accelerations, more recent studies have produced artificial enzymes with catalytic activities approaching those of natural enzymes [303]. It must be noted, however, that in general, the rate acceleration of abzymes are in the range of 10^3 to 10^5, though in some cases they may reach 10^6. The corresponding values for most enzymes are far better (>10^8).

The following pros and cons for the utilization of abzymes can be summarized as follows:
- They offer a unique access to artificial tailor-made enzyme-like catalysts and are able to catalyse reactions which have no equivalent counterpart within the arsenal of natural enzymes.
- The construction of enzymes having an opposite stereochemical preference is possible.
- The catalytic efficiency is low in comparison to that of natural enzymes, mainly because the transition state mimicks which are used to elicit the antibody is only a template for the true transition state.
- The production of abzymes on a large scale is tedious [304]. Furthermore, to avoid the occurrence of catalytically active enzyme impurities, the purity standards are very high.

References

1. Borgström B, Brockman HL (eds.) (1984) Lipases; Elsevier, Amsterdam
2. Dordick JS (1989) Enzyme Microb. Technol. 11: 194
3. Brink LES, Tramper J, Luyben KCAM, Vant't Riet K (1988) Enzyme Microb. Technol. 10: 736
4. Klibanov AM (1990) Acc. Chem. Res. 23: 114
5. Klibanov AM (1986) ChemTech 354
6. Klibanov AM (1989) Trends Biochem. Sci. 14: 141
7. Laane C, Tramper J, Lilly MD (eds) (1987) Biocatalysis in Organic Media, Studies in Organic Chemistry, vol 29, Elsevier, Amsterdam
8. Khmelnitsky YL, Levashov AV, Klyachko NL, Martinek K (1988) Enzyme Microb. Technol. 10: 710
9. Schultz GE, Schirmer RH (1979) Principles of Protein Structure; Springer, New York
10. Zaks A, Klibanov AM (1986) J. Am. Chem. Soc. 108: 2767
11. Zaks A, Klibanov AM (1988) J. Biol. Chem. 263: 3194
12. Kazandjian RZ, Klibanov AM (1985) J. Am. Chem. Soc. 107: 5448
13. Cooke R, Kuntz ID (1974) Ann. Rev. Biophys. Bioeng. 3: 95
14. Cotterill IC, Sutherland AG, Roberts SM, Grobbauer R, Spreitz J, Faber K (1991) J. Chem. Soc., Perkin Trans. I, 1365
15. Yamamoto Y, Yamamoto K, Nishioka T, Oda J (1989) Agric. Biol. Chem. 52: 3087
16. Kazandjian RZ, Klibanov AM (1985) J. Am. Chem. Soc. 107: 5448
17. Wang Y-F, Chen S-T, Liu K K-C, Wong C-H (1989) Tetrahedron Lett. 1917
18. Laumen K, Seemayer R, Schneider MP (1990) J. Chem. Soc., Chem. Commun. 49
19. Aldercreutz P, Mattiasson B (1987) Biocatalysis 1: 99
20. Zaks A, Klibanov AM (1984) Science 224: 1249
21. Rupley JA, Gratton E, Carieri G (1983) Trends Biochem. Sci. 8: 18
22. Zaks A, Klibanov AM (1986) J. Am. Chem. Soc. 108: 2767
23. Russell AJ, Klibanov AM (1988) J. Biol. Chem. 263: 11624
24. Gaertner H, Puigserver A (1989) Eur. J. Biochem. 181: 207
25. Ferjancic A, Puigserver A, Gaertner A (1990) Appl. Microbiol. Biotechnol. 32: 651
26. Ottolina G, Carrea G, Riva S (1991) Biocatalysis 5: 131
27. Sakurai T, Margolin AL, Russell AJ, Klibanov AM (1988) J. Am. Chem. Soc. 110: 7236

28. Kitaguchi H, Fitzpatrick PA, Huber JE, Klibanov AM (1989) J. Am. Chem. Soc. 111: 3094
29. Fitzpatrick PA, Klibanov AM (1991) J. Am. Chem. Soc. 113: 3166
30. Kise H, Hayakawa A, Noritomi H (1990) J. Biotechnol. 14: 239
31. Langrand G, Baratti J, Buono G, Triantaphylides C (1986) Tetrahedron Lett. 29
32. Koshiro S, Sonomoto K, Tanaka A, Fukui S (1985) J. Biotechnol. 2: 47
33. Inagaki T, Ueda H (1987) Agric. Biol. Chem. 51: 1345
34. Morrow CJ, Wallace JS (1990) Synthesis of Polyesters by Lipase-Catalysed Polycondensation in Organic Media; Abramowicz DA (ed) Biocatalysis, p 25, Van Nostrand Reinhold, New York
35. Margolin AL, Fitzpatrick PA, Klibanov AM (1991) J. Am. Chem. Soc. 113: 4693
36. Gutman AL, Bravdo T (1989) J. Org. Chem. 54: 4263
37. Makita A, Nihira T, Yamada Y (1987) Tetrahedron Lett. 805
38. Kitaguchi H, Fitzpatrick PA, Huber JE, Klibanov AM (1989) J. Am. Chem. Soc. 111: 3094
39. Gotor V, Brieva R, Rebolledo F (1988) Tetrahedron Lett. 6973
40. Kullmann W (1987) Enzymatic Peptide Synthesis; CRC Press, Boca Raton (FL)
41. Therisod M, Klibanov AM (1986) J. Am. Chem. Soc. 108: 5638
42. Riva S, Klibanov AM (1988) J. Am. Chem. Soc. 110: 3291
43. Riva S, Chopineau J, Kieboom APG, Klibanov AM (1988) J. Am. Chem. Soc. 110: 584
44. Douzou P (1977) Cryobiochemistry; Academic Press, London
45. Fink AL, Cartwright SJ (1981) CRC Crit. Rev. Biochem. 11: 145
46. Carrea G (1984) Trends Biotechnol. 2: 102
47. Anderson E, Hahn-Hägerdal B (1990) Enzyme Microb. Technol. 12: 242
48. Lilly MD (1982) J. Chem. Technol. Biotechnol. 32: 162
49. Antonini E, Carrea G, Cremonesi P (1981) Enzyme Microb. Technol. 3: 291
50. Kim KH, Kwon DY, Rhee JS (1984) Lipids 19: 975
51. Brink LES, Tramper J (1985) Biotechnol. Bioeng. 27: 1258
52. Zaks A, Klibanov AM (1988) J. Biol. Chem. 263: 8017
53. Kastle JH, Loevenhart AS (1900) Am. Chem. Soc. 24: 491
54. Valivety RH, Brown L, Halling PJ, Johnston GA, Suckling CJ (1990) Enzyme Reactions in Predominantly Organic Meida: Measurement and Changes of pH; In: Copping LG, Martin RE, Pickett JA, Bucke C, Bunch AW (eds) Opportunities in Biotransformations, p 81, Elsevier, London
55. Zaks A, Klibanov AM (1985) Proc. Natl. Acad. Sci. USA 82: 3192
56. Hsu S-H, Wu S-S, Wang Y-F, Wong C-H (1990) Tetrahedron Lett. 6403
57. Laane C, Boeren S, Hilhorst R, Veeger C (1987) Optimization of Biocatalysts in Organic Media; Laane C, Tramper J, Lilly MD (eds) Biocatalysis in Organic Media, p 65, Elsevier, Amsterdam
58. Laane C, Boeren S, Vos K, Veeger C (1987) Biotechnol. Bioeng. 30: 81
59. Rekker RF, de Kort HM (1979) Eur. J. Med., Chim. Ther. 14: 479
60. Faber K (1991) J. Mol: Catalysis 65: L49
61. Johnson LN (1984) Inclusion Compds. 3: 509
62. Nakamura K (1990) Tibtech 8: 288
63. Aaltonen O, Rantakylä M (1991) Chemtech 240
64. Marty A, Chulalaksananukul W, Condoret JS, Willemot RM, Durand G (1990) Biotechnol. Lett. 12: 11
65. Chi YM, Nakamura K, Yano T (1988) Agric. Biol. Chem. 52: 1541
66. Pasta P, Mazzola G, Carrea G, Riva S (1989) Biotechnol. Lett. 11: 643
67. van Eijs AMM, de Jong PJP (1989) Procestechniek 8: 50
68. Randolph TW, Blanch HW, Prausnitz JM, Wilke CR (1985) Biotechnol. Lett. 7: 325
69. Hammond DA, Karel M, Klibanov AM, Krukonis V (1985) J. Appl. Biochem. Biotechnol. 11: 393
70. Randolph TW, Clark DS, Blanch HW, Prausnitz JM (1988) Science 238: 387
71. Kasche V, Schlothauer R, Brunner G (1988) Biotechnol. Lett. 10: 569
72. Langrand G, Baratti J, Buono G, Triantaphylides C (1986) Tetrahedron Lett. 29
73. Koshiro S, Sonomoto K, Tanaka A, Fukui S (1985) J. Biotechnol. 2: 47

3.3 Modified and Artificial Enzymes

74. Inagaki T, Ueda H (1987) Agric. Biol. Chem. 51: 1345
75. Björkling F, Godtfredsen SE, Kirk O (1989) J. Chem. Soc., Chem. Commun. 934
76. Bell G, Blain JA, Paterson JDE, Shaw CEL, Todd RJ (1978) FEMS Microbiol. Lett. 3: 223
77. Paterson JDE, Blain JA, Shaw CEL, Todd RJ (1979) Biotechnol. Lett. 1: 211
78. Björkling F, Godtfredsen SE, Kirk O (1989) J. Chem. Soc., Chem. Commun. 934
79. Riva S, Faber K (1992) Synthesis, in press
80. Kirchner G, Scollar MP, Klibanov AM (1985) J. Am. Chem. Soc. 107: 7072
81. Mischitz M, Pöschl U, Faber K (1991) Biotechnol. Lett. 13: 653
82. Ghogare A, Kumar GS (1989) J. Chem. Soc., Chem. Commun. 1533
83. Degueil-Castaing M, De Jeso B, Drouillard S, Maillard B (1987) Tetrahedron Lett. 953
84. Wang Y-F, Wong C-H (1988) J. Org. Chem. 53: 3129
85. Wang Y-F, Lalonde JJ, Momongan M, Bergbreiter DE, Wong C-H (1988) J. Am. Chem. Soc. 110: 7200
86. Ledl F, Schleicher E (1990) Angew. Chem., Int. Ed. Engl. 29: 565
87. Donohue TM, Tuma DJ, Sorrell MF (1983) Arch. Biochem. Biophys. 220: 239
88. Berger B, Faber K (1991) J. Chem. Soc., Chem. Commun. 1198
89. Sugai T, Ohta H (1989) Agric. Biol. Chem. 53: 2009
90. Holla EW (1989) Angew. Chem., Int. Ed. Engl. 28: 220
91. Bianchi D, Cesti P, Battistel E (1988) J. Org. Chem. 53: 5531
92. Berger B, Rabiller CG, Königsberger K, Faber K, Griengl H (1990) Tetrahedron: Asymmetry 1: 541
93. Nicotra F, Riva S, Secundo F, Zucchelli L (1990) Synth. Commun. 20: 679
94. Fourneron J-D, Chiche M, Pieroni G, (1990) Tetrahedron Lett. 4875
95. Caer E, Kindler A (1962) Biochemistry 1: 518
96. Suemune H, Mizuhara Y, Akita H, Sakai K (1986) Chem. Pharm. Bull. 34: 3440
97. Morishima H, Koike Y, Nakano M, Atsuumi S, Tanaka S, Funabashi H, Hashimoto J, Sawasaki Y, Mino N, Nakano K, Matsushima K, Nakamichi K, Yano M (1989) Biochem. Biophys. Res. Commun. 159: 999
98. Santaniello E, Ferraboschi P, Grisenti P (1990) Tetrahedron Lett. 5657
99. Atsuumi S, Nakano M, Koike Y, Tanaka S, Ohkubo M (1990) Tetrahedron Lett. 1601
100. Baba N, Yoneda K, Tahara S, Iwase J, Kaneko T, Matsuo M (1990) J. Chem. Soc., Chem. Commun. 1281
101. Terao Y, Murata M, Achiwa K (1988) Tetrahedron Lett. 5173
102. Ader U, Breitgoff D, Laumen KE, Schneider MP (1989) Tetrahedron Lett. 1793
103. Ferraboschi P, Grisenti P, Manzocchi A, Santaniello E (1990) J. Org. Chem. 55: 6214
104. Laumen K, Breitgoff D, Schneider MP (1988) J. Chem. Soc., Chem. Commun. 1459
105. Finn MG, Sharpless KB (1985) Asymmetric Synthesis, Morrison JC (ed), vol. 5, p 247, Academic Press, New York
106. Carlier PR, Mungall WS, Schroder G, Sharpless KB (1988) J. Am. Chem. Soc. 110: 2978
107. Burgess K, Jennings LD (1990) J. Am. Chem. Soc. 112: 7434
108. Hayward RC, Overton CH, Witham GH (1976) J. Chem. Soc., Perkin Trans. I, 2413
109. Cunningham AF, Kündig EP (1988) J. Org. Chem. 53: 1823
110. Seemayer R, Schneider MP (1991) J. Chem. Soc., Chem. Commun. 49
111. Hemmerle H, Gais HJ (1987) Tetrahedron Lett. 3471
112. Xie ZF, Nakamura I, Suemune H, Sakai K (1988) J. Chem. Soc., Chem. Commun. 966
113. Laumen K, Seemayer R, Schneider MP (1990) J. Chem. Soc., Chem. Commun. 49
114. Chen C-S, Liu Y-C, Marsella M (1990) J. Chem. Soc., Perkin Trans I, 2559
115. Chen C-S, Liu Y-C (1989) Tetrahedron Lett. 7165
116. Hiratake J, Inagaki M, Nishioka T, Oda J (1988) J. Org. Chem. 53: 6130
117. Solladié-Cavallo A (1989) In: Advances in Metal-Organic Chemistry; Liebeskind LS (ed), vol. 1, JAI Press, p 99-131
118. Nakamura K, Ishihara K, Ohno A, Uemura M, Nishimura H, Hayashi Y (1990) Tetrahedron Lett. 3603
119. Yamazaki Y, Hosono K (1990) Tetrahedron Lett. 3895

120. Gokel GW, Marquarding D, Ugi IK (1972) J. Org. Chem. 37: 3052
121. Boaz NW (1989) Tetrahedron Lett. 2061
122. Mitsuda S, Nabeshima S, Hirohara H (1989) Appl. Microbiol. Biotechnol. 31: 334
123. Tinapp P (1971) Chem. Ber. 104: 2266
124. Satoh T, Suzuki S, Suzuki Y, Miyaji Y, Imai Z (1969) Tetrahedron Lett. 4555
125. Mitsuda S, Yamamoto H, Umemura T, Hirohara H, Nabeshima S (1990) Agric. Biol. Chem. 54: 2907
126. Effenberger F, Gutterer B, Ziegler T, Eckhardt E, Aichholz R (1991) Liebigs Ann. Chem. 47
127. Wang Y-F, Chen S-T, Liu K K-C, Wong C-H (1989) Tetrahedron Lett. 1917
128. Effenberger F, Ziegler T, Förster S (1987) Angew. Chem., Int. Ed. Engl. 26: 458
129. Feichter C, Faber K, Griengl H (1989) Tetrahedron Lett. 551
130. Feichter C, Faber K, Griengl H (1990) Biocatalysis 3: 145
131. Burgess K, Henderson I (1990) Tetrahedron: Asymmetry 1: 57
132. Burgess K, Cassidy J, Henderson I (1991) J. Org. Chem. 56: 2050
133. Burgess K, Jennings LD (1990) J. Org. Chem. 55: 1138
134. Pfenninger A (1986) Synthesis 89
135. Babiak KA, Ng JS, Dygos JH, Weyker CL, Wang Y-F, Wong C-H (1990) J. Org. Chem. 55: 3377
136. Baba N, Mimura M, Hiratake J, Uchida K, Oda J (1988) Agric. Biol. Chem. 52: 2685
137. Baba N, Tateno K, Iwasa J, Oda J (1990) Agric. Biol. Chem. 54: 3349
138. Greene TW (1981) Protective groups in Organic Chemistry, Wiley, New York
139. Hennen WJ, Sweers HM, Wang Y-F, Wong C-H (1988) J. Org. Chem. 53: 4939
140. Therisod M, Klibanov AM (1986) J. Am. Chem. Soc. 108: 5638
141. Riva S, Chopineau J, Kieboom APG, Klibanov AM (1988) J. Am. Chem. Soc. 110: 584
142. Riva S, Klibanov AM (1988) J. Am. Chem. Soc. 110: 3291
143. Riva S, Bovara R, Ottolina G, Secundo F, Carrea G (1989) J. Org. Chem. 54: 3161
144. Colombo D, Ronchetti F, Toma L (1991) Tetrahedron 47: 103
145. Therisod M, Klibanov AM (1987) J. Am. Chem. Soc. 109: 3977
146. Nicotra F, Riva S, Secundo F, Zucchelli L (1989) Tetrahedron Lett. 1703
147. Gruters RA, Neefjes JJ, Tersmette M, De Goede REJ, Tulp A, Huisman HG, Miedema F, Ploegh HL (1987) Nature 330: 74
148. Margolin AL, Delinck DL, Whalon MR (1990) J. Am. Chem. Soc. 112: 2849
149. Gatfield IL (1984) Ann. N. Y. Acad. Sci. 434: 569
150. Gutman AL, Oren D, Boltanski A, Bravdo T (1987) Tetrahedron Lett. 5367
151. Guo Z, Ngooi TK, Scilimati A, Fülling G, Sih CJ (1988) Tetrahedron Lett. 5583
152. Gutman AL, Zuobi K, Boltansky A (1987) Tetrahedron Lett. 3861
153. Gutman AL, Bravdo T (1989) J. Org. Chem. 54: 4263
154. Makita A, Nihira T, Yamada Y (1987) Tetrahedron Lett. 805
155. Guo Z, Sih CJ (1988) J. Am. Chem. Soc. 110: 1999
156. Ngooi TK, Scilimati A, Guo Z-W, Sih CJ (1989) J. Org. Chem. 54: 911
157. Rebolledo F, Brieva R, Gotor V (1989) Tetrahedron Lett. 5345
158. Fastrez J, Fersht AR (1973) Biochemistry 12: 2025
159. Yagisawa S (1981) J. Biochem. (Tokyo) 89: 491
160. Gotor V, Astorga C, Rebolledo F (1990) Synlett. 387
161. Gotor V, Garcia MJ, Rebolledo F (1990) Tetrahedron: Asymmetry 1: 277
162. Gotor V, Brieva R, Rebolledo F (1988) Tetrahedron Lett. 6973
163. Gotor V, Brieva R, Rebolledo F (1988) J. Chem. Soc., Chem. Commun. 957
164. Brieva R, Rebolledo F, Gotor V (1990) J. Chem. Soc., Chem. Commun. 1386
165. Kitaguchi H, Fitzpatrick PA, Huber JE, Klibanov AM (1989) J. Am. Chem. Soc. 111: 3094
166. Fruton JS (1982) Adv. Enzymol. 53: 239
167. Jakubke H-D, Kuhl P, Könnecke A (1985) Angew Chem., Int. Ed. Engl. 24: 85
168. Jakubke H-D (1987) The Peptides 9: 103
169. Kullmann (1987) Enzymatic Peptide Synthesis, CRC Press, Boca Raton, FL
170. Morihara K (1987) Trends Biotechnol. 5: 164
171. Glass JD (1981) Enzyme Microb. Technol. 3: 2

172. Chaiken IM, Komoriya A, Ojno M, Widmer F (1982) Appl. Biochem. Biotechnol. 7: 385
173. Bergmann M, Fraenkel-Conrat H (1938) J. Biol. Chem. 124: 1
174. Kuhl P, Wilsdorf A, Jakubke H-D (1983) Monatsh. Chem. 114: 571
175. Homandberg GA, Komoriya A, Chaiken IM (1982) Biochemistry 21: 3385
176. Inouye K, Watanabe K, Tochino Y, Kobayashi M, Shigeta Y (1981) Biopolymers 20: 1845
177. Margolin AL, Klibanov AM (1987) J. Am. Chem. Soc. 109: 3802
178. West JB, Wong C-H (1987) Tetrahedron Lett. 1629
179. Matos JR, West JB, Wong C-H (1987) Biotechnol. Lett. 9: 233
180. Isowa Y, Ohmori M, Ichikawa T, Mori K, Nonaka Y, Kihara K, Oyama K (1979) Tetrahedron Lett. 2611
181. Stoineva IB, Petkov DD (1985) FEBS Lett. 183: 103
182. Di Maio J, Nguyen TM-D, Lemieux C, Schiller PW (1982) J. Med. Chem. 25: 1432
183. Inouye K, Watanabe K, Morihara K, Tochino Y, Kanaya T, Emura J, Sakakibara S (1979) J. Am. Chem. Soc. 101: 751
184. Rose K, Gladstone J, Offord RE (1984) Biochem. J. 220: 189
185. Obermeier R, Seipke G (1984) In: Voelter W, Bayer E, Ovchinnikov YA, Wünsch E (eds) Chemistry of Peptides and proteins, vol 2, p 3, de Gruyter, Berlin
186. Masaki T, Nakamura K, Isono M, Soejima M (1978) Agric. Biol. Chem. 42: 1443
187. Morihara K, Oka T, Tsuzuki H, Tochino Y, Kanaya T (1980) Biochem. Biophys. Res. Commun. 92: 396
188. Björkling F, Godtfredsen SE, Kirk O (1990) J. Chem. Soc., Chem. Commun. 1301
189. Grunwald J, Wirz B, Scollar MP, Klibanov AM (1986) J. Am. Chem. Soc. 108: 6732
190. Malmstrom BG, Ryden L (1968) In: Biological Oxidations; Singer TP (ed) p 419, Wiley, New York
191. Wood BJB, Ingraham LL (1965) Nature 205: 291
192. Kazandjian RZ, Klibanov AM (1985) 107: 5448
193. Siegel SM, Roberts K (1968) Space Life Sci. 1: 131
194. Suckling CJ, Suckling KE (1974) Chem. Soc. Rev. 3: 387
195. Sharma BP, Bailey LF, Messing RA (1982) Angew. Chem., Int. Ed. Engl. 21, 837
196. Rosevaer A (1984) J. Chem. Technol. Biotechnol. 34: 127
197. Zaborsky OR (1973) Immobilized Enzymes, CRC Press, Cleveland, Ohio
198. Trevan MD (1980) Immobilized Enzymes: Introduction and Applications in Biotechnology, Wiley, New York
199. Hartmeier W (1986) Immobilisierte Biokatalysatoren, Springer, Berlin
200. Suckling CJ (1977) Chem. Soc. Rev. 6: 215
201. Christen M, Crout DHG (1987) Enzymatic Reduction of β-Ketoesters using Immobilized Yeast, In Bioreactors and Biotransformations, p 213, Moody GW, Baker PB (eds) Elsevier, London
202. Martinek K, Klibanov AM, Goldmacher VS, Tchernysheva AV, Mozhaev VV, Berezin IV, Glotov BO (1977) Biochim. Biophys. Acta 485: 13
203. Klibanov AM (1983) Science 219: 722
204. Miyawaki O, Wingard jr LB (1984) Biotechnol. Bioeng. 26: 1364
205. Krakowiak W, Jach M, Korona J, Sugier H (1984) Starch 36: 396
206. Bianchi D, Cesti P, Battistel E (1988) J. Org. Chem. 53: 5531
207. Wiegel J, Dykstra M (1984) Appl. Microbiol. Biotechnol. 20: 59
208. Kato T, Horikoshi K (1984) Biotechnol. Bioeng. 26: 595
209. Sugiura M, Isobe M (1976) Chem. Pharm. Bull. 24: 72
210. Hsu S-H, Wu S-S, Wang Y-F, Wong C-H (1990) Tetrahedron Lett. 6403
211. Lavayre J, Baratti J (1982) Biotechnol. Bioeng. 24: 1007
212. Boudrant J, Ceheftel C (1975) Biotechnol. Bioeng. 17: 827
213. Tosa T, Mori T, Fuse N, Chibata I (1967) Enzymologia 31: 214
214. Tosa T, Mori T, Chibata I (1969) Agric. Biol. Chem. 33: 1053
215. Weetall HH, Mason RD (1973) Biotechnol. Bioeng. 15: 455
216. Cannon JJ, Chen L-F, Flickinger MC, Tsao GT (1984) Biotechnol. Bioeng. 26: 167
217. Ibrahim M, Hubert P, Dellacherie E, Magdalou J, Muller J, Siest G (1985) Enzyme Microb. Technol. 7: 66

218. Monsan P, Combes D (1984) Biotechnol. Bioeng. 26: 347
219. Chipley JR (1974) Microbios 10: 115
220. Marek M, Valentova O, Kas J (1984) Biotechnol. Bioeng. 26: 1223
221. Vilanova E, Manjon A, Iborra JL (1984) Biotechnol. Bioeng. 26: 1306
222. Miyama H, Kobayashi T, Nosaka Y (1984) Biotechnol. Bioeng. 26: 1390
223. Burg K, Mauz S, Noetzel S, Sauber K (1988) Angew. Makromol. Chem. 157: 105
224. Khan SS, Siddiqui AM (1985) Biotechnol. Bioeng. 27: 415
225. Kaul R, D'Souza SF, Nadkarni GB (1984) Biotechnol. Bioeng. 26: 901
226. Karube I, Kawarai M, Matsuoka H, Suzuki S (1985) Appl. Microbiol. Biotechnol. 21: 270
227. Qureshi N, Tamhane DV (1985) Appl. Microbiol. Biotechnol. 21: 280
228. Umemura I, Takamatsu S, Sato T, Tosa T, Chibata I (1984) Appl. Microbiol. Biotechnol. 20: 291
229. Fukui S, Tanaka A (1984) Adv. Biochem. Eng. Biotechnol. 29: 1
230. Mori T, Sato T, Tosa T, Chibata I (1972) Enzymologia 43: 213
231. Martinek K, Klibanov AM, Goldmacher VS, Berezin IV (1977) Biochim. Biophys. Acta 485: 1
232. Hoar TP, Schulman JH (1943) Nature 152: 102
233. Bonner FJ, Wolf R, Luisi PL (1980) Solid Phase Biochem. 5: 255
234. Poon PH, Wells MA (1974) Biochemistry 13: 4928
235. Wells MA (1974) Biochemistry 13: 4937
236. Martinek K, Levashov AV, Klyachko NL, Khmelnitsky YL, Berezin IV (1986) Eur. J. Biochem. 155: 453
237. Luisi PL, Laane C (1986) Trends Biotechnol. 4: 153
238. Meier P, Luisi PL (1980) Solid Phase Biochem. 5: 269
239. Barbaric S, Luisi PL (1981) J. Am. Chem. Soc. 103: 4239
240. Grandi C, Smith RE, Luisi PL (1981) J. Biol. Chem. 256: 837
241. Martinek K, Semenov AN, Berezin IV (1981) Biochim. Biophys. Acta 658: 76
242. Martinek K, Levashov AV, Khmelnitsky YL, Klyachko NL, Berezin IV (1982) Science 218: 889
243. Hilhorst R, Spruijt R, Laane C, Veeger C (1984) Eur. J. Biochem. 144: 459
244. Flaschel E, Wandrey C, Kula M-R (1983) Adv. Biochem. Eng./Biotechnol. 26: 73
245. Bednarski MD, Chenault HK, Simon ES, Whitesides GM (1987) J. Am. Chem. Soc. 109: 1283
246. Thiem J, Stangier P (1990) Liebigs Ann. Chem. 1101
247. Furukawa S, Katayama N, Iizuka T, Urabe I, Okada H (1980) FEBS Lett. 121: 239
248. Bückmann AF, Kula M-R, Wichmann R, Wandrey C (1981) J. Appl. Biochem. 3: 301
249. Vasic-Racki DJ, Jonas M, Wandrey C, Hummel W, Kula M-R (1989) Appl. Microbiol. Biotechnol. 31: 215
250. Wykes JR, Dunnill P, Lilly MD (1972) Biochim. Biophys. Acta 286: 260
251. Malinauskas AA, Kulis JJ (1978) Appl. Biochem. Microbiol. 14: 706
252. Lundblad RL (1991) Chemical Reagents for Protein Modification, 2nd Edn, CRC Press, London
253. Glazer AN (1976) The Proteins, Neurath H, Hill RL (eds) vol II, p 1, Academic Press, London
254. Inada Y, Takahashi K, Yoshimoto T, Ajima A, Matsushima A, Saito Y (1986) Trends Biotechnol. 4: 190
255. Kodera Y, Takahashi K, Nishimura H, Matsushima A, Saito Y, Inada Y (1986) Biotechnol. Lett. 8: 881
256. Takahashi K, Ajima A, Yoshimoto T, Okada M, Matsushima A, Tamaura Y, Inada Y (1985) J. Org. Chem. 50: 3414
257. Takahashi K, Ajima A, Yoshimoto T, Inada Y (1984) Biochem. Biophys. Res. Commun. 125: 761
258. Uemura T, Fujimori M, Lee H-H, Ikeda S, Aso K (1990) Agric. Biol. Chem. 54: 2277
259. Ferjancic A, Puigserver A, Gaertner H (1988) Biotechnol. Lett. 10: 101
260. Pasta P, Riva S, Carrea G (1988) FEBS Lett. 236: 329

261. Yoshimoto T, Ritani A, Ohwada K, Takahashi K, Kodera Y, Matsushima A, Saito Y, Inada Y (1987) Biochem. Biophys. Res. Commun. 148: 876
262. Bückmann AF, Morr M, Johansson G (1981) Makromol. Chem. 182: 1379
263. Takahashi K, Nishimura H, Yoshimoto T, Saito Y, Inada Y (1984) Biochem. Biophys. Res. Commun. 121: 261
264. Beecher JE, Andrews AT, Vulfson EN (1990) Enzyme Microb. Technol. 12. 955
265. Yoshimoto T, Takahashi K, Nishimura H, Ajima A, Tamaura Y, Inada Y (1984) Biotechnol. Lett. 6: 337
266. Yoshimoto T, Mihama T, Takahashi K, Saito Y, Tamaura Y, Inada Y (1987) Biochem. Biophys. Res. Commun. 145: 908
267. Nishio T, Takahashi K, Yoshimoto T, Kodera Y, Saito Y, Inada Y (1987) Biotechnol. Lett. 9: 187
268. Matsushima A, Kodera Y, Takahashi K, Saito Y, Inada Y (1986) Biotechnol. Lett. 8: 73
269. Ajima A,. Yoshimoto T, Takahashi K, Tamaura Y, Saito Y, Inada Y (1985) Biotechnol. Lett. 7: 303
270. Lee H, Takahashi K, Kodera Y, Ohwada K, Tsuzuki T, Matsushima A, Inada Y (1988) Biotechnol. Lett. 10: 403
271. Babonneau M-T, Jaquier R, Lazaro R, Viallefont P (1989) Tetrahedron Lett. 2787
272. Matsushima A, Okada M, Inada Y 81984) FEBS Lett. 178: 275
273. Kaiser ET (1988) Angew. Chem., Int. Ed. Engl. 27: 902
274. Nakatsuka T, Sasaki T, Kaiser ET (1987) J. Am. Chem. Soc. 109: 3808
275. Polgár L, Bender MC (1966) J. Am. Chem. Soc. 88: 3153
276. Neet KE, Koshland DE (1966) Proc. Natl. Acad. Sci. (USA) 56: 1606
277. Hilvert D (1989) Design of Enzymatic Catalysts, In Biocatalysis in Agricultural Biotechnology, Whitaker JR, Sonnet PE (eds) ACS Symposium Series 389, p 14, Am. Chem. Soc., Washington
278. Kaiser ET, Lawrence DS (1984) Science 226: 505
279. Levine HL, Kaiser ET (1980) J. Am. Chem. Soc. 102: 343
280. Slama JT, Radziejewski C, Oruganti SR, Kaiser ET (1984) J. Am. Chem. Soc. 106: 6778
281. Radziejewski C, Ballou DP, Kaiser ET (1985) J. Am. Chem. Soc. 107: 3352
282. Hilvert D, Hatanaka Y, Kaiser ET (1988) J. Am. Chem. Soc. 110: 682
283. Schultz PG, Lerner RA, Benkovic SJ (1990) Chem. Eng. News, May 28, 26
284. Schultz PG (1989) Angew. Chem., Int. Ed. Engl. 28: 1283
285. Lerner RA (1990) Chemtracts - Org. Chem. 3: 1
286. French DL, Laskov R, Scharff MD (1989) Science 244: 1152
287. Pollack SJ, Schultz PG (1989) J. Am. Chem. Soc. 111: 1929
288. Janjic N, Tramontano A (1989) J. Am. Chem. Soc. 111: 9109
289. Janda KD, Ashley JA, Jones TM, McLeod DA, Schloeder DM, Weinhouse MI (1990) J. Am. Chem. Soc. 112: 8886
290. Janda KD, Schloeder D, Benkovic SJ, Lerner RA (1988) Science 241: 1188
291. Tramontano A, Janda KD, Lerner RA (1986) Science 234: 1566
292. Pollack SJ, Jacobs JW, Schultz PG (1986) Science 234: 1570
293. Janda KD et al. (1991) J. Am. Chem. Soc. 113: 291
294. Napper AD, Benkovic SJ, Tramontano A, Lerner RA (1987) Science 237: 1041
295. Janda KD, Benkovic SJ, Lerner RA (1989) Science 244: 437
296. Shokat KM, Leumann CJ, Sugasawara R, Schultz PG (1989) Nature 338: 269
297. Cochran AG, Sugasawara R, Schultz PG (1988) J. Am. Chem. Soc. 110: 7888
298. Hilvert D, Nared KD (1988) J. Am. Chem. Soc. 110: 5593
299. Jackson DY, Jackson DY, Jacobs JW, Sugasawara R, Reich SH, Bartlett PA, Schultz PG (1988) J. Am. Chem. Soc. 110: 4841
300. Braisted AC, Schultz PG (1990) J. Am. Chem. Soc. 112: 7430
301. Iverson BL, Cameron KE, Jahangiri GK, Pasternak DS (1990) J. Am. Chem. Soc. 112: 5320
302. Jackson DY, Schultz PG (1991) J. Am. Chem. Soc. 113: 2319
303. Tramontano A, Ammann AA, Lerner RA (1988) J. Am. Chem. Soc. 110: 2282
304. Kitazume T, Lin JT, Takeda M, Yamazaki T (1991) J. Am. Chem. Soc. 113: 2123

4 State of the Art and Outlook

The biotransformations described in this book show that the area is in an active state of development. This is reflected by the fact that in 1991 8% of all papers published in the area of synthetic organic chemistry contained elements of biotransformations. Attention has been focused on those methods that are most useful for and accessible to synthetic chemists. In the following paragraphs a brief summary on the state of the art of biotransformations is given. An outlook on future developments, however, should be taken with some circumspection as it bears a resemblance to an Austrian weather forecast.

Hydrolytic enzymes such as proteases, esterases and lipases are ready-to-use catalysts for the preparation of optically active carboxylic acids, amino acids, alcohols and amines. The area is sufficiently well researched to be of potential use for a wide range of synthetic problems. About two thirds of the reported research on biotransformations involves these areas. Since there is a considerable arsenal of proteases and lipases from which to choose the right biocatalyst, development of techniques for the improvement of selectivities will be the subject of future studies as will be the development of simple models aimed at the prediction of the stereochemical outcome of a given reaction. A search for novel esterases to enrich the limited number of available enzymes from this group would be a worthwhile endeavour.

The synthesis of optically active phosphate esters is now possible and this strategy should be seriously considered by chemists entering this area of work.

The asymmetric hydrolysis of epoxides is hampered by the lack of a readily available source of microbial enzymes. Once this drawback is surmounted, the method may constitute a valuable alternative to the asymmetric epoxidation of olefins, particularly for those substrates where directing functional groups are absent.

The enzymatic or microbial hydrolysis of nitriles to amides or carboxylic acids often does not proceed with high enantioselectivity, but it offers a valuable mild alternative method to the harsh reaction conditions usually required for this conversion.

Isolated dehydrogenases and/or whole microbial (usually fungal) cells can be used for stereo- and enantioselective reduction of ketones to furnish the corresponding secondary alcohols on a laboratory scale. Due to the fact that the majority of isolated dehydrogenases follow Prelog´s rule, a search for enzymes with the opposite specificity would be an interesting goal. On the other hand, the asymmetric reduction of activated carbon-carbon double bonds is restricted due to the fact that, in general, these transformations must be carried out using whole cell systems; this is due to the sensitivity involved, i.e. the enoate reductases. With the exception of NADH, further research on coenzyme recycling has to be done before these methods can be used for large-scale processes. About a quarter of the of research in biotransformations involves the latter areas.

Biocatalytic oxygenation reactions are becoming increasingly important since traditional methodology is either not feasible or makes use of hypervalent metal oxides, which are ecologically undesirable when used on a large scale. As the use of isolated oxygenases will always be hampered due their requirement of NAD(P)H-recycling, many useful oxygenation reactions such as mono- and di-hydroxylation, epoxidation, sulphoxidation and Baeyer-Villiger reactions will continue to be performed using whole cell systems. Considering the lack of cofactor-dependence, hydrogen peroxide dependent oxidases are likely to gain more importance.

The formation of carbon-carbon bonds in an asymmetric manner by means of aldolases and transketolases is a well researched method, which is certain to gain prominence as a standard technique for carbohydrate synthesis once more enzymes of this group are made available at a reasonable cost using genetic engineering. The same is true for glycosylation reactions using glycosidases and glycosylases.

The synthesis of optically active (R)-cyanohydrins by oxynitrilase is well established and the publication of techniques for using the corresponding (S)-specific enzymes is to be expected in the near future.

Although halogenation and dehalogenation reactions can be catalysed by enzymes, it is doubtful whether these reactions will supplement other biocatalytic methods, mainly due to the fact that the corresponding conventional chemical methods are highly competitive.

The methodology concerning the employment of enzymes in non-aqueous solvents (as well as merits and limits) is well understood so as to be useful for organic chemists. Thus, the synthesis of esters, lactones, amides, peptides and peracids is possible using enzymes. The understanding of the influence of the nature of the solvent on an enzyme's selectivity is just beginning to be understood and research on this topic will yield additional methods for the selectivity-enhancement of biocatalytic methods.

Despite tremendous efforts in the field of genetic engineering (such as site-directed mutagenesis) aimed at the development of novel enzymes (e.g. enzymes possessing altered specificities) it is clear that at this stage screening for novel microorganisms and enzymes is much more likely to lead to a solution of a particular problem than painstaking modification of the enzyme. After all, enzymes have been optimizing their skills for more than 3×10^9 years so as to develop a lot of sophisticated chemistry, whereas organic chemists have a track-record of less than a century. Fortunately there is some evidence that the number of distinctly different enzyme mechanisms is finite, since we already know examples of enzyme molecules that are unrelated by evolution but possess almost identically arranged functional groups. In these cases Nature has obviously faced the same biochemical problem and has found the same optimum solution. Of more immediate use is the cloning and over-expression of enzymes in sturdy and easy-to-cultivate host organisms (for those cases where the enzymes are found only in trace amounts in sensitive microorganisms) thus providing sufficient quantities of biocatalysts for biotransformations on a synthetic scale.

Finally, microorganisms can synthesize extremely complicated optically active molecules such as penicillins and steroids from cheap sources. Although the initial investment in screening, selecting and the development of mutant strains or cloning techniques are very high, many valuable compounds can be synthesized very easily where simple chemical methods will fail.

5 Appendix

5.1 Abbreviations

ACE	acetylcholine esterase	LDH	lactate dehydrogenase
ADH	alcohol dehydrogenase	LG	leaving group
ADP	adenosine diphosphate	MEEC	membrane enclosed enzymatic catalysis
AMP	adenosine monophosphate		
ATP	adenosine triphosphate	MEH	microsomal epoxide hydrolase
Bn	benzyl	MSL	*Mucor* sp. lipase
Cbz	benzyloxycarbonyl	NAD^+/ NADH	nicotinamide adenine dinucleotide
CCL	*Candida cylindracea* lipase		
CEH	cytosolic epoxide hydrolase	$NADP^+$/ NADPH	nicotinamide adenine dinucleotide phosphate
CTP	cytosine triphosphate		
Cyt P-450	cytochrome P-450	NDP	nucleoside diphosphate
DAHP	3-deoxy-D-*arabino*-heptulosonate-7-phosphate	NeuAc	*N*-acetylneuraminic acid
		Nu	nucleophile
DER	2-deoxyribose-5-phosphate	O-5-P	orotidine-5-phosphate
DH	dehydrogenase	PEG	polyethylene glycol
DHAP	dihydroxyacetone phosphate	PEP	phosphoenol pyruvate
DOPA	3,4-dihydroxyphenyl alanine	PLE	porcine liver esterase
E	enantiomeric ratio	PPL	porcine pancreatic lipase
e.e.	enantiomeric excess	PQQ	pyrroloquinoline quinone
EH	epoxide hydrolase	PRPP	5-phospho-D-ribosyl-α-1-pyrophosphate
Enz	enzyme		
FAD	flavine adenine dinucleotide	PSL	*Pseudomonas* sp. lipase
FDH	formate dehydrogenase	PYR	pyruvate
FDP	fructose-1,6-diphosphate	RAMA	rabbit muscle aldolase
FMN	flavine mononucleotide	Sub	substrate
Gal	galactose	TBADH	*Thermoanaerobium brockii* alcohol dehydrogenase
GDH	glucose dehydrogenase		
Glc	glucose	TEPP	tetraethyl pyrophosphate
GluDG	glutamate dehydrogenase	TPP	thiamine pyrophosphate
G6P	glucose-6-phosphate	TTN	total turnover number
G6PDH	glucose-6-phosphate dehydrogenase	UDP	uridine diphosphate
		UMP	uridine monophosphate
GTP	guanosine triphosphate	UTP	uridine triphosphate
HLADH	horse liver alcohol dehydrogenase	XDP	nucleoside diphosphate
		XTP	nucleoside triphosphate
HLE	horse liver esterase	YADH	yeast alcohol dehydrogenase
HSDH	hydroxysteroid dehydrogenase	Z	*tert*-butyloxycarbonyl
KDO	3-deoxy-D-*manno*-2-octulosonate-8-phosphate		

5.2 Suppliers of Enzymes

Amano	Japan	Miles Laboratories	FRG
Biocatalysts	UK	Novo Industri	Denmark
Biozyme	UK	Oriental Yeast	Japan
Boehringer	FRG	Plant Genetic Systems	Belgium
Calbiochem	USA	Recordati	Italy
Enzymatix	UK	Röhm	FRG
Fluka	Switzerland	Rhône-Poulenc	France
Genzyme	UK	Sigma	USA
Gist Brocades	Netherlands	Toyo Yozo	Japan
ICN National Biochemicals	UK	US Pharmaceuticals	USA
International Bio-Synthetics	Netherlands	Worthington	USA
Meito Sangyo	Japan		

Index

Abzymes 297
Acetaldehyde 212
Acetate kinase 101
Acetobacter sp. 169
Acetyl phosphate 101
Acetyl-CoA 20
Acetylcholine esterase 48, 58
Acetylglucosamine 231
Acetyllactosamine 231
Acetylmannosamine 211
Acetylneuraminic acid 211
Acid anhydride 260
Acid phosphatase 97
Acinetobacter sp. 53, 190
Acremonium sp. 119
Acrylamide 116
Acrylonitrile 116
Activated charcoal 284
Activated esters 257
Activation energy 15
Active site model 72, 86
Acyl donor 257
Acyl migration 60, 250
Acyl-enzyme intermediate 24, 114
Acyl transfer 256
Acylase-method 45
Acyloin 214, 214, 221
Addition reaction 221
Adenylate kinase 105
Adsorption 250, 283
Aerosol OT 290
Agar gel 288
Agarose 285
Aglycon 236
Agrochemicals 4
AIDS 208, 271
Alcohol dehydrogenase 139, 141
Aldehyde dehydrogenase 170
Aldehydes 207
Aldol-reactions 204
Aldolases 204
Aldose 210
Alginate gel 288
Alkaline phosphatase 98, 208
   protease 43
Alkaloids 96, 198
Allene 63, 158, 240

Allergies 7
Allylic alcohols 87
Alternative fit 52, 149
Amidase 44, 113
Amidase-method 44
Amide synthesis 273
Amination 225
Amine oxidases 114
Amino acid 40
   synthesis 41
   racemases 42
Aminoacylases 45
Aminoadipate 140
Aminolysis of esters 275
Aminonitrile 119
Aminopeptidases 44
Ammonia 100, 221
AMP-ADP-ATP interconversion 102
Anisaldehyde 215
Anomeric centre 232
Anti-inflammatory agents 63
Antigen 298
Arene oxide 181
Aristeromycin 98
Armentomycin 158
Arsenate 207
Arthrobacter sp. 53
   lipase 92
Artificial enzyme 293
Asparagine 9
Aspartame 276
Aspartic acid 65
Aspergillus sp. 44, 45, 49, 65, 76
   lipase 90
Asymmetric synthesis 5
Asymmetrisation
   prochiral substrate 26, 52
   meso-substrate 28
ATP 20, 100
Autooxidation 283
Auxiliary reagent 5
Aziridines 112

Bacillus sp. 60, 148
Bacteridium sp. 116
Baeyer-Villiger reaction 189
Baker´s yeast 43, 214, 216, 225

enzyme inhibition 152, 153
hydrolysis of esters 61, 62
reduction of carbonyl groups 150, 159, 160
    model 154
reduction of C=C-bonds 157
sulphoxidation by 187
Barbituric acid 241
Beauveria sulfurescens 157
Benzaldehyde 214
Bichiral esters 94
Bidentate ligands 263
Biomass 8
Biotechnology 7
Biotin 20, 57
Biphasic system 73, 79, 87, 94, 222, 251
Brevibacterium sp. 60, 116
Brevicomin 208
Bromoperoxidase 238
Bücherer-Bergs reaction 47
Butyl hydroperoxide 188
C-Synthon 79, 115, 204
C=C-Bond
    reduction of 157
Caldariomyces fumago 188, 238
Camphor hydroxylase 176
Cancerogenicity 107
Candida sp. 76
    lipase 82, 95, 259, 276, 280
        model 85
Carbamate kinase 100
Carbamic acid 100
Carbamoyl phosphate 100
Carbohydrate 7, 77
Carbon dioxide 254
Carbon monoxide dehydrogenase 162
Carboxyl esterase NP 48, 63
Carotenoid 150, 161
κ-Carragenan 288
Carrier 281
Castanospermine 271
Catalase 170, 174
Catalytic antibodies 297
Catalytic triad 24, 90, 92
Cathode 162
Celite 250, 253, 284
Cellulose 285
Chelation 112
Chemical energy 19
Chemical operator 11, 23, 295
Chemoselectivity 4, 66
Chiral pool 5
Chirality, tpes of 12
Chloroethyl ester 43, 258
Chlorohydrins 160
Chloroketones 160
Chloroperoxidase 188, 238
Chloropropionic acid 245
Cholesterol 179

Cholesterol esterase 48, 77
Chromobacterium sp. lipase 92, 272
Chymotrypsin 249
    model 43
Cineole 179
Cinnamaldehyde 215
Citronellol 159
Claisen-rearrangement 3
Clostridium sp. 157
Coenzyme 19
Cofactor 7, 8, 19, 100, 135, 193, 291
    electrochemical regeneration 136
    enzymatic regeneration 136
        coupled enzyme regeneration 136
        coupled substrate regeneration 136
    photochemical regeneration 136
Conglomerate 5
Coordinated metals 20
Cope-rearrangement 3
Copolymer 293
Cosolvent 69, 222, 236, 250
Cosubstrate 135
Covalent attachment 285
Criegee-intermediate 189
Critical micellar concentration 73
Cross-linking 287
Crotonaldehyde 225
Crown-ether 263
Cryoenzymology 6, 251
Crystallography 70, 92
Cytidin triphosphate 20, 102
Curtius-rearrangement 52
Curvularia falcata 141
Cutinase 93
Cyanide 112
Cyanoglucosides 113, 124
Cyanohydrin 92, 221, 258, 266, 267
Cyanolipids 113
Cyanomethyl ester 257
Cyanopyrazine 119
Cyanopyridines 117
Cyclohexadienediols 197
Cyclohexanone mono-oxygenase 190
Cyclohexene oxides 110
Cytochrome P-450 176

DAHP synthetase 212
Decarboxylation 64
Dehalogenase 244
Dehalogenation 238, 243
Dehydrase 224
Dehydrogenase 41, 135, 193, 218, 280
Dehydroglutamate 64
Deactivation 6, 9, 251
Denaturation 283
Deoxyribose-5-phosphate 212
Deoxyribose-5-phosphate aldolase 212
Deoxysugar 206
Dephosphorylation 97

Deracemisation 170
Desolvation theory 11
Detergent industry 20
Detoxification 107
Di-oxygenase 174, 186, 193, 196
Diabetes 278
Diastereomeric esters 16
Diastereoselectivity 4, 155
Diclofop 63
Dielectric constant 253
Diels-Alder reactions 3, 197
Diffusion 254
Dihydrodiol dehydrogenase 196
Dihydropyrimidinases 47
Dihydroxyacetone 104, 148, 207
    phosphate 97, 105, 206
Dihydroxylation of aromatic compounds
    196
Diketone, reduction of 155
Diketopiperazines 43
Diene, reduction of 158
Diolide 272
Diol 156, 205
Dioxetan 194
Dipole moment 253
Discrimination
    enantioface 15
    enantiomer 31
    enantiotopos 14
Distal region 196
Distomer 4
Disulfide bond 9, 297
Disulphide 243
DOPA 182
Double-sieving method 37

Ecosystem 243
Electronic effects 79
Elimination reaction 221, 299
Enantiodivergent reaction 193
Enantioface differentiation 25, 157
Enantiomer
    differentiation 13, 31, 32
    biological effects 5
Enantiomeric ratio 32, 34
Enantioselective inhibition 95
Enantioselective phosphorylation 105
Enantioselectivity 4, 25, 34, 250
Enantiotopos 26, 27
    face 16
    group 16
Endo-peroxide 193
Enkephalin amide 276
Enoate reductase 157, 162
Enol ester 258
Entrapment 152, 288
Entropy 12
Environment 3, 112
Enzymatic resolution

racemate of 31
    sequential 39
Enzymation 7
Enzyme 7, 8
    classification 17, 18
    Commission 18
    conformation 11
    deactivation 9
    model 10
    structure 9
    Source 20
    surface 9
    transition-state 16
Enzyme-substrate complex 16
Epichlorohydrin 286
Epoxidation 135, 185
    microbial 107
    alkene 183
Epoxide 107, 287
Epoxide hydrolase 107
    hepatic 107
    mechanism 108
    microbial 112
    substrate-type 109
Epoxy alcohols 80
Equilibrium constant 35
Ester
    aminolysis 25
    hydrolysis 48
    synthesis 255
Esterase-method 42
Esterases 41, 48
    microbial 60
Esterification 38, 255, 261
Eudismic ratio 4
Eutomer 4

Fermentation 7
    conditions 150
Ferrocenylethanol 266
Flavine 20, 136, 178, 190
    mononucleotide 136, 170
    dinucleotide 136, 222
Flavours 191
Fluorine 239, 245
Fluorohydrins 244
Fluorosugars 206
Fluvalinate 40
Food industry 20
Formate dehydrogenase 138, 162
Formation of carbon-carbon bonds 204
Fosfomycin 186
Fragrance 179, 261
Free energy 17
Fructose-1,6-diphosphate aldolase 206
Fuculose-1-phosphate aldolase 212
Fumarase 224
Fumaric acid 224
Fusarium solani pisi 93

Galactosidase 234
Galactosyl transferase 230
Geotrichum candidum 71, 148
    lipase 92
Geraniol 159, 225
Gluconolactone 138
Glucopyranosides 255
Glucose dehydrogenase 138
Glucose oxidase 174
Glucose-6-phosphate 97, 103, 138, 230
    dehydrogenase 138
Glutamate 140
    dehydrogenase 140
Glutamic acid 65
Glutamine 9
Glutardialdehyde 287
Glyceraldehyde-3-phosphate 206, 212
Glycerol 105, 148
    dehydrogenase 148
    kinase 105, 207
        model 106
Glycine 9
Glycohydrolases 232
Glycol 170, 194, 197
Glycolysis 204
Glyconeogenesis 204
Glycosidases 232
Glycosyl Transferases 228
Glycosylation 233
Grahamimycin 88
Guanosin triphosphate 20

Haemoprotein 177
Halido-hydrolase 244
Haloacid dehalogenase 245
Haloamines 243
Haloethyl ester 257
Halogenation 238
    alkene 239
    alkyne 240
    aromatic 241
    C-H group 241
    N- and S-atom 243
Halohydrin 239, 265
    epoxidase 244
Haloketone 240, 242
Halolactone 239
Halometabolites 238
Haloperoxidase 238, 241
Helical chirality 58
Helminthosporium sativum 112
Hexokinase 103, 138
Hexose 204
Hildebrandt solubility parameter 253
Horse liver alcohol dehydrogenase 143, 171, 280
    model 145
Horse liver esterase 48

Horseradish peroxidase 189, 239
Hydantoin 47
Hydantoinase 41
Hydantoinase-method 47
Hydrazide 25, 277
Hydrazinolysis of esters 25, 273
Hydrocarbon 251
Hydrogen cyanide 92, 221
Hydrogen peroxide 170
Hydrogen-bonding 298
Hydrogenase 140, 162
Hydrolase 19
Hydrolysis
    amide 23, 40
    epoxide 107
    ester 23, 48
        mechanism 23
    nitrile 112
    peptide 9
Hydroperoxide 178, 193, 194, 268
Hydroxamic acid 25
Hydroxyketone 148
Hydroxylamine 25
Hydroxylase 184
Hydroxylation 135, 184, 281
    Alkanes 178
    Aromatic Compounds 181
    Beauveria sulfurescens by 180
Hydroxysteroid dehydrogenase 147
Hypohalous acid 239

Immobilization 8, 150, 250, 283
    cells of 8
    covalent 259
Induced-fit mechanism 10
Inducer 116
Industrial effluents 116
Inhibition 6, 95, 137, 150, 151, 169, 181, 249, 251
Inhibitors 190
Inositol 198
nsulin 278
Interesterification 25, 256
Interface 73
Iodide 245
Iodoperoxidase 238
Ionic binding 285
Isoenzyme 85
Isomerase 19
Isomerisation 300
Iso-propenyl acetate 258, 262

Katal 18
KDO synthetase 212
Ketoacid, reduction of 153
Ketoadipate 140
Ketoester, reduction of 153
Ketoglutarate 140
Ketorolac 66

Ketose 210
Kinase 100, 139
Kinetic control 210
Kinetics 4
  double step 30
  irreversible reaction 33
  Michaelis-Menten 33
  reversible reaction 35
  single step 29
K_m-value 283

Lactam 43, 180
Lactate 155
  dehydrogenases 148
Lactols 171, 173
Lactone 171, 178, 190, 299
  hydrolysis 82
  synthesis 272
Leloir pathway 228
Leuconostoc mesenteroides 138
Leukotriene 81
Ligase 19
Linoleic acid 195
Lipase 11, 43, 72, 251, 279
  mechanism 73
    steric requirements 76
    substrate-type 75
Lipoxygenase 193, 196
Lithiocholic acid 179
Liver microsomes 107
Lock and Key mechanism 10
Log P 253
Lyase 19, 41, 221, 105
Lyophilization 9
Lysine 259, 279

Macrolide 273
Magnetite 295
Maillard-reaction 259
Malic acid 224
Malonic diesters 52
Mandelate 155
Mannosamine 211
Mass-transfer 73
  coefficient 251
Matrix 288
Mediator 140, 162
Medium engineering 69, 94
Membrane 289
Membrane-enclosed-enzymatic-catalysis 291
Meso-dicarboxylates 54, 55
  diols, oxidation of 172
Meso-substrate 27, 28, 50
Meso-trick 52, 55, 74
Mesophilic organism 9
Metal-porphyrin 20
Metallo-enzyme 112
Methanotroph 183

Methioninol 97
Methoxycarbonyl phosphate 102
Methylaspartase 224
Micelle 290
Michael-type addition 197, 216
Michaelis-Menten kinetics 73, 149
Micro-pH 255, 260
Micrococcus sp. 116
Microsomal epoxide hydrolase 107
Mild conditions 3, 6, 51, 98
Mixed anhydride 45
Model 70, 106
  hydroxylation 181
Modified enzyme 293
Molecular modelling 71
Molecular sieve 259
Mono-oxygenase 107, 174, 176, 183, 186
Monophasic solvent 94, 250, 252
Mosher´s acid 66
Mucor sp. 71, 76
  lipase 90
Mutant 178, 190

NAD(P)+/NAD(P)H 20, 103, 135, 183, 281
Naproxen 63
Nerolidol 225
Nicotinamide 117
  cofactor 136
Nitrilase 114
  mechanism 115
Nitrile hydrolysis 114, 116, 121
  aliphatic 113
  aromatic 113
  enantioselectivity 119
Nitrile hydratase 113
  mechanism 115
Nitro-olefin 162
Nitrous acid 47
Nocardia corallina 184
Nojirimycin 208
Nomenclature
  enantiotopos, enantioface 14
Nucleophilicity 257
Nucleoside triphosphate 100

Oligosaccharide 228
Optimisation of selectivity 67, 94
Organic solvent 3, 73, 248
Organometallic compound 265
Orotidine-5´-pyrophosphorylase 105
Oxidase 135, 174, 280
  haeme-protein 174
  metallo-flavin 174
Oxidation 169
  alcohols and aldehydes 169
  polyols 169
Oxidoreductase 19, 149
Oxime ester 257, 258

Oxygen 174
　activation 176
Oxygenase 20, 135, 174
Oxygenation reactions 174, 175
Oxynitrilase 221, 266

π-π Stacking 9
PAF-antagonists 262
Pantolactone 62, 169
Papain 49, 63, 65, 297
Paraoxon 190
Partition coefficient 253
Penicillin acylase 45, 49, 65
Penicillin G 65
Penicillium sp. 45, 186
Pentose pathway 213
Pepsin 49
Peptide 40
　synthesis 274
Peracid
Peracid 25, 189
　synthesis 279
Peroxidase 174, 282
Peroxide 194
Pesticides 4
pH 70, 222, 252, 255, 266, 283
Pharmaceuticals 4
Phase-distribution 251
Phenols 182, 281
Phenylacetate ester 65
Pheromone 150, 208
Phosphatase 97
Phosphate donor 100
Phosphate ester 97
　formation 99
　hydrolysis 97, 98
Phospho-lipases 77
Phosphoenol pyruvate 101, 231
Phosphoglucomutase 230
Phospholipids 105
Phosphonic ester 299
Phosphorylation 97, 100, 207
　chemoselective 105
Photochemical reaction 299
Phytol 159
Pichia miso 60
Picolinamide 117
Pig liver esterase 43, 50, 54, 58, 68, 90, 276
　acetone-powder 58
　substrate model 71
Pinitol 198
Platelet-activating-factor 90
Polyethylene glycol 293
　polyhydroxy compounds, regioselective protection 269
Polymerisation 181, 249
Polyol dehydrogenase 211
Polyphenol oxidase 181, 249, 281

Polyprenyl pyrophosphate 97
Polysaccharides 228
Polyvinyl acetate 286
Porcine pancreatic lipase 77, 90, 250, 257, 262, 269, 276
Prelog´s rule 141, 148, 150, 155, 160, 214, 218
Prenalterol 183
Primary structure 8
Prochiral centre 53
Product inhibition 116
Productivity number 136
Progesterone 178
Propionitriles 120
Prostaglandin 51, 56, 79
Prostethic group 114, 222
Protease 41, 48
　esterase-activity 63
Protein purification 21
Proteus sp. 157
Proximal region 196
Pseudomonas sp. 44, 76, 116, 169, 184
　lipase 86, 95, 259, 262, 265, 267, 280
Pyrenophorin 273
Pyrethroid 91, 221
Pyridoxal-phosphate 20
Pyrroloquinoline quinone 114, 136
Pyruvate 141, 211
　decarboxylase 214
　kinase 101

Quasi-irreversible 257
Quinones 181, 281

Rabbit muscle aldolase 206
Racemisation 48, 249
Radical 162
Rate acceleration 2, 15
Rearrangement 158
Redox cofactor 135
Reduction 135
　aldehyde of 141
　ketones of 141, 146
Reductive dehalogenation 243
Regioselective reaction 4, 77, 169, 250, 269, 270
　hydrolysis 51, 64
　hydroxylation 181
　oxidation 170
　phosphorylation 103
Remote centre 87
Residual water 9
Resolution of racemic ester 58
Resting cell 8
Retro-fat 65
Reverse micelle 290
Rhizopus nigricans 60
Rhodococcus sp. 44, 116, 117, 120, 169
Ribokinase 105

Ribose-5-phosphate 105
Ricinine 113
Rubredoxin 184

Salt bridge 9
SAM 20
Schiff's base 41, 113, 205, 259, 287
Selectivity 6
Selenium 296
Semisynthetic enzyme 295
Sequential resolution 36
Serine hydrolase 23
Serine-hydroxymethyl transferase 212
Sharpless epoxidation 80, 107, 263, 268
Sialic acid 211
Sialyl Aldolase 211
Side-reaction 249
Silica 250
Sodium dithionite 136
Solid protein 254
Solvation-substitution theory 12
Soybean-lipoxygenase 194
SPAC-reaction 87
Space-time yield 283
Staphylococcus epidermidis 148
Starch 285
Steroid 48, 178, 240, 241, 252, 290
Strecker synthesis 120
Streptomyces sp 66, 239
Structural water 9
Substrate
    inhibition 116
    model 71, 85
    modification 67, 88, 94, 150, 181
    specificity 250
    tolerance 3
Subtilisin 43, 49, 65, 70, 249, 251, 254, 295
Sulfenyl halides 243
Sulfhydryl residue 114
Sulphone 186
Sulphoxidation 186
Sulphoxides 186, 243
Supercritical gas 254
Surfactant 255, 290
Sweetener 274

Thermoanaerobium brockii 141, 145
    dehydrogenase 141
Temperature 6, 70, 262, 283
Teratogenicity 107
Terpene alcohol 225
Tetrahedral intermediate 178
Thermodynamic control 210
Thermolysin 43
Thermophile organism 9
Thermostable enzymes 148
Thiamine pyrophosphate 213, 214
Thiirane 112

Thioacetal 188
Thioester 74
Thioether 187
Three-point attachment rule 12
Threonine 279
Tocopherol 216
Total turnover number 136
Tranexamic acid 119
Trans-cyanation 223
Transesterification 256
Transferase 19, 105
Transglycosylation 234
Transition-state 15
Transketolase 205, 213
Transpeptidation 275
Trehalose 231
    phosphorylase 231
Trichloroethyl ester 43, 258
Trifluoroacetaldehyde 218
Trifluoroethanol 218
Triglyceride 72
Triolein 73
Trypsin 49
Tryptophan 45
Turnover numbers 100
Twistanone 143
Two-step resolution 34
Tyrian purple dye 238

Uridin monophosphate 104
Unfolding 250
Unit, enzyme activity 18
Unnatural substances 3
Unsaturated carboxaldehydes 158
    carboxylic acids 157
    esters 157
    nitro compounds 158
Uridin triphosphate 20, 102, 229
Ursodeoxycholic acid 179

Van-der-Waals 9, 298
Vanadate 208
Vesicle 290
Vinyl acetate 258, 262
Viologen 162
Vitamin 62, 159, 216

Water 6, 249
    activity 276
    bound water 249, 254, 256
    bulk water 9, 249
    scavengers 256
Wax ester 93
Whole cell systems 7

Xenobiotic 107

Yeast alcohol dehydrogenase 141

P. Gacesa, University College, Cardiff, UK;
J. Hubble, University of Bath, UK

Enzymtechnologie

Aus dem Englischen übersetzt von B. Vollert-Schmid
Überarbeitet von G. Hummel
1992. Etwa 220 S. 68 Abb. 19 Tab. (Springer-Lehrbuch)
Broch. DM 54,- ISBN 3-540-55183-2

Die Entwicklung der Immobilisierungstechniken für Enzyme seit den 50er Jahren hat zu einem Boom bei der industriellen Anwendung von Enzymen geführt und hat an Wissenschaftler mit verschiedenstem akademischen Hintergrund die Anforderung gestellt, in der Enzymtechnologie zu arbeiten.

Dieses Buch wendet sich an Studenten sowie in der Industrie tätige Wissenschaftler, die aus den Fachbereichen Chemieingenieurwesen, Chemie, Biochemie, Mikrobiologie/Biologie, Technische Biologie sowie Biotechnologie kommen und ein allgemeines Interesse an Enzymen haben. Es bringt dem Ingenieur die Feinheiten der Enzyme sowie das Potential der Techniken in der Molekulargenetik nahe, mit denen diese Katalysatoren für spezifische Anwendungen geschneidert werden können.
Für jene mit einem chemisch/biochemischen oder biologischen Hintergrund werden in diesem Buch vor allem die biochemisch-technischen Beschreibungen, wie kinetische Eigenschaften und Reaktorkonstruktion, von Nutzen sein.

Springer-Verlag
Berlin
Heidelberg
New York
London
Paris
Tokyo
Hong Kong
Barcelona
Budapest